# INEXHAUSTIBILITY

A non-exhaustive treatment

LECTURE NOTES IN LOGIC

A Publication of
THE ASSOCIATION FOR SYMBOLIC LOGIC

Lecture Notes in Logic 16

# INEXHAUSTIBILITY

## A non-exhaustive treatment

by

Torkel Franzén
Department of Computer Science and Electrical Engineering
Luleå University of Technology

CRC Press
Taylor & Francis Group
Boca Raton London New York

CRC Press is an imprint of the
Taylor & Francis Group, an informa business

Addresses of the Editors of Lecture Notes in Logic and a Statement of Editorial Policy may be found at the back of this book.

**Sales and Customer Service:**
A K Peters, Ltd.
888 Worcester Street, Suite 230
Wellesley, Massachusetts 02482, USA

**Association for Symbolic Logic:**
C. Ward Henson, Publisher
Mathematics Department
University of Illinois
1409 West Green Street
Urbana, Illinois 61801, USA



**Library of Congress Cataloging-in-Publication Data**

Franzén, Torkel.
Inexhaustibility : a non-exhaustive treatment / by Torkel Franzén.
p. cm. – (Lecture notes in logic ; 16)
Includes bibliographical references and index.
ISBN 1-56881-174-8 – ISBN 1-56881-175-6 (pbk.)
1. Incompleteness theorems. 2. Logic, Symbolic and mathematical. 3. Mathematics–Philosophy. I. Title. II. Series.

QA9.56 .F73 2004
511.3–dc21 2002025765

CRC Press
6000 Broken Sound Parkway, NW
Suite 300, Boca Raton, FL 33487
270 Madison Avenue
New York, NY 10016
2 Park Square, Milton Park
Abingdon, Oxon OX14 4RN, UK

Publisher's note: This book was typeset in LaTeX, by the ASL Typesetting Office, from electronic files produced by the author, using the ASL documentclass `asl.cls`. The fonts are Monotype Times Roman. This book was printed by Friesens, of Manitoba, Canada, on acid-free paper. The cover design is by Richard Hannus, Hannus Design Associates, Boston, Massachusetts.

A Druidic myth relates how Lucanor, coming upon the other gods as they sat at the banquet table, found them drinking mead in grand style, to the effect that several were drunk, while others remained inexplicably sober; could some be slyly swilling down more than their share? The disparity led to bickering, and it seemed that a serious quarrel was brewing. Lucanor bade the group to serenity, stating that the controversy no doubt could be settled without recourse either to blows or to bitterness. Then and there Lucanor formulated the concept of numbers and enumeration, which heretofore had not existed. The gods henceforth could tally with precision the number of horns each had consumed and, by this novel method, ensure general equity and further, explain why some were drunk and others not. "The answer, once the new method is mastered, becomes simple!" explained Lucanor. "It is that the drunken gods have taken a greater number of horns than the sober gods, and the mystery is resolved." For this, the invention of mathematics, Lucanor was given much honor.

Jack Vance, *Madouc*

# PREFACE

The logical and mathematical results presented in this book are not at all due to the author, except perhaps for some minor simplifications and observations. There are some references in the text to the people who have created and developed the subject of Gödelian extensions of theories, but I'm aware that there are many others who could have been mentioned, and I ask their indulgence for my casual style of presentation. The contribution of the book consists in a sustained presentation, starting from basic concepts, of a particular view of the topic of Gödelian extensions. It is my optimistic hope that philosophers, mathematicians, and others with an interest in this topic will find the book helpful in acquring a basic grasp of the philosophical and logical results and issues.

Writing this book, over the past couple of years, was made possible through the facilities and pleasant working environment offered by the computer science department at Luleå University of Technology. Several people have commented on the text, and I am grateful in particular to Sam Buss, Calvin Ostrum, Charles Silver, Walter Felscher, Leon Horsten, Damjan Bojadziev, and Alex Blum for useful comments. My thanks also to C. Ward Henson and Cathryn Young for their work on producing the book.

Marcia, my wife, hasn't commented on the book except to say "That's very nice, dear", but must nevertheless be held partly responsible for its contents, in that she, through a series of strange and wonderful events, separated me from a sloth-like bachelor existence, causing me to do all sorts of things, like finding a job and writing a book. I also choose to conveniently hold her responsible for any remaining mistakes in the book.

Torkel Franzén
*Luleå, Sweden*

# CONTENTS

# CHAPTER 1

# INTRODUCTION

## 1.1. Inexhaustibility: the positive side of incompleteness

In 1931, Kurt Gödel presented his famous incompleteness theorem, which has since had a great and continuing impact on logic and the philosophy of mathematics, and has also like no other result in formal logic caught the interest and imagination of the general public.

Gödel presented two results in his 1931 paper, usually referred to as the first and the second incompleteness theorem. The proof of the first incompleteness theorem shows that for every consistent formal axiomatic theory in a wide class of such theories, there is at least one statement which can be formulated in the language of the theory but can neither be proved nor disproved in the theory. Such a statement is said to be *undecidable* in the theory. By a consistent formal theory is meant one in which no logical contradiction—both a statement $A$ and its negation not-$A$—can be proved. The second theorem states that for a wide class of such theories $T$, if $T$ is consistent, the consistency of $T$ cannot be proved using only the axioms of the theory $T$ itself.

In discussions of the meaning and implications of these theorems, their negative or limiting aspects are most often kept in the foreground: a formal theory of arithmetic cannot be complete, a theory cannot be proved consistent using only the resources of that theory. But the second incompleteness theorem also has a positive aspect, which was emphasized by Gödel (*Collected Works*, vol. III, p. 309, italics in the original):

> It is this theorem [the second incompleteness theorem] which makes the incompletability of mathematics particularly evident. For, *it makes it impossible that someone should set up a certain well-defined system of axioms and rules and consistently make the following assertion about it: All of these axioms and rules I perceive (with mathematical certitude) to be correct, and moreover I believe that they contain all of mathematics.* If somebody makes such a statement he contradicts himself. For if he perceives the axioms under consideration to be correct, he also perceives (with the same certainty) that they are consistent. Hence he has a mathematical insight not derivable from his axioms.

This positive implication of the incompleteness theorem is that we have a way of extending any system of axioms for mathematics that we recognize as correct to a logically stronger system of axioms that we will also recognize as correct, namely by adding the statement that the old system is consistent as a new axiom. That the resulting system is logically stronger than the old system means that it proves everything that the old system proves, and more besides. Thus our mathematical knowledge would appear to be *inexhaustible* in the sense that it cannot be pinned down in any one formal axiomatic theory.

This implication, or apparent implication, of the incompleteness theorem is neither uncontroversial nor unproblematic. There is something rather strange about the impossibility of pinning down which axioms we recognize as correct. Shouldn't it be possible to somehow incorporate in those axioms the very principle that any collection of valid axioms can be extended to a stronger collection, and thereby set down in axiomatic form all of our current mathematical knowledge? And if this really can't be done, do we have some sort of ineffable or unformalizable insight that cannot be fully expressed in rules? If so, what are the limits of what can be proved using this insight, and how can those limits be described? In particular, just what is it that happens when we try to systematically and ad infinitum extend a theory by adding successive consistency statements as new axioms?

These are the questions to be pursued in this book, along with some important side issues. Although the questions are philosophical rather than mathematical, we need to understand the relevant mathematical concepts and theorems in order to arrive at any good answers to the questions, and so much of the book consists of mathematical definitions and proofs.

The book is aimed at readers who wish to understand the inexhaustibility phenomenon pointed out by Gödel, whatever their level of logical expertise. No knowledge of logic or mathematics will be presupposed except some basic school arithmetic and some slight acquaintance with sets and functions. The relevant basics of predicate logic, recursion theory, and the theory of ordinals and ordinal notations will be introduced and explained in an ad hoc fashion as the exposition proceeds. It may be that the presentation is sometimes too brief to suffice as an introduction, in which case the non-expert reader will need to consult other sources for supplementary explanations or for a more extended and systematic treatment.

In a few places a mathematical illustration may be used which assumes some further mathematical background. In particular, this is true of the presentation and application of Euler's product formula, which presupposes some calculus. Nothing essential will be lost by skipping lightly over such passages.

In the philosophy of mathematics, we need a good choice of mathematical examples to work with. This doesn't mean that the examples must be difficult or advanced mathematics. It's not the case that the philosophical problems become more difficult or sophisticated as more difficult mathematics is introduced. The mathematician G. H. Hardy went so far as to proclaim (in Hardy [1929]) that "Philosophy proper is a subject, on the one hand so hopelessly obscure, on the other so astonishingly elementary, that there knowledge hardly counts." In a philosophical discussion of the apparent phenomenon of inexhaustibility, it's clear enough that some knowledge of Gödel's theorem, the properties of sequences of Gödelian extensions of theories (here based on autonomous recursive progressions of theories as conceived by Turing and Feferman) and some other technical topics treated in the book is needed, and it's also a good idea to have some understanding of mathematical proof as it exists in mathematics. However, the precise choice of definitions and results from logic and mathematics in this book has no profound justification, and to a considerable extent only reflects the author's preferences and limitations.

The book begins in Chapter 2 with a bit of arithmetic, both because some arithmetic will be used in later chapters and in order to provide some examples for the philosophical and logical discussion to refer to. There is an emphasis on some logical aspects of elementary arithmetical reasoning that will be described more formally in later chapters. Chapter 3 introduces some further examples of mathematics in the form of some classical results about primes, and also explains the informal distinction between arithmetical, elementary, and analytic proofs of arithmetical theorems. The formal development of the logical results used or pondered in the philosophical argument of the book begins in Chapter 4. The material in chapters 4 to 11 covers some basic predicate logic, set theory, and recursion theory leading up to the proof of the incompleteness theorems in Chapter 12. The last four chapters, beginning with Chapter 12, form the core of the argument of the book concerning inexhaustibility. A reader with a good background in logic can go straight to Chapter 12 or 13 and consult the index and earlier chapters as needed. A reader with a not-so-good background in logic who finds some of the earlier chapters heavy going may also choose to skip forward to the later chapters, and judge on the basis of the argument presented in §15.3 whether it would be worthwhile to go to the trouble of studying the earlier chapters.

The remainder of this introductory chapter will be devoted to some general comments about the aims and claims of the book, and about the difficulties a reader may encounter in coming to grips with a rather odd mixture of technical results and philosophical argument of this sort.

## 1.2. Two Gödelian traditions

In translation, the title of Gödel's 1931 paper is "On formally undecidable propositions of *Principia Mathematica* and related systems I". In the paper, Gödel proved the incompleteness theorem for a particular formal system P, which was not that of Russell's and Whitehead's monumental work *Principia*

*Mathematica*, but was indeed related to that system. At the end of §2 of the paper, he remarks:

> In the proof of Theorem IV [the first incompleteness theorem] no properties of the system P were used besides the following:
>
> 1. The class of axioms and the rules of inference (that is, the relation "immediate consequence") are recursively definable (as soon as we replace the primitive signs in some way by natural numbers);
>
> 2. Every recursive relation is definable (in the sense of theorem V) in the system P.
>
> Therefore, in every formal system that satisfies the assumptions 1 and 2 and is $\omega$-consistent there are undecidable propositions of the form (x)F(x), where F is a recursively defined property of natural numbers, and likewise in every extension of such a system by a recursively definable $\omega$-consistent class of axioms.

Gödel goes on to remark that these conditions are clearly satisfied by various formal theories used in logic and set theory.

This is a common pattern in logic, probably more so than in other formal sciences. A result is proved or a concept defined for some particular formalism, and it's "clear" how to extend or adapt it to various other related formalisms. The formalisms are typically axiomatic theories, or systematizations of the rules of logical reasoning, or formalisms for expressing algorithms. It is "clear" how to extend or adapt the result or concept only on the basis of experience with these different formalisms, together with an appreciation of what is central to the definition or result in question. One might think that it should be feasible to formulate the definitions or results and their proofs in sufficient generality to cover these different formalisms. To some extent this can be done, but on the whole it is probably correct to say that Gödel's theorem and many other results in logic are best understood by seeing the proof carried through in detail for some particular formalism, rather than by attempting to achieve any great generality in the proof itself.

This feature carries over to philosophical discussions connected with Gödel's theorem. In this book, inexhaustibility will be discussed in connection with arithmetic and extensions of arithmetic, but it will be clear how the discussion applies, for example, in the case of set theory and extensions of set theory. Also, the consequences of extending a theory by adding consistency statements or reflection principles ad infinitum will be formally worked out for one particular approach, that of transfinite recursive progressions, but with the understanding that a similar treatment can be carried out for other approaches to be found in the literature. The philosophical and informal conclusions and arguments in the book are independent of the particular formal approach taken.

Gödel also initiated in his paper another tradition that presents a considerable difficulty for the beginning student. Gödel's proof of the second incompleteness theorem, which states that a formalization of the statement

"$T$ is consistent" is not provable in $T$ itself, is based on the fact that the (first half of the) first incompleteness theorem for $T$ *is* provable in $T$ itself. Gödel only presented this fact as a plausible claim, pointing out that the proof of the first incompleteness theorem only used arithmetical reasoning of a kind formalizable in $T$. He intended to give a rigorous proof of the claim in part II of the paper, which was also to substantiate the earlier observations regarding the wide applicability of the incompleteness theorem. However, part II never appeared. One reason for this was that the argument Gödel gave for the second incompleteness theorem was in fact quite convincing to his readers, another reason was that a detailed formal proof of the claim made in connection with the second theorem was given by Hilbert and Bernays in their two-volume *Grundlagen der Mathematik* (1939). The formal details turned out to require quite a lot of rather tedious work. In most proofs of Gödel's theorem it is customary even today to skip these formal details when it comes to proving the second incompleteness theorem.

If one looks today at the logical literature, one finds that there is a great deal of handwaving going on of the same kind as that used by Gödel in his paper. That is, informal arguments are given to show that a formalization of a certain statement is formally provable in a certain theory. These informal arguments can be quite difficult to grasp for a non-expert, but they are convincing and clear enough to the experts, who have no doubt that they could be expanded to any level of detail required to establish the assertion formally. So why the handwaving? Because the full formal proofs would in most cases be very long and boring and not add anything to our understanding of the proof or the result proved. Of course this situation is not found only in logic. In mathematical proofs in general, many steps are often skipped or sketched when it is clear to an expert reader how they could be carried out if it were necessary to do so. But in logic we come upon this kind of reasoning at a rather early stage in our studies, and applied in non-trivial cases. Also, because of such phenomena as $\omega$-incompleteness (to be dealt with later in the book), there are often some subtleties involved in understanding why a proof can be formalized in one theory but not in another.

In the later chapters of this book, there will be quite a lot of handwaving of this traditional kind, in connection with the step from an ordinary or informal proof of a mathematical statement to the conclusion that there is a formal derivation in some particular theory of a formalization of the statement. However, the earlier chapters present and illustrate methods that should allow a determined reader to replace the handwaving with detailed arguments.

## 1.3. Truth and provability

Any difficulties that a reader may have with the mathematical or logical reasoning in a book such as this can be cleared up, by wrestling with proofs and definitions and by consulting other sources. The philosophical reasoning is another matter. It's common for philosophical writers to complain that they fail to recognize their own opinions or arguments in the critical remarks of others, and for readers to find numerous unclear points in and possible interpretations of those opinions and arguments. The comments on truth and provability in this introductory section are intended to help clarify the main argument of the book and forestall some possible misunderstandings.

## Truth

Throughout the book, the exposition and argument will rely on the notion of *truth* (in the form of the predicate "is true") applied to mathematical (in particular arithmetical) statements. Thus for example it will be argued that the property of a theory $T$ that is chiefly of interest in connection with inexhaustibility is that the theory is *sound*, meaning that all theorems of $T$ are true.

A common reaction to philosophical arguments involving truth is to ask just what is meant by "true". Indeed this question is often asked even in a purely mathematical context, as when it is stated that Gödel's theorem proves, for a wide class of theories $T$, that if $T$ is consistent there is a true arithmetical statement $\phi_T$ not provable in $T$. The question then arises, if $\phi_T$ isn't provable in $T$, in what sense is it true? Is it provable in some other theory, and if so which one? Or is it true in the sense that it can be "seen to be true"? Or is it perhaps true in the sense of there being a correspondence between $\phi_T$ and some mathematical reality that is independent of our knowledge?

Underlying such questions is a tendency to think that whenever we speak of the *truth* of mathematical statements (whether these statements represent axioms or theorems or open questions), we are no longer talking about mathematical matters, but have entered a realm of epistemological or metaphysical argument or speculation. Mathematicians and others who shy away from such argument and speculation will therefore often surround "true" with scare quotes or avoid using the word altogether, preferring to speak about provability or derivability rather than truth.

This association of the concept of truth with philosophical issues is a natural one, since the question "What is mathematical truth?" is often used as a label for a set of fuzzily defined but distinctly philosophical questions about the nature of mathematics. But there is also a formal and mathematical use of "true" in logic, and it would be a mistake to assume that references to truth automatically take us into the realm of philosophy. In particular, in this book the reply to all questions about what is meant by "true arithmetical statement", in any mathematical or non-mathematical context, is that being true is a *mathematically* defined property of statements in the formal language of arithmetic. Thus the statement "$\phi_T$ is true" is itself a mathematical statement.

When we unravel the mathematical definition of "true arithmetical statement" we find that for any arithmetical statement $\phi$, "$\phi$ is a true arithmetical statement" is mathematically equivalent to $\phi$ itself. Similarly for references to the truth or falsity of other formalized mathematical statements.

For a reader who is unfamiliar with formal logic, it will probably be unclear at this point what a mathematical definition of "true arithmetical statement" might look like. The latter part of Chapter 7 shows how to define truth mathematically, after the necessary mathematical and logical concepts have been introduced. Basically, what such definitions amount to is a mathematical spelling out of the explanation that a statement is true if and only if things are as they are said to be in the statement. This was called "the semantic conception of truth" by Tarski, who was the first to introduce such mathematical definitions of truth (see Tarski [1944]). The essential point regarding this use of "true", as understood in this book, can be appreciated without knowledge of any of the technical details: "$\phi$ is true", said of an arithmetical statement $\phi$, is not a statement about what can be proved or known, and nor is it a statement about any correspondence between $\phi$ and a mathematical reality. It is a mathematical statement, mathematically equivalent to $\phi$ itself.

This stipulation concerning the meaning of "true" applied to arithmetical and other mathematical statements is basic to the presentation in the book. Note that it is only a stipulation. I am not arguing that this is what *should* be meant by "true" in a discussion of inexhaustibility, and I am not expressing any theory about mathematical truth or about anything else by using "true" in this way. In particular, the various philosophical problems and misgivings commonly associated with the concept of mathematical truth do not go away when we define "true" mathematically so as to make $\phi$ and "$\phi$ is true" equivalent. The mathematical definitions of "true statement" (for various formally defined classes of mathematical statements) to be given in later chapters will not serve to allay any doubts or misgivings that a reader may have concerning the meaning of mathematical statements or the justification of basic mathematical principles, since the definitions themselves freely use ordinary mathematics. Although there will be some discussion of the interpretation and justification of mathematical axioms in later chapters, there is no attempt to systematically introduce and justify mathematical or logical concepts and principles from a philosophical or foundational point of view. Rather, the argument of the book introduces and uses whatever parts of logic and mathematics are seen as relevant—arithmetic, inductive definitions and proofs, ordinals, functions, sets, and so on. Readers who regard some or all of the logical and mathematical apparatus invoked in the discussion as standing in need of further explanation or justification (either generally, or in the context of a philosophical discussion of this kind) will need to decide for themselves how this affects the significance of the conclusions put forward in the book.

## Provability

The question what "provable" means is less often asked than the corresponding question for "true", but is less easily answered. For the discussion in this book, it is essential to distinguish between what will be called formal provability and actual provability. These terms will next be explained.

Gödel's theorem and the extensions of that theorem considered in this book are about provability in axiomatic theories, that is to say about the *existence*, in the mathematical sense, of formal proofs in such theories. A *formal proof* of a statement $A$ in a theory $T$ is a mathematical structure, such as a tree or sequence, built up from statements in the formally defined language of the theory $T$ using axioms and rules of reasoning associated with $T$, and with a formalization of the statement $A$ as its designated conclusion. For emphasis, the term *formal provability* will sometimes be used when we are talking about a formalized statement being provable in a formal theory. Formal provability is always relative to some such theory.

That a formalization of a mathematical statement—say the statement "there are infinitely many twin primes"—is provable in a formal theory does not in itself imply that the statement can be *proved* in the ordinary mathematical sense, that is, that an argument establishing the statement as a mathematical theorem can be given. As an extreme instance, any statement is provable in a theory in which it is taken as an axiom, but this tells us nothing about whether or not the statement can be proved in the ordinary sense. Less trivially, even if the statement is provable in a formal theory embodying axioms and rules of reasoning recognized as correct in ordinary mathematics, it doesn't follow that the statement can be proved in any practical sense, simply because there is a limit on the length and complexity of proofs that human beings can produce or understand.

The term *actually provable* will be used for statements that can in fact be proved to the satisfaction of the mathematical community, that is, for which some actual argument or series of arguments establishing the statement as a theorem can be produced, studied, printed, and so on. Such arguments will be referred to as *actual proofs*. When we ask where the proof of some theorem first appeared, or ask if somebody has studied the proof, or if anybody has checked the whole proof, we are asking about the actual proof—the existing text, figures, reports of the results of computations, and so on, on the basis of which the theorem is accepted as such in the mathematical community.

While formal provability is a precise and indeed mathematically defined notion, (given some formal theory or class of formal theories), actual provability is neither precise, nor mathematically definable. Rather, it is a vague concept which depends on the consensus of some mathematical community and on the resources available to that community. The only thing that is clear about actual provability is that if a statement has in fact been proved, then it is actually provable. Even so, there are in some cases disagreements over whether a statement has been proved or not, either because of uncertainties regarding the correctness of steps (or computations) in the proof, or more fundamentally because of disagreement over which axioms or methods of reasoning are admissible in mathematical proofs.

There are two important connections between formal provability and actual provability. The first lies in the foundational role played by formal systems in mathematics. It is accepted in the mainstream mathematical community that if a statement has actually been proved, there exists a formal derivation of a formalization of that statement in some standard formal system such as the system of axiomatic set theory ZFC. This serves two useful purposes. First, it provides a standard for what we may call *local rigor* in proofs: any uncertainty about whether or not a particular step or argument in a proof is valid can be resolved by making that part of the proof more explicit and detailed, if necessary to the point where there is no doubt that it can be formalized in some particular formal theory whose axioms and rules of inference are accepted as correct. (Note that this does not mean that it would be at all helpful or feasible to replace the entire proof by a formal proof.) The second useful purpose is to allow people to set aside, in a purely mathematical context, any disagreements over what axioms or methods of reasoning are admissible. For example, a published proof of a particular theorem can be recognized as having established that (a formalization of) the theorem is provable in ZFC, even while it remains an open question just how far that proof can be simplified and whether the theorem or its proof makes sense from some particular foundational point of view (such as finitism or intuitionism).

The second significant connection between formal and actual provability consists in the existence of *negative* results about formal provability with implications for actual provability. If it is known that there is no formal proof in a theory $T$ of a statement $A$, and if the rules and axioms of $T$ include formalized versions of some particular rules of mathematical reasoning, we can conclude that there can be no actual proof of $A$ that uses only those rules. In particular, if $A$ is known to be unprovable in a formal system such as ZFC, it can be concluded that $A$ cannot be proved using the methods of "ordinary mathematics", since those methods are known to be formalizable in ZFC. Thus most mathematicians will cease to regard it as a mathematical problem to prove or disprove a statement $A$ if $A$ is shown to be undecidable in ZFC. Of course this does not mean that $A$ has been shown to be undecidable in any absolute sense. On the contrary, it is perfectly possible that "ordinary mathematics" will in the future come to encompass axioms or rules that do make it possible to prove or disprove $A$. But finding such new rules and axioms is not part of ordinary mathematical activity, and it remains a (mathematically) highly significant fact that $A$ cannot be settled by mathematical proof as now understood.

The comments above were made in terms of *known* negative results about formal provability. If we know that $A$ is not provable in ZFC, we know that $A$ cannot be proved by ordinary mathematical reasoning. But of course we can also say that whether or not we know that $A$ is not provable in ZFC, if it is in fact not provable in ZFC, then it cannot in fact be proved by ordinary mathematical reasoning, even though we may never come to realize this. Thus it is sometimes suggested that the reason why some famous mathematical conjecture has so far withstood all attempts to prove or disprove it may be that it is in fact undecidable in ZFC. Although such suggestions usually have nothing to support them, they cannot be dismissed on general grounds.

Now consider *positive* results about formal provability. If we *know* that $A$ is provable in some theory $T$ whose rules and axioms we accept as valid, we will not necessarily thereby be able to produce an actual proof of $A$ that can be formalized in $T$ (for reasons already touched on, and elaborated below). We will however be able to prove $A$ on the basis of the knowledge that it is provable in $T$ by invoking the principle that every statement provable in $T$ is true. This is an application of a *reflection principle* for $T$ to produce an actual proof of a theorem which is formally provable in $T$ without the reflection principle. We may come to know that $A$ is provable in $T$ for example by using computers to search for a formal proof of $A$ in $T$. The formal proof found by the computer may be too long or complex to be at all intelligible to us, but by using the reflection principle we can conclude that $A$ is true. Such arguments, which include reports of the results of computations that go far beyond what has traditionally been used in mathematical proofs, are actual proofs in the sense defined, although to some extent controversial, since proofs that do not appeal to the results of computations that cannot be carried out by hand are generally considered more satisfactory. But reflection principles for $T$ can also be used to prove theorems that are *not* provable in $T$ itself—in particular, every reflection principle for $T$ has the consequence that $T$ is consistent—and for this reason they will play a central role in later chapters.

While the mere fact of $A$ being unprovable in ZFC had consequences for actual provability (whether or not we are able to establish that $A$ is unprovable in ZFC), the mere fact of a statement $A$ being provable in ZFC, or in any theory $T$ whose rules and axioms we recognize as valid, does not imply that $A$ is actually provable. This is most simply illustrated by numerical statements of the form "$p$ is a prime", "$p$ has exactly two prime factors", and so on. All true statements of this form are formally provable using the ordinary rules of arithmetic, and indeed in a logically very weak formal theory, but as far as actual mathematics is concerned, there is no guarantee whatever, for a given such statement with $p$ a large integer expressed in some standard notation, that it is possible to decide whether the statement is true or false. Specialists in factoring techniques may or may not succeed in doing so, using various mathematical ideas and results in combination with computer calculations.

Once we have established the truth of a statement of this kind, by whatever means, we know that it is provable in a logically weak axiomatic theory $T$, but this does not mean that we could in fact have used only $T$ to arrive at the truth of the statement.

The main significance of these distinctions in the present context is that the questions and arguments in this book concern only formal provability, not actual provability. (Thus in particular there will be no discussion of the use of reflection principles to make possible or simplify actual proofs; for this topic, see Harrison [1995].) More precisely, they concern formal provability in arithmetic and certain extensions of arithmetic by axioms that we recognize as valid. Thus one may say that in speaking of provability in a discussion of this kind, we are speaking about what truths are implicit in axioms and rules that we recognize as valid, rather than about methods by which we actually prove or could prove mathematical theorems. The apparent inexhaustibility of our mathematical knowledge, as brought out by Gödel, is based on the discovery that the truth of certain mathematical statements ("the theory $T$ is consistent") is *not* formally implicit in the axioms and rules of a theory, even though it is implicit in the theory being valid. Our accepting the theory as valid means that we are equally justified in accepting the formally undecidable statement as true, and so we are faced with the questions outlined in the introductory section.

# CHAPTER 2

# ARITHMETICAL PRELIMINARIES

## 2.1. Numbers

Counting things is an extraordinarily useful and important part of our everyday thinking. Suppose you are taking the ducklings Quacky, Beaky, Tailfeather, Downhead, and Webfoot on an outing. At several points—getting off the bus, distributing worms, returning home—you're anxious to make sure that none of the ducklings has slipped away or been inadvertently left behind. You can do this by looking for and locating each of the ducklings, or by having them respond when you call out their names. However, if you don't know the ducklings personally and don't have a list of their names it may be quite difficult to keep track of them this way. You can then adopt the expedient of simply counting the ducklings at each check, making sure there are five of them. This way, you don't need to look for Quacky, Beaky, etc. individually, and it doesn't matter if you don't know them or their names.

There is a price to pay for this greater convenience. Counting the ducklings means abstracting from, or in other words disregarding, their particular identities. The only information noted is their number. Thus it is quite compatible with there being at all times five ducklings in your charge that Quacky has absconded and his place been taken by another duckling. If, as is to be assumed, such an event is undesirable, counting will replace individual inspection only when the risk that a substitution will occur is thought to be negligible.

In other cases, it doesn't matter if a substitution takes place, as long as the item substituted is recognizably of the proper kind. This is the case since ancient times when people count coins, sacks of grain or other commercial items, and when generals count soldiers.

The great usefulness of counting amply justifies and explains the occurrence in language of names and notations for the counting numbers 1, 2, 3, ... and the very concept of "number" associated with them. "Three soldiers", "three sacks of grain", "three days", are recognized as having in common that there are three of them. When counting, the items counted are tallied against the names of the numbers in a memorized order, and thus a total is arrived at.

These totals can be checked and compared with other totals, giving useful information of various kinds.

Counting soldiers, coins, and sacks of grain quickly leads to adding and multiplying the numbers counted. The practical utility of addition and multiplication, and of their associated inverse operations of subtraction and division, is easily appreciated. These operations are defined as operations on *numbers*, that is, in abstraction from what is actually being counted. When we learn to add and multiply, to subtract and divide, we carry out these operations using special symbols or particular objects, and accept the methods and results as applicable to whatever objects we may later wish to count.

The use of numbers, then, may reasonably be assumed to go back to the dawn of human language, and can be understood in terms of the great utility of numbers. But the realm of numbers and the use of numbers have also been greatly extended and expanded, through considerations turning partly on utility, but partly on internal mathematical factors.

Thus in algebra a further step is taken in that individual numbers are also abstracted from, and rules and computations are introduced that operate on "symbolic quantities" and apply to any specific numbers that may be chosen. Equations such as

$$(a+b)^2 = a^2 + 2ab + b^2$$

are justified as valid for any numbers $a, b$ through symbolic computation, and we are naturally led to such generalizations as the binomial theorem, which gives a similar expansion of $(a+b)^n$ for numbers $n > 2$.

Indeed once we start pushing symbols around, the formalism seems to take on a life of its own. To solve the equation $b + x = a$ we can subtract $b$ from both sides of the equation to get $x = a - b$. Of course if we are thinking about positive integers, we must assume that $a > b$, so that the subtraction can be carried out. If it can't, the equation has no solution. But why not just subtract $b$ from $a$ whatever $a$ and $b$ may be, and introduce a number $x$ as a solution of the equation $b + x = a$ if it didn't exist before? Carrying out this scheme, we come up with such numbers as $0 = 5 - 5$, $-2 = 8 - 10$. As we become used to these negative numbers, we find numerous applications of them, for example in keeping track of debts as well as assets.

And so it goes on, as mathematics develops in an intricate interplay between the needs of practical activities like counting and surveying, the needs of scientific theory, and the forces of mathematical generalization. At least for the past few thousand years, an important part of this process has been the interest aroused in people by numbers and their properties for their own sake, regardless of any practical utility they may have. In particular, Western mathematics, science, and philosophy have been greatly influenced by the mathematical work in geometry and arithmetic compiled in Euclid's *Elements* (ca. 300 BC). As will emerge, much of the arithmetic actually used in this book can be found in the *Elements*.

In this chapter, which introduces some mathematical, logical and philosophical concepts and results that will be used or referred to in the remainder of the book, we will be concerned with the *integers*, the numbers $0, 1, -1, 2, -2, \ldots$. The positive integers are $1, 2, 3, \ldots$ and the non-negative integers or *natural numbers* are $0, 1, 2, 3 \ldots$.

## 2.2. Mathematical induction

The proofs of basic theorems about the integers combine algebraic manipulations with the use of the *principle of mathematical induction*: whatever is true of 0, and if true of any natural number $n$ is also true of $n + 1$, is true of all natural numbers. That the assertion is true of 0 is known as the *base case*, and the assumption that it is true of $n$, which we use as a hypothesis in showing that it is true of $n + 1$, is the *induction hypothesis*. A variant principle uses the ordering relation between the natural numbers and is called the principle of *well-founded induction* (or *course-of-values induction*): whatever is true of a natural number $n$ whenever it is true of every natural number strictly smaller than $n$, is true of all natural numbers. Here the induction hypothesis is the assumption that the assertion is true of every natural number strictly smaller than $n$, and there is no base case. Yet another principle is the *smallest number principle*: if there is a natural number having a certain property, then there is a *smallest natural number* having that property.

These three principles are equivalent, in the sense that we can deduce the others given one of them. To show this, let's assume the principle of mathematical induction and deduce the principle of well-founded induction. Suppose that $F(n)$ holds for $n$ whenever $F(m)$ holds for every $m < n$. We prove by mathematical induction that every natural number $n$ has the following property $P(n)$: for every natural number $k < n$, $F(k)$. Clearly if $P(n)$ holds for every $n$, then since $n < n + 1$ and $P(n + 1)$, it follows that $F(n)$ holds for every $n$. $P(0)$ holds vacuously (meaning that since there is no natural number $< 0$, anything is true of "every natural number $< 0$"). Assuming $P(n)$ we get $P(n + 1)$, since any number $k$ which is smaller than $n + 1$ is either $n$, in which case $F(k)$ holds since $F(m)$ holds for every $m < n$, or smaller than $n$, in which $F(k)$ holds by the assumption $P(n)$. So the principle of mathematical induction implies the principle of well-founded induction.

The argument above illustrates a common feature of proofs by induction: in order to prove by induction that every natural number has some property $P$, it is often necessary to replace $P$ by some property $P'$, often a stronger property, which makes it easier to draw conclusions from the induction hypothesis. Choosing the right property is often the crucial step in a proof by induction.

The principle of well-founded induction in turn implies the smallest number principle, as we see by proving by well-founded induction on $n$ that if there is a $k < n$ such that $P(k)$, then there is a smallest natural number $k$ such that $P(k)$.

The details of this proof are left to the reader. Finally, the smallest number principle implies the principle of mathematical induction: suppose $F(0)$ is true and furthermore $F(n+1)$ is true whenever $F(n)$ is. If there is a natural number $k$ such that $F(k)$ is not true, let $m$ be the smallest such. $m$ is not 0, so it must be $n+1$ for some $n$. But since $n < m$ and $m$ is the smallest $k$ such that $F(k)$ is not true, $F(n)$ must be true, so $F(m)$ is also true by our assumption. This is a contradiction, so there can be no $k$ such that $F(k)$ is not true.

The arguments just given use without explicit mention some properties of the ordering relation between natural numbers, and they also use several general logical principles of reasoning. In later chapters we will see how to formulate explicit axioms and rules for arithmetic and logical reasoning that allow us to *formalize* the arguments above, that is to formulate them as formal derivations in axiomatic theories. Such a set of rules also gives an unambiguous sense to the statement that the three principles are equivalent. As long as we don't specify just what we are allowed to use for example in deducing the principle of well-founded induction from the principle of mathematical induction, there is some uncertainty as to what is at issue. In a logical context, when we speak of two axioms or basic principles as equivalent, we usually have in mind some background theory within which neither axiom or principle is provable, but each can be derived from the other.

In proving by induction that some statement involving several variables is true for all values of those variables, we often have a choice which variable to use in the induction argument. We then say that the proof is by induction *on* that variable. Thus to prove that $A(n,k,m)$ holds for all natural numbers $n,k,m$ by induction on $k$ means to prove by induction that every $k$ satisfies the condition "for every $n$ and every $m$, $A(n,k,m)$".

Another common phrase in induction proofs is "by induction on $f(x)$", where $f$ is some function with natural numbers as values. In proving by induction on $f(x)$ that every $x$ in some set $M$ satisfies a condition $A(x)$, the induction hypothesis is that $A(y)$ holds for every $y$ such that $f(y) < f(x)$. This in fact is a special case of well-founded induction, as can be seen by casting the argument as a proof by well-founded induction of the statement "for every $k$, $B(k)$", where $B(k)$ is the condition "for every $x$ in $M$, if $f(x) = k$ then $A(x)$". Another way of looking at induction on $f(x)$ is to justify it using the smallest number principle: if there is an $x$ such that $A(x)$ does not hold, there is a smallest number $f(x)$ such that $A(x)$ is false.

## 2.3. Division and Euclid's algorithm

### Division

For any natural number $k$ and positive integer $m$, there is a unique *quotient* $q$ and a similarly unique *remainder* $r$ obtained by dividing $k$ by $m$. $q$ and $r$ are defined by the equation $k = qm + r$ and the condition that $r$ is greater than or equal to 0, but strictly smaller than $m$. If $r = 0$, $k$ is (evenly) *divisible* by $m$, and we write $m \mid k$ ("$m$ divides $k$" or "$m$ is a divisor of $k$" or "$k$ is a multiple of $m$"). That the quotient and remainder are unique means that given $k$ and $m$ there is exactly one pair $q, r$ satisfying the stated conditions.

For negative $k$, $-k$ is positive, and by the above we have $-k = qm + r$ for some non-negative $q$ and $r$, so that $k = -qm - r$. In the case when $r > 0$, the remainder when dividing the negative number $k$ by $m$ may be taken to be the negative $-r$, or the non-negative $m - r$. In the former case, the quotient will be $-q$, in the latter case it will be $-(q + 1)$. Here the remainder will always be taken to be non-negative.

That the quotient and the remainder *exist* ("there is a unique quotient") would perhaps by most people who are not mathematicians automatically be taken to imply or mean that they can be *computed*, given the numbers $k$ and $m$. And indeed, through familiarity with long-hand division or electronic calculators, we know that this is the case. In general, though, existence and computability are not necessarily the same in what is known as "classical" arithmetic, where a "pure existence theorem" may assert the existence of a number $k$ satisfying some condition $P(k)$ without giving any means of computing such a $k$.

If we disregard—as we will consistently do in this book—all questions of efficiency or practicality and consider long division from a theoretical point of view, we see that for positive $k$, the quotient and remainder when dividing $k$ by $m$ can be obtained by subtracting $m$ from $k$ if $m \leq k$, then again subtracting $m$ from the result, repeating this until the number that remains is smaller than $m$. This remaining number is the remainder $r$, and the quotient $q$ counts how many times $m$ was subtracted.

It is only in a purely mathematical context that one would set out to *prove* the existence of a unique quotient and remainder. For the uniqueness part, we use algebraic manipulation: if $k = qm + r$ and $k = q'm + r'$, we get $m(q - q') = r' - r$, where we may assume $r \leq r'$. Since $r'$ and $r$ are both greater than or equal to 0 and strictly smaller than $m$, the same is true of $r' - r$. Since $m$ and $r' - r$ are non-negative, $q - q'$ is non-negative, but must then be 0, since otherwise $m(q - q') \geq m$. So $r = r'$, but then $q = q'$ also follows.

For the existence part, we use well-founded induction. Suppose, for given $m$ and $k$, that the quotient and remainder exist for every $k' < k$. If $k = m$, the quotient is one and the remainder 0. If $k < m$, the quotient is 0 and the remainder $k$. If $k > m$, let $k' = k - m$. By the induction hypothesis (since $k' < k$), there are $q'$ and $r'$ with $k' = mq' + r'$, so $k = k' + m = m(q' + 1) + r'$.

Another way of looking at this argument is to see it as proving by induction that the procedure of repeatedly subtracting $m$ described above will eventually terminate. Let's consider another computational procedure from this point of view.

## Euclid's algorithm

For any two positive integers $k$ and $m$ there is a *greatest common divisor* (g.c.d.) of $k$ and $m$, which we denote by $(k,m)$. This is the greatest positive integer which divides both $k$ and $m$. Thus for example $(39,26)$ is 13, while $(39,25)$ is 1. In school we used greatest common divisors to reduce fractions: the fraction 39/26 reduces to 3/2, while the fraction 39/25 is irreducible.

Euclid's *Elements* presents an algorithm for computing $(k,m)$. It goes as follows:

> We start with the two numbers $k$ and $m$. If $k$ and $m$ are not equal, we replace the larger of the two numbers with the result of subtracting the smaller number from the larger. We continue in this way until we have equal numbers, and that one number is then the g.c.d. of the original two numbers.

Some readers may be more familiar with a computationally more efficient version of Euclid's algorithm which uses division instead of subtraction. When division in turn is described in terms of repeated subtraction we get the above formulation of the algorithm.

Actually, $(k,m)$ also exists if $k$ or $m$ is 0—since every number divides 0, $(k,0)$ is $k$. The algorithm formulated above does not work if we allow $m = 0$, since we will then just subtract 0 from $m$ forever. A somewhat different formulation covers this case as well:

> We start with the two numbers $k$ and $m$. If neither of $k$ and $m$ is 0, we replace the larger of the two numbers with the result of subtracting the smaller number from the larger, or replace one of the numbers with 0 if they are equal. We continue in this way until at least one of the numbers is 0. The other number is then the g.c.d. of the original two numbers.

Another way of explaining Euclid's algorithm (in its first version), familiar to users of programming languages, is the following:

$$\begin{aligned}&\text{If } k = m \text{ then } (k,m) = k;\\ &\text{otherwise, if } k > m \text{ then } (k,m) = (k-m,m);\\ &\text{otherwise } (k,m) = (k,m-k).\end{aligned}$$

To say that the formulations above describe an *algorithm* for computing the g.c.d. is to say that they give a set of explicit instructions for how to proceed in the computation, when to stop, and what to present as the result of the computation. A function is *computable* if and only if there is at least one algorithm for computing it. When we formulate algorithms in ordinary language, "explicit" is always a matter of degree, and each of the three formulations above might be more or less effective as a set of instructions, some people considering them clear and unambiguous, others finding them all but unintelligible.

In mathematics and computing there are various ways of formally defining algorithms, all of them leading to the same class of computable functions. One way of formally defining the computable functions will be introduced in Chapter 9.

An algorithm which is to produce any result must eventually terminate, that is, carrying out the instructions must eventually lead to a situation in which we are instructed to stop. (This is sometimes taken to be part of what is meant by an algorithm.) Usually, when we present an algorithm for solving a certain problem—adding two numbers, finding the greatest common divisor of two numbers, sorting a list of words alphabetically, and so on—there are two things we need to verify: that the algorithm terminates, and that any result that it may yield is in fact a solution of the problem, or of a given instance of the problem. In programming, the two conditions together are referred to as "correctness", while the second condition, somewhat confusingly, is known as "partial correctness".

To see that any final result yielded by Euclid's algorithm is in fact the g.c.d. of $k$ and $m$ we need to show that

$$(k,m) = (m,k), \text{ and if } k = m \text{ then } (k,m) = k, \text{ and if } k > m \text{ then } (k,m) = (k-m,m)$$

The first two parts of this assertion are obvious, and the third part is true because if $a \mid k$ and $a \mid m$, then $a \mid k - m$, and conversely if $a \mid k - m$ and $a \mid m$ then $a \mid k$.

Why do the above observations establish that any final result yielded by Euclid's algorithm is in fact the g.c.d. of $k$ and $m$? Implicit in the presentation of the observations is a proof by mathematical induction on $n$ that if Euclid's algorithm terminates after $n$ steps, then the result returned by the algorithm is the g.c.d. of $k$ and $m$. The observation that $(k,k) = k$ is used in the base case, and the remaining observations in the inductive step.

We also see that the algorithm must eventually yield a result, since at each step one of the two numbers involved is decreased while the other is unchanged. This means that the sum of the two numbers keeps getting smaller, so the procedure must stop, by the smallest number principle applied to the set of sums of the numbers in a pair.

Let's take a closer look at this argument. We write $\langle k,m\rangle$ for the ordered pair consisting of the two numbers $k$ and $m$. If $\langle k,m\rangle$ and $\langle k',m'\rangle$ are successive pairs of numbers in an application of Euclid's algorithm, the following relation holds between the two pairs:

$$k' = k \text{ and } m' = m - k, \text{ or } k' = k - m \text{ and } m' = m$$

We write $R(\langle k',m'\rangle, \langle k,m\rangle)$ to express the above relation. That Euclid's algorithm terminates for every pair $\langle k,m\rangle$ of positive integers follows if we prove that there is no infinite sequence $\langle k_0,m_0\rangle, \langle k_1,m_1\rangle, \ldots$ of pairs of positive integers such that $R(\langle k_{i+1}, m_{i+1}\rangle, \langle k_i, m_i\rangle)$ holds for every $i$. The argument suggested above is the following. Given any such sequence, finite or infinite, let $M$ be the set of all sums of the form $k_i + m_i$. This set has a smallest member $n$, which is the sum of the numbers in the pair $\langle k_i, m_i\rangle$ for some $i$. But then $\langle k_i, m_i\rangle$ must be the last element of the series, for if $R(\langle k_{i+1}, m_{i+1}\rangle, \langle k_i, m_i\rangle)$, the sum $k_{i+1} + m_{i+1}$ is smaller than $n$, which is impossible.

While this argument is perfectly correct, one may wonder why it should be necessary to speak of infinite sequences at all. And indeed we can formulate instead the assertion that Euclid's algorithm terminates for every $k, m$ as the statement "For every pair $\langle k, m\rangle$ of positive integers, there is a finite sequence $\langle k_0, m_0\rangle, \langle k_1, m_1\rangle, \ldots$ beginning with $\langle k, m\rangle$ and ending with a pair $\langle k_n, m_n\rangle$ such that $k_n = m_n$, where every pair $\langle k_{i+1}, m_{i+1}\rangle$ is obtained from the preceding pair $\langle k_i, m_i\rangle$ as in Euclid's algorithm." To prove this, instead of invoking the smallest number principle, we show by well-founded induction on $n$ that for every pair $\langle k, m\rangle$ such that $k + m = n$, a finite sequence of the kind indicated exists. For if $\langle k, m\rangle$ is any pair such that $k + m = n$, either $k = m$, in which case we have a finite sequence of length 1, or an application of Euclid's algorithm yields a next pair $\langle k', m'\rangle$ such that $R(\langle k', m'\rangle, \langle k, m\rangle)$, so that a finite sequence starting from $\langle k', m'\rangle$ exists by the induction hypothesis.

We can look at the argument from yet another point of view, as a proof by well-founded induction on the relation $R$. This point of view is prominent in the part of logic known as proof theory, and will play an important role in later chapters, so it merits some further comments.

## Induction on well-founded relations

The relation $R(\langle k', m'\rangle, \langle k, m\rangle)$ between ordered pairs of positive integers defined by

$$k' = k \text{ and } m' = m - k, \text{ or } k' = k - m \text{ and } m' = m$$

has the property of being *well-founded*. This means that there is no infinite sequence $\langle k_0, m_0\rangle, \langle k_1, m_1\rangle, \ldots$, called an *infinite descending sequence* for $R$, of pairs of positive integers such that $R(\langle k_{i+1}, m_{i+1}\rangle, \langle k_i, m_i\rangle)$ holds for every $i$. In this case $R$ is a relation between pairs of positive integers, but in general a well-founded relation $R$ can be a relation between objects of any kind.

That a relation $R$ is well-founded entails that a principle of *well-founded induction* applies with respect to that relation: if $F(a)$ is true for every $a$ such that $F(b)$ is true for every $b$ for which $R(b, a)$, then $F(a)$ is true for every $a$. For if there is some $a_0$ for which $F(a_0)$ is not true, it follows that there is an $a_1$ such that $R(a_1, a_0)$ and $F(a_1)$ is not true. Continuing, we get an infinite descending sequence for $R$. The ordinary principle of well-founded induction over the natural numbers is a special case of this, with the usual ordering relation $<$ between numbers as the relation $R$. The relation $<$ is indeed well-founded as a relation between natural numbers, although not of course as a relation between integers in general.

Conversely, if the principle of well-founded induction holds for a relation $R$, then $R$ is well-founded, since it follows by well-founded induction on $a$ that there is no infinite descending sequence beginning with $a$.

As in the case of the ordering of the natural numbers, the principle of well-founded induction for an arbitrary well-founded relation $R$ can equivalently be formulated as a minimality principle: if $R$ is a well-founded relation on a set $M$, and $K$ is any nonempty subset of $M$, there is an $a$ in $K$ such that $b$ is not in $K$ for any $b$ such that $R(b, a)$. If $R$ is the order relation between natural numbers, this is the same as the smallest number principle.

The designation "minimality principle" does not imply that $R$ is necessarily an *order relation*. Since orderings will appear in many contexts in the book, it is as well to introduce a cluster of terms associated with them. A relation $R$ between elements in some set $D$ is called a *reflexive partial order* if it is *transitive*—whenever $R(a,b)$ and $R(b,c)$ hold, $R(a,c)$ holds—and *antisymmetric*—if $R(a,b)$ and $R(b,a)$ then $a = b$—and *reflexive*—$R(a,a)$ holds for every $a$ in $D$. If $R$ is instead *irreflexive*—meaning that $R(a,a)$ never holds—it is called a *strict* partial order. (In the case of strict partial orders we don't need to require antisymmetry, since by irreflexivity and transitivity, it is never true that $R(a,b)$ and $R(b,a)$.) $R$ is a *total* or *linear* order (whether strict or reflexive) if for every $a,b$ in $D$, $R(a,b)$ or $R(b,a)$ or $a = b$. Finally, $R$ is called a *well-ordering* if $R$ is a well-founded total order. In these terms, then, a well-founded relation need not be a partial order, and indeed the well-founded relation $R$ associated with Euclid's algorithm is not transitive. It is however irreflexive, as every well-founded relation must be, since if $R(a,a)$, there is an infinite descending sequence $a, a, a, \ldots$.

The ordering relation $<$ between integers is an example of a strict total order which is not a well-ordering, but it becomes a well-ordering when restricted to the natural numbers. The relation "$A$ is a subset of $B$" between sets of natural numbers is an example of a reflexive partial order which is not total. The *lexicographic* ordering of pairs of natural numbers is defined by

$$R(\langle k', m'\rangle, \langle k, m\rangle) \text{ if and only if } k' < k, \text{ or } k' = k \text{ and } m' < m$$

This is a strict total order, since a comparison, first between $k'$ and $k$ and then, if $k = k'$, between $m'$ and $m$, will always put one of two different pairs before the other. By an argument to be given below, it follows that this ordering is in fact a well-ordering.

The principle of well-founded induction for $<$ can be proved, as we have seen, using ordinary mathematical induction. In the case of the particular relation $R$ associated with Euclid's algorithm, we have seen that the principle of well-founded induction can be proved for this relation too using ordinary mathematical induction. The essential step in establishing this was to note that on the assumption $R(\langle k_{i+1}, m_{i+1}\rangle, \langle k_i, m_i\rangle)$, it follows that the sum $k_{i+1} + m_{i+1}$ is smaller than the sum $k_i + m_i$, so that no infinite descent is possible.

Another way of describing the fact that the principle of well-founded induction for $R$ can be proved using ordinary mathematical induction is to say that any proof carried out using well-founded induction on $R$ can instead be carried out using ordinary mathematical induction.

The relation $R$ above made its appearance in connection with Euclid's algorithm for finding the greatest common divisor of two positive integers. Now let's consider a different algorithm, operating on a pair of natural numbers $\langle k, m\rangle$:

If $k = m = 0$ then the algorithm terminates with value 0;

otherwise, if $k > 0$ we replace the pair $\langle k, m\rangle$ with the pair $\langle k-1, m\rangle$;

otherwise we replace the pair $\langle k, m\rangle$ with the pair $\langle m^m, m-1\rangle$.

Obviously this algorithm, if it computes any value at all, computes the value 0, in a highly roundabout way. If we try out the algorithm in practice we find that it seems to require a great many steps before it terminates, in most cases. For example, if the initial pair is 5, 10, it will take almost two billion steps before the algorithm terminates. That it does terminate for every pair follows by well-founded induction on the relation $S$ between pairs $\langle k', m'\rangle$ and $\langle k, m\rangle$ defined by

$$m' = m \text{ and } k' < k, \text{ or } m' < m$$

$S$, it will be noted, is like the lexicographic ordering defined above, except that the roles of the first and second elements are reversed. It is therefore known as the reverse lexicographic ordering of pairs. Since $S(\langle k', m'\rangle, \langle k, m\rangle)$ holds for successive pairs in the algorithm, it follows that the algorithm always terminates provided $S$ is well-founded.

That $S$ is well-founded can again be proved using ordinary induction. Equivalently, any proof by well-founded induction on $S$ can be carried through instead using ordinary well-founded induction. For assume $F(\langle k, m\rangle)$ holds for every pair $\langle k, m\rangle$ such that $F(\langle k', m'\rangle)$ holds for every $\langle k', m'\rangle$ with $S(\langle k', m'\rangle, \langle k, m\rangle)$. We prove by ordinary well-founded induction on $k$ that every $k$ has the following property $P(k)$: for every $m$, $F(\langle k, m\rangle)$. So assume that $P(k')$ holds for every $k' < k$. To show that for every $m$, $F(\langle k, m\rangle)$, we use ordinary well-founded induction on $m$. Assume $F(\langle k, m'\rangle)$ holds for every $m' < m$. Then $F(\langle k', m'\rangle)$ holds for every $\langle k', m'\rangle$ where $k = k'$ and $m' < m$ (by the "inner" induction hypothesis), and also for every $\langle k', m'\rangle$ with $k < k'$ (by the "outer" induction hypothesis), and therefore (by the assumption on $F$) it follows that $F(\langle k, m\rangle)$ for every $m$, or in other words $P(k)$. (The proof also shows that the lexicographic ordering is well-founded.)

The inductive argument showing that $S$ is well-founded appears more complicated than the inductive argument showing that the relation $R$ associated with Euclid's algorithm is well-founded, in that the former argument but not the latter uses a *nested* induction, meaning an inductive argument within an inductive argument. This appearance is not misleading. Indeed, in proof theory, what is called the proof-theoretic strength of theories is in a sense measured by which well-founded relations they can prove to be well-founded. In the context of this book, the extent to which arithmetical theories are able to prove relations to be well-founded will feature in connection with constructive ordinal notations and autonomous progressions of theories, topics to be introduced later.

## Set theory and arithmetic

The argument above showing that well-founded induction holds for any well-founded relation uses what is known as the *axiom of dependent choice*: "If for every $x$ in $M$ there is a $y$ in $M$ such that $S(x, y)$, then there is an infinite sequence $x_0, x_1, \ldots$ of elements in $M$ such that $S(x_i, x_{i+1})$ for every $i$." (The argument applies this principle to the *converse* of $R$, that is, the relation $S$ for which $S(x, y)$ holds if and only if $R(y, x)$, and the set $M$ of $x$ such that $F(x)$ does not hold.) The reason why we need to invoke such a principle in the argument is that we have no way of defining, when we are talking about relations in general, any particular $a_{i+1}$ such that $R(a_{i+1}, a_i)$, given the assumption that well-founded induction does not hold for $R$. However, if we are talking about relations defined, say, on the set of pairs of natural numbers, we don't need to invoke the axiom of dependent choice, since given that for every $x$ in $M$ there is a $y$ in $M$ such that $S(x, y)$, we can explicitly define a sequence $x_0, x_1, \ldots$ by saying that $x_{i+1}$ is the first pair in the lexicographic order for which $S(x_i, x_{i+1})$ holds. This situation is not untypical: in arithmetical reasoning, we have a use for general set-theoretical principles, but we can in many cases more or less easily replace the use of such principles by more specific arithmetical arguments. In fact, in the particular case of the axiom of dependent choice, we know (from work in logic) that its use can always be eliminated from proofs of arithmetical theorems. That other uses of set theory in arriving at arithmetical conclusions cannot always be replaced by purely arithmetical reasoning will be seen to follow from the incompleteness theorem.

The explanation given above of why a special axiom is needed to prove that well-founded induction applies to any well-founded relation is in a way misleading. Suppose we wish to prove that for any $R$, and any $n$, if well-founded induction does not hold for $R$, there is a descending sequence $a_0, a_1, \ldots, a_{n-1}$ of length $n$. It may seem that we will still need to invoke some special choice principle, since it is still true that we cannot in the general argument specify any particular $a_{i+1}$ such that $R(a_{i+1}, a_i)$. In fact we do not need to invoke any special principle, since this argument can be handled by mathematical induction and general logical reasoning. It is only when an infinite number of choices must be made that we may need some special set-theoretical principle. This distinction will be clarified in Chapter 8, in connection with the formal rules of logical reasoning.

## Euclid's lemma

Returning to traditional arithmetic, we can characterize $(k, m)$ in a different way. Let $I$ be the set of integers obtainable by adding a multiple of $k$ and a multiple of $m$. That is, $I$ contains every integer $kx + my$, where $x$ and $y$ are any integers.

$I$ has the following properties:

(1) If $a$ and $b$ are in $I$, then $a + b$ is also in $I$.
(2) If $a$ is in $I$, then $ax$ is in $I$, for any integer $x$.

We verify the properties (1) and (2) by noting that

$$(kx + my) + (kx_1 + my_1) = k(x + x_1) + m(y + y_1)$$

and

$$(kx + my)x_1 = kxx_1 + myx_1$$

By the smallest number principle, $I$ contains a smallest positive integer, $a$. This $a$ must in fact divide every member of $I$. For if $b$ is in $I$ and $b = ax + r$, the remainder $r$ also belongs to $I$ by (1) and (2), and so cannot be positive and smaller than $a$, but must be 0.

We next note that $(k, m)$ also belongs to $I$. For in Euclid's algorithm, we start with two numbers ($k$ and $m$) that belong to $I$, and the subtraction operation always yields another member of $I$ by (1) and (2). Also, $(k, m)$ divides both $k$ and $m$ and therefore divides every number $kx + my$ in $I$.

Together these two observations imply an assertion that we will refer to as

> **Euclid's lemma:** $(k, m)$ is the smallest positive number in $I$. It follows that $(k, m) = 1$ if and only if there are integers $x$ and $y$ such that $kx + my = 1$. (The "if" part of this assertion is Proposition 1 of Book VII of the *Elements*.)

Euclid's lemma can be used to show that $(k, m)$ can be characterized as the number $n$ which divides both $k$ and $m$, and is a multiple of every number that divides both $k$ and $m$. This characterization is used to define "greatest common divisor" in other mathematical contexts, for example when talking about polynomial division.

## 2.4. Primes and congruence classes

If $(k, m) = 1$, where $k$ and $m$ are positive integers, we say that $k$ and $m$ are *relatively prime*. If $p$ is an integer greater than 1 such that $(p, m) = 1$ for every positive integer $m$ smaller than $p$, we say that $p$ is a *prime number*.

An implication of two numbers being relatively prime that we will use in several later arguments is expressed in the

**Division lemma:** if $k$ divides $ab$, and $k$ and $a$ are relatively prime, then $k$ divides $b$.

The division lemma follows since by Euclid's lemma there are integers $x$ and $y$ such that $kx + ay = 1$, which implies that $bkx + bay = b$, and since $k$ divides both $ba$ and $bkx$, $k$ divides $b$.

### Congruence classes

Suppose $k$ and $n$ are relatively prime. Then, as we have seen, there are integers $x$ and $y$ such that $kx + ny = 1$. (Since $k$ and $n$ are positive, one of $x$ and $y$ will be negative.) Another way of putting this is to say that there is an $x$ such that $kx - 1$ is divisible by $n$. In other words, $kx$ leaves the remainder 1 when divided by $n$, and we say that $kx$ is *congruent to 1 modulo n*, written $kx \equiv 1(n)$.

More generally, $a$ and $b$ are congruent modulo $n$, where $n$ is a positive integer, if $n$ divides the difference $a - b$. By the division algorithm, for every $a$ and positive $n$ there is a unique number $r$ such that $a$ is congruent to $r$ modulo $n$ and $r$ satisfies the further condition $0 \leq r < n$.

Congruence modulo $n$ is an *equivalence relation*, which is to say that it is reflexive and transitive and *symmetric*, the last condition meaning that the relation holds between $a$ and $b$ if it holds between $b$ and $a$. Whenever we have an equivalence relation, we may in some circumstances choose to disregard (abstract from) the differences between equivalent $a$ and $b$, as we usually (although not in numismatics) disregard differences between coins of the same denomination. In mathematics, we say that we *identify* $a$ and $b$, which in formal terms means that we think in terms of *equivalence classes*. An equivalence class is a class containing those and only those $b$ that are equivalent to some particular $a$. Thus all $a$ and $b$ in an equivalence class $A$ are equivalent, and any $a$ equivalent to any (and therefore to every) element in $A$ is itself in $A$. The equivalence class containing an element $a$ is sometimes written $[a]$, where it is assumed to be clear from the context which equivalence relation is intended.

Tradition dictates that equivalence classes are called "equivalence classes" rather than "equivalence sets", but there is no difference between "class" and "set" in this context. More generally, "class" does have a use in set theory as a technical term distinct from "set" (most often in the form "proper class"), but in this book the terms "class" and "set" are interchangeable, stylistic considerations apart.

In the case of the congruence relation, the equivalence classes are called *congruence classes* (modulo $n$). Thus for any positive integer $n$, there are precisely $n$ congruence classes, $[0], [1], \ldots, [n-1]$. Note that $[n] = [0]$.

Having identified equivalent elements in a domain, everything we wish to say about them can be expressed as statements about equivalence classes. But it is often still convenient to refer to an equivalence class using some particular element $a$, as the equivalence class containing $a$. In defining properties of and relations between equivalence classes, we must then ensure that it doesn't matter which particular element we use to identify a particular equivalence class.

For an example of this, suppose we wish to define a multiplication operation on congruence classes. A natural way of doing this is by means of the equation $[a] * [b] = [ab]$. The meaning of this is that for any two equivalence classes $A$ and $B$, we define $A * B$ as the equivalence class $C$ which is identical with $[ab]$ whenever $A = [a]$ and $B = [b]$. But of course speaking of *the* equivalence class $C$ is justified only if $[ab]$ is in fact independent of which particular $a$ and $b$ we have chosen to represent the equivalence classes $A$ and $B$. In other words, we must show that

$$\text{if } [a] = [a'] \text{ and } [b] = [b'] \text{ then } [ab] = [a'b']$$

To verify this, we show that that if $n$ divides both $a - a'$ and $b - b'$, then $n$ divides $ab - a'b' = b(a - a') + a'(b - b')$.

In contrast, if we try to define an exponentiation operation on congruence classes by saying that $[a]^{[b]} = [a^b]$ for every $a$ and $b$ we find that this doesn't work, since for example 5 and 2 are congruent modulo 3, as are 2 and 2, but $2^5$ and $2^2$ are not congruent modulo 3.

The situation illustrated above using congruence classes will appear in later contexts in the book, for example in defining functions by recursion on the ordinals in Chapter 6, and in the construction used in the proof of the completeness theorem in Chapter 8.

## 2.5. Two classical theorems

### The fundamental theorem of arithmetic

This is the traditional name for the theorem, also first proved in the *Elements*, that every positive integer greater than 1 has an essentially unique factorization into primes. "Essentially unique" means that if $n$ can be written in two ways as a product of primes, $n = q_1 \ldots q_k$ and $n = r_1 \ldots r_m$, we must have $k = m$, and the $q_i$ are a rearrangement of the $r_i$. Thus, for example, however 120 is written as a product of primes, there are 5 prime factors,

$120 = q_1 \ldots q_5$, and the sequence $q_1, \ldots, q_5$ is a rearrangement of 2, 2, 2, 3, 5. Another way of putting this is that for every natural number $n > 1$ there are unique natural numbers $n_0, \ldots, n_k$ such that,

$$n = p_0^{n_0} \ldots p_k^{n_k}$$

where $n_k > 0$ and $p_0, p_1, \ldots$ is an enumeration of the primes in order: $2, 3, \ldots$.

For the proof, we first need to show that every $n > 2$ can be factored into a product of primes in at least one way. If $n$ is not itself a prime, we can find numbers $a$ and $b$ greater than 1 and smaller than $n$ such that $n = ab$. By well-founded induction, $a$ and $b$ are products of primes, and therefore $ab$ is also such a product.

To show that the prime factorization of $n$ is essentially unique, suppose we have two such factorizations $q_1 \ldots q_k$ and $r_1 \ldots r_m$. $q_1$ divides $r_1 \ldots r_m$, and since all the numbers involved are primes, we can conclude from the division lemma that $q_1$ must divide one of $r_1, \ldots, r_m$, say $r_j$. (Here we are implicitly using induction on $m$.) But again since $q_1$ and $r_j$ are primes, it follows that $q_1 = r_j$. So we can divide both sides of the equation by $q_1$. Again arguing by well-founded induction, we conclude by the induction hypothesis that $k - 1 = m - 1$ and $q_2, \ldots, q_k$ is a rearrangement of the $r_i$ for $i \neq j$, and so the conclusion follows.

### The Chinese remainder theorem

Suppose $k$ and $n$ are relatively prime. By Euclid's lemma, this implies that there are integers $x$ and $y$ such that $kx + ny = 1$. Another way of putting this is that the equation $kx \equiv 1$ has a solution modulo $n$, or in other words there is an integer $x$ such that $kx - 1$ is divisible by $n$.

Let us generalize this. Suppose we have positive integers $n_1, \ldots, n_m$ which are pairwise relatively prime, that is, $(n_i, n_j) = 1$ whenever $i \neq j$. Also suppose $k_1$ and $n_1$ are relatively prime, and let $a_1$ be any integer. Then there is an $x$ for which $k_1 x \equiv a_1$ modulo $n_1$, while $x \equiv 0$ modulo $n_j$ for every $j \neq 1$.

To find such an $x$, we need to find a $y$ for which $k_1 y n_2 \ldots n_m$ is congruent to 1 modulo $n_1$. $x = a_1 y n_2 \ldots n_m$ will then satisfy the conditions stated. By Euclid's lemma, there is such a $y$ if $(k_1 n_2 \ldots n_m, n_1) = 1$. That this is in fact the case follows by the division lemma, for if $w$ divides $k_1 n_2 \ldots n_m$ and also divides $n_1$, $w$ and each of $k_1, n_2, \ldots, n_m$ must all be relatively prime, so repeated application of the division lemma yields that $w$ must divide $n_1$ and one of $k_1, n_2, \ldots, n_m$ and therefore be equal to 1.

We can generalize a bit further. Suppose we have $k_1, n_1, \ldots, n_m$ as above, but also $k_2, \ldots, k_m$ such that $(k_i, n_i) = 1$ for every $i$, and let $a_1, \ldots, a_m$ be any integers. We can then find an $x$ such that $k_i x$ is congruent to $a_i$ modulo $n_i$ for every index $i$. For by the argument above there is for each $i$ an $x_i$ such that $k_i x_i$ is congruent to $a_i$ modulo $n_i$ while $x_i$ is congruent to 0 modulo $n_j$

for every $j$ different from $i$. We get the solution $x$ by adding all these $x_i$. This classical result is known as the Chinese Remainder Theorem.

Finally an observation that will be used later: if $x$ is a solution to the system of congruences $k_i x \equiv a_i(n_i)$ $(i = 1, \ldots, m)$ then an integer $y$ is also a solution if $x \equiv y(n_i)$ for $i = 1, \ldots, m$, and in particular $x - n_1 \ldots n_m$ is also a solution. It follows that the smallest positive solution of the system is no greater than the product $n_1 \ldots n_m$.

## 2.6. Sequences of numbers

The two classical results presented above have a rather special application in mathematical logic. In arithmetical reasoning, one argues using not only numbers but also finite (and sometimes infinite) sets and sequences of numbers, as has been illustrated in several of the arguments above. In logic one often wants to reduce talk of finite sequences of numbers to talk of single numbers. This is done by encoding or representing a sequence $n_1, \ldots, n_k$ of natural numbers as a single natural number $n$, from which the elements of the sequence can then be retrieved by means of some arithmetical operation.

One way of achieving this relies on the fundamental theorem of arithmetic. Let $p_0, p_1, p_2, \ldots$ enumerate the primes in order. Thus $p_0$ is 2, $p_1$ is 3, $p_2$ is 5, and so on. We can represent the sequence $n_0, \ldots, n_k$ of natural numbers by the number $p_0^{n_0+1} \ldots p_k^{n_k+1}$. This representation satisfies two essential conditions. First, given a sequence of natural numbers, we can compute the number representing this sequence. Second, given any natural number it can be decided whether or not it represents any finite sequence, and if it does represent such a sequence we can read off from the number which sequence that is. Here we invoke the fundamental theorem of arithmetic, which ensures that $n_0, \ldots, n_k$ can be computed from $n$, and establishes that $n$ represents a sequence if and only if it has the property that for every $p_i$ that divides $n$, $p_j$ also divides $n$ for every $j < i$.

Note that the formulations "we can compute", "it can be decided", "we can read off" are not to be taken as stating that anything can actually be computed, decided or read off. Factoring a large number may be a task well beyond the combined computational resources of all nations. These formulations are to be understood, rather, as referring to what can "in principle" be done. The justification for this simplification is that we are concerned with certain theoretical (logical, mathematical and philosophical) aspects of arithmetic rather than with achieving any computational results.

Gödel (for reasons that will become clear in later chapters) needed to show that finite sequences can be represented as numbers using only multiplication and addition. (Note that the description of the correspondence above made use of the exponentiation function.) For this purpose he introduced the $\beta$-*function*, defined for natural numbers $k, m, i$ by the stipulation that $\beta(k, m, i)$

is the remainder when dividing $k$ by $1+(i+1)m$. The point of this definition is to be found in the following result:

> **$\beta$-lemma:** For every finite sequence $a_0, \dots, a_n$ of natural numbers there are natural numbers $k$ and $m$ such that $\beta(k, m, i) = a_i$ for $i = 0, \dots, n$.

For the proof, let $m$ be the product of all positive integers smaller than or equal to the largest of $n, a_0, \dots, a_n$. Then $(1+(i+1)m, 1+(j+1)m) = 1$ if $i$ and $j$ are different numbers from 0 to $n$. For if $p$ is a prime that divides both these numbers, it divides their difference $(i-j)m$ (where we assume that $i > j$). By the division lemma, $p$ must then divide $i-j$ or $m$. But $p$ does not divide $m$, for that would mean, since $p$ divides $1+(i+1)m$, that $p$ divides 1. And since $i-j$ divides $m$, $p$ cannot divide $i-j$ either. Hence the numbers $e_i = 1+(i+1)m$ are pairwise relatively prime. By the Chinese Remainder Theorem there is a $k$ such that $k$ is congruent to $a_i$ modulo $e_i$ for $i$ from 0 to $n$. By the definition of the $\beta$-function, $\beta(k, m, i) = a_i$, since $0 \leq a_i < 1+(1+i)m$. Note that $\beta(k, m, i) \leq k$ for all $k$, $m$, $i$. Also note that by the definition of $m$ and the observation made earlier about the smallest positive solution of a system of congruences, we can construct an algebraic expression involving $n, a_0, \dots, a_n$ which yields an upper bound for both $k$ and $m$. Specifically, if all of $n, a_0, \dots, a_n$ are smaller than $b$, both $k$ and $m$ are smaller than $(1+(b+1)b!)^b$, where $b!$ ($b$ factorial) is the product of all positive integers smaller than or equal to $b$. This expression will later be referred to as "the upper bound for $k$ and $m$ in the $\beta$-lemma".

How the $\beta$-lemma is used will become clear in later chapters.

CHAPTER 3

# PRIMES AND PROOFS

## 3.1. The infinity of primes

### Euclid's proof

Euclid proved in Proposition 20, book IX of the *Elements* that "prime numbers are more than any assigned multitude of prime numbers". His proof went as follows: Let $A$, $B$, $C$ be prime numbers. Then there is another prime number. For let $D$ be the number $ABC + 1$. $D$ is not divisible by any of $A$, $B$, $C$, since division by one of these primes yields a remainder 1. Thus either $D$ is itself a prime number or it is divisible by some prime number other than $A$, $B$, $C$.

It was implicit in Euclid's proof, but clear to his readers, that the argument applies to any enumeration of primes $q_1, \ldots, q_n$. The product $q_1 \ldots q_n + 1$ is either itself a prime different from all of $q_1, \ldots, q_n$, or it has a prime factor different from all of $q_1, \ldots, q_n$. So for every finite sequence of primes, there is a prime not in the sequence.

The result is usually expressed as "there are infinitely many primes". This is not to be understood as implying that an infinite totality exists, in any philosophically controversial sense, but only that there is no upper bound to the primes, that for every natural number $n$ there is a prime greater than $n$. And indeed a common variant of Euclid's argument starts with an arbitrary $n$, and notes that the number $n! + 1$ is either a prime or divisible by a prime greater than $n$.

Thus the statement "there are infinitely many primes", although containing the phrase "infinitely many" is an *arithmetical statement*, in the sense of a statement that refers only to integers and to finite constructions based on the integers. This concept of arithmetical statement is an informal one, which will be replaced by a formally defined concept in later chapters.

It is often said that Euclid's proof is *indirect*, that is, proceeds by assuming the negation of what is to be proved and deriving a contradiction. As can be seen from the formulation above, this is not the case. Some presentations do however turn the argument into an indirect one, quite unnecessarily. This point will be taken up again in connection with the rules of logic in Chapter 8.

## Euler's proof

No essentially new proof of the inexhaustibility of the primes was found until Leonhard Euler's proof in the 18th century. Euler introduced an interesting formula, now known as *Euler's product formula*:

$$\sum_{k=1}^{\infty} \frac{1}{k^s} = \prod_{p} \frac{1}{\left(1 - \left(\frac{1}{p}\right)^s\right)}$$

In this formula, $s$ is any real number greater than 1, and the product is taken over all *primes* $p$. We know from calculus (by an integral test) that the infinite sum on the left converges for $s$ greater than 1 but diverges for $s \leq 1$. For $s = 1$, the series on the left is known as the *harmonic* series. To prove the formula, we use the fundamental theorem of arithmetic and the formula for the sum of a geometric series. Let $A_m$ be the set of positive integers all of whose prime factors are among the first $m$ primes $p_1, \ldots, p_m$. We have that

$$\sum_{k=1}^{\infty} \frac{1}{k^s} = lim_{m\to\infty} \sum_{n\in A_m} \frac{1}{n^s} = lim_{m\to\infty} \sum_{k_1,\ldots,k_m=0}^{\infty} 1\Big/\left(p_1^{k_1} \ldots p_m^{k_m}\right)^s$$

$$= lim_{m\to\infty} \prod_{i=1}^{m} \left(1 + \frac{1}{p_i^s} + \frac{1}{p_i^{2s}} + \ldots\right) = \prod_{i=1}^{\infty} \frac{1}{\left(1 - \left(\frac{1}{p_i}\right)^s\right)}$$

Here we have used, first the fundamental theorem of arithmetic, then a basic result from calculus about rearranging absolutely convergent infinite series, and finally the formula for the sum of a geometric series.

Using Euler's formula, we get a new proof that there are infinitely many primes. For as $s$ tends to 1, the left side in the equation

$$\sum_{k=1}^{\infty} \frac{1}{k^s} = \prod_{p \text{ a prime}} 1\Big/\left(1 - \left(\frac{1}{p}\right)^s\right)$$

tends to infinity, while if there are only finitely many primes, the product on the right side will have as its limit a finite product.

## Euler's stronger result

Although simple as mathematical proofs go, Euler's proof that there are infinitely many primes is not nearly as simple and perspicuous as Euclid's. At the same time it seems clear that Euler has introduced a new idea, one that should give us some new information about the primes. And indeed

Euler could prove, using his formula, a stronger statement than that proved by Euclid's argument, namely that the infinite sum

$$\sum_{p \text{ a prime}} \frac{1}{p}$$

is divergent, that is, there is no upper bound to the sums of reciprocals of primes. This clearly implies that the primes are infinite in number, but it also tells us that the primes are more dense among the natural numbers than for example the perfect squares, since the sum

$$\sum_{k=1}^{\infty} \frac{1}{k^2}$$

is convergent.

To prove Euler's result, we take logarithms:

$$\begin{aligned}\log \sum_{k=1}^{\infty} \frac{1}{k^s} &= \log \prod_p 1 \Big/ \left(1 - \left(\frac{1}{p}\right)^s\right) = \sum_p -\log\left(1 - \left(\frac{1}{p}\right)^s\right) \\ &= \sum_p \left(p^{-s} + O\left(p^{-2s}\right)\right) = \sum_p p^{-s} + O\left(\sum_p p^{-2s}\right) \\ &= \sum_p p^{-s} + O(1)\end{aligned}$$

Here the $O$-notation indicates a value smaller than or equal to (in absolute value) the expression within parentheses, multiplied by some constant. We get the estimate $O(p^{-2s})$ from a Taylor expansion of the function $-\log(1-x)$, and the $O(1)$ from the convergence of $\sum_p p^{-2s}$. Since $\log \sum_{k=1}^{\infty} \frac{1}{k^s}$ tends to infinity as $s$ tends to 1, it follows that the same is true of the sum $\sum_p p^{-s}$, and so the sum of the reciprocals of the primes diverges. (Refining this argument yields that $\sum_{p \leq n} p^{-1}$ is approximately $\log \log n$.)

## Dirichlet's theorem

This line of argument was taken very much further by Dirichlet, who proved in 1837 that whenever $(a, k) = 1$, there are infinitely many primes of the form $kn + a$. Thus there are infinitely many primes in every arithmetical series, with the obvious exception of those where every term is a multiple of some particular number. To prove this he introduced a generalization of Euler's formula:

$$\sum_{k=1}^{\infty} \frac{\chi(n)}{n^s} = \prod_{p \text{ a prime}} 1 \Big/ \left(1 - \frac{\chi(p)}{p^s}\right)$$

Here $\chi$ is any so-called Dirichlet character—essentially, a multiplicative homomorphism from the congruence classes modulo $k$ to the complex numbers. Dirichlet then applied a complex logarithm function to both sides of the above formula, and using information about the existence and number of Dirichlet characters obtained an expression involving the sum of reciprocals of the primes of the form $kn + a$. By means of a fairly involved argument using further estimates, integrals and series, he obtained the result that this sum of reciprocals diverges.

In present-day mathematics, there are many refinements and improvements of these results. But I shall stop here, to consider these proofs from the point of view of this book.

## 3.2. Arithmetical, elementary, and analytic proofs

Comparing Euclid's and Euler's proofs of the inexhaustibility of the primes, there is an obvious difference in that Euler's proof invokes real numbers and limits, not just integers and finite sequences or sets of integers. But of course, as Euclid's proof shows, this invocation is not essential. The assertion "for every integer $k$, there is a prime greater than $k$" can be proved without mentioning real numbers or limits. We will call Euclid's proof an *arithmetical proof*, or *purely arithmetical proof* for emphasis. This is not a formally defined notion, but an informal term meaning that the proof reasons about integers and finite constructions based on the integers (such as finite sets and sequences) but does not introduce concepts from calculus, such as limits, integrals, derivatives, or any theory of infinite sets.

### Rational numbers and arithmetical proofs

What of the stronger statement proved by Euler, that the sum of the reciprocals of primes diverges? This is a more informative statement than "there are infinitely many primes", and it is formulated using not only integers, but rational numbers. However, the statement "the sum of the reciprocals of primes diverges" can be "translated" into a statement about integers. Let $p_1, p_2, \ldots$ be an enumeration of the primes in order. To say that

$$\frac{1}{p_1} + \frac{1}{p_2} + \cdots + \frac{1}{p_k} > n$$

is, by the rules for computing with rational numbers, the same as saying that

$$m_1 + m_2 + \cdots + m_k > nm$$

where $m$ is the product of the first $k$ primes, and $m_i$ is the product of those primes except $p_i$. That the sum diverges means that for every $n$ there is a $k$ such that the above inequality holds, and thus this statement can be formulated as a statement about natural numbers and finite sequences of natural numbers.

More generally, every mathematical statement involving sums and products of rational numbers, comparisons of rational numbers, sequences of rational numbers, and so on, can be expressed as a statement about integers. Indeed, in the foundations of mathematics (both in philosophy and as presented in introductory chapters of mathematics texts) one usually identifies a rational number $a/b$ with the (equivalence) class of all ordered pairs of integers $\langle m, n\rangle$ such that $an = bm$. The rational numbers are thus the equivalence classes of the equivalence relation $\sim$ between ordered pairs of integers defined by $\langle m, n\rangle \sim \langle m', n'\rangle$ if and only if $mn' = m'n$. Addition and other arithmetical operations on rational numbers are then defined in terms of arithmetical operations on integers. For example we define addition (writing $[m, n]$ for the equivalence class containing $\langle m, n\rangle$) by

$$[m, n] + [m_1, n_1] = [mn_1 + m_1 n, nn_1]$$

where, as in §2.3, we must verify that the equation above does define an operation on equivalence classes by checking that the value on the right is independent of the choice of representatives of the equivalence classes involved.

We will look at this reduction of the rational numbers to the integers somewhat differently. In purely arithmetical proofs, we introduce only the integers and finite constructions based on the integers. The equivalence classes of ordered pairs of integers are infinite, since there are infinitely many pairs $\langle k, l\rangle$ satisfying the equation $ml = nk$. But references to these classes can be understood in an arithmetical context as a figure of speech, since they can always be replaced by references to representatives of a class. Thus, for example, to say that for any two positive rational numbers $r, r'$ where $r < r'$ there is a rational $s$ such that $r < s < r'$ is to say that for any positive integers $p$, $q$ and $m$, $n$ with $pn < mq$ there are positive integers $i, j$ such that $pj < iq$ and $in < mj$. Similarly a statement about finite sequences $r_1, \ldots, r_n$ of rational number can be formulated as a statement about finite sequences of pairs of integers. So we don't need to actually introduce any objects corresponding to the rational numbers. Instead we just translate every statement about rational numbers into an equivalent statement about integers. The equivalence holds on the basis of general properties of the rational numbers, independently of the particular statement translated.

Thus rational numbers are in a strong sense reducible to the integers and finite sets and sequences of integers: everything we say about rational numbers can equivalently be expressed as a statement about integers, and proofs that reason about rational numbers can be translated into proofs that reason about integers. This reduction doesn't change the fact that many questions posed in terms of rational numbers—for example the question about the sum of the reciprocals of primes—quite possibly would never have occurred to us if we had only talked about integers. Also, proofs using rational numbers may be much more easily found, and understood once they've been found, than the corresponding proofs that use only integers. This is typical of foundational reductions in general. In a logical context, such as the discussion in this book, we are usually concerned only with problems and arguments for which the distinction between rational numbers and the corresponding equivalence classes of pairs of integers is not important, but it shouldn't be assumed that the discussion presupposes that there is no difference.

## An arithmetical proof of Euler's result

So can we prove that the reciprocal prime series diverges by reasoning only about rational numbers and finite sequences of such, without invoking limits or real numbers? Yes we can. To begin with, the argument will be formulated in a style that speaks of infinite series. Suppose the infinite sum

$$\sum_{i=1}^{\infty} \frac{1}{p_i}$$

converges. We can then choose $N$ so large that

$$\sum_{i=N}^{\infty} \frac{1}{p_i} < 1$$

Let $A$ be the set containing 1 together with those positive integers all of whose prime factors are among $p_1, \ldots, p_{N-1}$, and let $B$ be the set containing 1 together with those positive integers all of whose prime factors are among the $p_i$ for $i \geq N$. We then have

$$\sum_n \frac{1}{n} = \sum_{m \in A, k \in B} \frac{1}{mk} = \left( \sum_{m \in A} \frac{1}{m} \right) \left( \sum_{k \in B} \frac{1}{k} \right)$$

In this last product, we observe that the first factor is bounded, since by the argument used to derive Euler's formula it is a finite product. The second factor is bounded as well, as we shall see. Let $P(k_1, \ldots, k_n, j_1, \ldots, j_n, m)$ abbreviate the condition $k_1 + \cdots + k_n = m$, $k_1, \ldots, k_n > 0$, $N \leq j_1 < j_2 < \cdots < j_n$. Then

$$\sum_{k \in B} \frac{1}{k} = \sum_m \sum_{P(k_1,\ldots,k_n,j_1,\ldots,j_n,m)} \frac{1}{p_{j_1}^{k_1} \cdots p_{j_n}^{k_n}} = \sum_m \left( \sum_{j=N}^{\infty} \frac{1}{p_i} \right)^m$$

and the last sum ($m$ summed over all natural numbers) converges by the condition on $N$. Thus we get that the harmonic series converges, and since this is not the case, it follows that $\sum_{i=1}^{\infty} \frac{1}{p_i}$ doesn't converge either.

The proof just given isn't strictly speaking an arithmetical proof, since it doesn't refer only to integers and finite sets or sequences of integers. Instead it invokes rational numbers and infinite sums of rational numbers. But talk of rational numbers can, as noted above, be replaced by talk of integers, and the manipulations involving infinite series in the argument can be replaced by purely arithmetical arguments. Thus, for the divergence of the harmonic series, we prove by induction on $k$ that

$$1/1 + 1/2 + \cdots + 1/(2^k - 1) \geq k$$

The assumption that $\sum_{i=1}^{\infty} \frac{1}{p_i}$ converges means, arithmetically expressed, that there is some $m$ such that $\sum_{i=1}^{n} \frac{1}{p_i} < m$ for every $n$. To get $N$ as above, let $m$ be the smallest natural number such that $2\sum_{i=1}^{n} \frac{1}{p_i} < m$ for all $n$. There is then an $N$ such that

$$2\sum_{i=1}^{n} \frac{1}{p_i} \leq m < 2\sum_{i=1}^{N-1} \frac{1}{p_i} + 1$$

for all $n$, which implies that

$$\sum_{i=N}^{n} \frac{1}{p_i} < \frac{1}{2}$$

for all $n > N$. The boundedness of the partial sums in the geometric series based on $\sum_{i=N}^{n} \frac{1}{p_i}$ can then be shown using an arithmetical inequality, and the argument goes through as a purely arithmetical proof. Similarly, the use of the argument proving Euler's formula to show that the first factor $\sum_{m \in A} \frac{1}{m}$ above is bounded can be expanded into a purely arithmetical proof. Thus one will ordinarily present the argument above as arithmetical, taking for granted that it can be expanded into purely arithmetical form, at the cost of making it longer and less perspicuous.

### Arithmetical proofs of Dirichlet's theorem

What then of Dirichlet's theorem? Can we give an arithmetical proof of the assertion that if $(k, n) = 1$, there are infinitely many primes of the form $kn + a$? I don't know of any such arithmetical proof. However, even if nobody has in fact produced any purely arithmetical proof of the assertion, there is no doubt that such a proof can be produced. Dirichlet's own proof, and modern variants of that proof, are not purely arithmetical, but they are what in mathematics is called *elementary proofs*. This means that they don't rely on complex analysis or results from algebraic topology or other fields that use a lot of mathematical apparatus not found in arithmetic. In positive terms, it means that even though logarithms, limits, integrals, complex numbers are used in the proof, it is clear to the expert that they could—at least "in principle"—be replaced by rational approximations, finite sums, arithmetical inequalities. Thus it is seen that a purely arithmetical proof exists at least in a mathematical sense, that is, one could in principle write down an arithmetical proof of the theorem, derived from Dirichlet's proof. In other words, the theorem is formally provable on the basis of purely arithmetical axioms and rules of reasoning. A proof resulting from this procedure would probably be long, boring, difficult or impossible to follow. There may or may not be an arithmetical proof of the result that is reasonably easy to follow and not too long to write down, but for the purposes of this book it is enough that there is at least one arithmetical proof, however long and however useless in practice.

## Analytical versus elementary proofs

Dirichlet's proof contrasts with the original proof of another classical theorem, the *prime number theorem.* In one version, conjectured by Gauss in 1792, this theorem states that the number of primes smaller than $n$ (which we denote by $\pi(n)$) is asymptotically $n/\log n$, meaning that

$$\frac{\pi(n)}{n/\log n} \to 1 \text{ as } n \to \infty$$

This statement involves logarithms and limits and thus isn't even an arithmetical statement, so it doesn't on the face of it make sense to ask if it has an arithmetical proof. However, using an infinite series expansion of the logarithm function we can reformulate it as an arithmetical statement, and it makes sense to ask whether the resulting arithmetical (although hardly intelligible) statement has an arithmetical proof.

The prime number theorem was first proved at the end of the 19th century using complex analysis, and in this case it was not at all clear how an elementary (and thereby arithmetical) proof could be extracted from the proof. The proof was a typical *analytical* proof. This again is an informal term, referring to a proof that freely uses methods from different mathematical fields, in the classical case typically complex analysis (contour integrals). Indeed when an elementary proof of the prime number theorem was presented in 1950 by Selberg and Erdös, this was seen as a great achievement in the mathematical community. Since the proof found by Selberg and Erdös was elementary (in the sense explained), a purely arithmetical proof is known to exist. Again, as in the case of Dirichlet's theorem, this doesn't imply that there is any purely arithmetical proof that is intelligible to human beings. Also, the reformulation of the theorem in arithmetical terms can hardly be understood without reference to the statement about limits and logarithms that it reformulates. Still, in theory the arithmetical version of the theorem has a purely arithmetical proof.

Today, through extensive work done by logicians, it is in most cases pretty clear to experts (in proof theory) that non-elementary proofs of arithmetical statements can ("in principle") be converted into elementary proofs. This is not to say that it is always clear. At the moment, the proof by Andrew Wiles of (a conjecture known to imply) Fermat's last theorem, that the equation $x^n + y^n = z^n$ has no solution in positive integers for $n > 2$, which is quite involved and uses heavy mathematical machinery, is not known to have any elementary counterpart. On the basis of general experience however, it is reasonable to expect that an elementary proof will eventually be seen to exist, again "in principle".

### Does every arithmetical theorem have an arithmetical proof?

On the basis of experience, then, it would be a not unreasonable guess that every arithmetical statement has a purely arithmetical proof if it can be proved at all in mathematics as we know it. This is not to say that it necessarily has any purely arithmetical proof that we could in fact have found, or could in fact use to convince ourselves of the truth of the statement. It may be that any purely arithmetical proof will simply be too long and too complicated to be useful to us as a proof. But it may yet exist from a mathematical point of view, as an idealized argument securing the truth of the statement.

We know that this is in fact the situation in the case of computational statements, statements whose truth or falsity could in principle be settled by a computation determined by the form of the statement. Consider for example a statement of the form "there is a prime between $n$ and $2n$", where $n$ is an expression in some standard notation denoting a very large natural number. This is a computational statement: an exhaustive testing of the numbers between $n$ and $2n$ will show whether or not there is any such prime. In practice, if we choose $n$ large enough, it may well be that we can no more carry out such a computation than we can dismantle the moon and rebuild it outside our solar system. It may be "possible in principle", but it is quite irrelevant to anything people actually do. We know that there is a prime between $n$ and $2n$ only because we have a proof of this fact which reasons in general terms about hypothetical computations on arbitrary numbers. Still, from a mathematical rather than practical point of view we can say that the truth of the statement "there is a prime between $n$ and $2n$" entails that such a computation could in principle be carried out, and if carried out would verify the statement.

Similarly it may seem that an arithmetical statement, if it can be shown to be true at all, should in principle admit a proof that does not involve any other objects than those involved in the statement itself: the integers, and finite constructions based on the integers. Of course there is nothing compelling about this idea: why shouldn't there be statements about the integers that can only be seen to be true (even "in principle") by going beyond the integers? But at least it should be agreed that it would be interesting to see an example of such a statement, to see *why* it can only be shown, even in principle, to be true by going beyond the integers.

## 3.3. The need for formalization

In order to show that an arithmetical statement has an arithmetical proof it is sufficient to present such a proof, or to indicate how an arithmetical proof of the statement could in principle be extracted from a known proof. But in order to show that some arithmetical statement which has been proved by other means, or at least is not known to be false, does not have an arithmetical proof, it is necessary to come up with some formal characterization of arithmetical proofs, or at least of arithmetical provability, so as to make possible a demonstration that a certain mathematical theorem or conjecture has no arithmetical proof.

This is analogous to the case of algorithmically solvable problems. To show that there is an algorithm for computing some function, it is sufficient to exhibit an algorithm or show how one can be found. But to show that there is no algorithm for computing a particular function, it is necessary to have some formal characterization of the algorithmically computable functions.

In the case of computability, such a formal characterization of the algorithmically computable functions was found around the same time as Gödel proved his incompleteness theorem, and since then several equivalent characterizations have been given. These formal characterizations are generally accepted as definite: a function can be computed by an algorithm if and only if it is computable according to the definition, which states that a function is computable if there is an algorithm in some particular class of formally defined algorithms which can be used to compute it. The most famous formally defined type of algorithm is the *Turing machine*, a mathematical idealization of a digital computer, introduced by Alan Turing in 1936. The claim that every function that can be computed by an algorithm can also be computed by a Turing machine, or by an algorithm in one of the equivalent classes of algorithms, is known as Church's Thesis, or the Church-Turing thesis, after its two most prominent proponents.

In the case of arithmetical provability, the situation is much less clear-cut, as we shall see. But a reasonable way of characterizing the arithmetically provable statements emerged from the work in mathematical logic in the early part of the twentieth century, in the form of formal axiomatic theories—usually called formal systems—such that all ordinary theorems of mathematics known to have arithmetical proofs can be seen to be provable in those systems. So if there is an arithmetical statement which cannot be proved or disproved in such a system, we have a candidate for a true arithmetical statement that is not arithmetically provable. Whether such a candidate is provable by non-arithmetical proof is another matter, which will have to be decided from case to case.

Note the distinction above between proofs and provability, algorithms and computability. There are algorithms of many different kinds, and there is no formal description that covers every type of algorithm that exists or may be introduced. But according to the generally accepted characterization of computable functions, if there is any algorithm at all for computing a function, then there is a Turing machine which computes the function.

Similarly, in the case of arithmetical proofs, it is not to be expected that any formal description could cover everything that we would want to call an arithmetical proof. But the formal theories referred to above are reasonable candidates for characterizing arithmetical *provability*, that is, if an arithmetical statement has anything that we would call an arithmetical proof, it also has a formal derivation in the theory.

## 3.4. Gödel's theorem

So suppose $T$ is a formal theory in which arithmetical proofs from ordinary mathematics can be seen to have formal counterparts. Just what such a $T$ can look like will become clearer in the following chapters. Gödel's second incompleteness theorem then yields an arithmetical statement which although true cannot be proved in $T$, namely a formalization of the statement that $T$ is consistent. The first incompleteness theorem yields instead the so-called Gödel statement for $T$, but this statement is equivalent in $T$ to the consistency assertion, for the relevant $T$. The Gödel statement formalizes a self-referential statement which "says of itself that it is not provable in $T$", and discussions of the Gödel statement sometimes give the impression that statements undecidable in $T$ must be in this sense self-referential. This is not the case, since the formalization of "$T$ is consistent" is also undecidable, by the second incompleteness theorem, and is not a self-referential statement. But nor is it an ordinary arithmetical statement. Instead it is a statement about the formalism of $T$ itself, which is expressed as an arithmetical statement by using the fact that the sentences and terms of $T$ can be represented as numbers, and their syntactic properties (such as being provable using certain formal rules) defined in arithmetical terms.

As a candidate for a true arithmetical statement that is not arithmetically provable, "$T$ is consistent" is rather startling, or perhaps disappointing. From experience in number theory, one might expect such a candidate to take the form of a theorem—about the distribution of primes, say—that uses analytical or topological methods in some essential way that just can't be replaced by purely arithmetical methods. But "$T$ is consistent" goes off in a completely different direction. As an arithmetical statement, it's an oddity: it is arithmetical only in virtue of a representation of syntactic objects like terms and statements as numbers, and it is not on the face of it connected with any traditional arithmetical problems.

During the past twenty-five years, much work has been done on finding arithmetical statements which can be shown to be unprovable in $T$ on the basis of Gödel's theorem, but which are closer to mathematical practice, in the sense of being more "normal" arithmetical statements, with implications in ordinary mathematics. Quite a lot has been achieved in this direction (beginning with a result by Paris and Harrington on a combinatorial principle not provable in elementary arithmetic—see Paris and Harrington [1977]) but this line of development will not be considered in this book. Instead, as explained in the introduction, we will be concerned with just those aspects of the consistency statement that make it untypical of the arithmetical statements that feature in ordinary mathematics. First we need to put the whole discussion on a formal footing, by introducing the formal languages, axioms and rules of inference dear to the hearts of logicians, and this will be the topic of the next several chapters.

CHAPTER 4

# THE LANGUAGE OF ARITHMETIC

## 4.1. Integers, rationals, and natural numbers

The informal term "arithmetical proof" has been explained as referring to proofs that invoke only the integers and finite structures (such as sets and sequences) based on the integers. As explained in §3.2, this term can be taken to cover also proofs that make use of the rational numbers, since statements about rational numbers can be reduced to statements about integers (and pairs of integers). The basis for this reduction is the fact that we can represent a rational number by a pair $\langle m, n\rangle$ of integers ($n \neq 0$), interpreted as the rational number $m/n$, so that two pairs $\langle m, n\rangle$ and $\langle m', n'\rangle$ represent the same rational number if and only if $mn' = m'n$. We can then define addition, multiplication, comparisons and other operations on rational numbers in terms of operations on the pairs of integers representing them. Thus we can formulate our arithmetical statements about arbitrary rational numbers, sequences of rational numbers, and so on, as statements about arbitrary pairs of integers, sequences of such pairs, and so on.

It is possible to similarly reduce the integers to the natural numbers. We can, for example, represent an integer as a pair $\langle m, n\rangle$ of natural numbers, interpreted as the integer $m - n$, so that two such pairs $\langle m, n\rangle$ and $\langle m', n'\rangle$ represent the same integer if and only if $m+n' = m'+n$. As with the rationals, we can define addition, multiplication, comparisons and other operations on integers in terms of operations on the pairs of natural numbers representing them. Thus for example the equality

$$(m_1 - m_2)(n_1 - n_2) = m_1n_1 + m_2n_2 - (m_1n_2 + m_2n_1)$$

shows that we should define the result of multiplying the pairs $\langle m_1, m_2\rangle$ and $\langle n_1, n_2\rangle$ as the pair $\langle m_1n_1 + m_2n_2, m_1n_2 + m_2n_1\rangle$. A similar but simpler rule gives addition of pairs. A natural number $n$ regarded as a non-negative integer is represented by $\langle n, 0\rangle$. We can combine this interpretation of integers as pairs of natural numbers with the interpretation of rational numbers as pairs of integers to obtain an interpretation of rational numbers in terms of finite structures (pairs of pairs) based on the natural numbers. This gives us a way of systematically translating statements about rational numbers into statements about natural numbers.

Consider as an example how the statement "there is a rational number $r$ such that $r^2 - r = 1/2$" translates into a statement about the natural numbers. It is equivalent to "there are integers $m$ and $n$ such that $(m/n)^2 - (m/n) = 1/2$", which reduces to "there are integers $m$ and $n$ such that $2m^2 = n^2 + 2mn$". Eliminating the reference to integers as indicated above, this becomes

> there are pairs $\langle m_1, m_2\rangle$ and $\langle n_1, n_2\rangle$ of natural numbers such that $2\langle m_1, m_2\rangle\langle m_1, m_2\rangle$ represents the same integer as $\langle n_1, n_2\rangle\langle n_1, n_2\rangle + 2\langle m_1, m_2\rangle\langle n_1, n_2\rangle$,

which becomes

> there are natural numbers $m_1, m_2, n_1, n_2$ such that $\langle 2m_1^2+2m_2^2, 4m_1m_2\rangle$ and $\langle n_1^2+n_2^2+2m_1n_1+2m_2n_2, 2n_1n_2+2m_1n_2+2m_2n_1\rangle$ represent the same integer,

which finally becomes

> there are natural numbers $m_1, m_2, n_1, n_2$ such that $2m_1^2+2m_2^2+2n_1n_2+2m_1n_2+2m_2n_1 = n_1^2+n_2^2+2m_1n_1+2m_2n_2+4m_1m_2$.

This method of translation is clearly cumbersome and impractical, but it shows that from a theoretical point of view, in considering statements about rational numbers and various philosophical and logical questions connected with them, we can restrict our attention to the natural numbers. (In practice, many statements about integers or rational numbers have much more intelligible and easily obtained translations into statements about the natural numbers than the one yielded by the general translation described.)

In the following, "number" will always mean "natural number" unless otherwise stated, and arithmetical statements will be taken to invoke only natural numbers and finite structures based on the natural numbers. By a *number-theoretic function* will be meant a function taking natural numbers as arguments and values, and similarly for relations.

## 4.2. Quantifiers and connectives

Let's consider how to define something that we might call the formal language of arithmetic, a formally defined language in which reasoning about the natural numbers and finite constructions based on the natural numbers can be expressed.

An example of a statement that we want to be able to formulate in the formal language of arithmetic is "There are infinitely many primes". As noted earlier, we can formulate this as an arithmetical statement in various equivalent ways. One formulation is "For every sequence $p_1, \ldots, p_n$ of primes, there is a prime not in the sequence", another is "For every $n$, there is a prime greater than $n$". The second formulation appears simpler than the first in that it doesn't invoke sequences of natural numbers, but only individual numbers. So we'll consider the second formulation first.

"$p$ is a prime" is defined in arithmetic to mean "$p > 1$ and for every $m$, if $m$ divides $p$ then $m = 1$ or $m = p$". "$m$ divides $p$" is defined in turn as "there is a $k$ such that $p = mk$". So in our arithmetical language we need to be able to express:

> For every $n$ there is a $p > 1$ which satisfies the following two conditions: $p > n$, and for every $m$, if there is a $k$ such that $p = mk$, then $m = 1$ or $m = p$.

This statement uses certain general logical constructs: the phrases "for every $n$", "there is a $k$", "if", "or", "and". Let us introduce special symbols for these phrases:

"For every" becomes $\forall$, and $\forall k$ is read "for every $k$ (it is the case that)". We call $\forall$ the *universal quantifier*. In an arithmetical statement, it is understood that "for every $k$" means "for every natural number $k$", and similarly for the other letters $n, m, \ldots$ that we use as variables for natural numbers.

"There is" becomes $\exists$, and $\exists k$ is read "there is a $k$ (meaning: at least one $k$) such that". We call $\exists$ the *existential quantifier*. Again it is understood in an arithmetical statement that "there is a $k$" means "there is a natural number $k$".

"Or" is written as $\vee$, and "and" is written as $\wedge$. "If" is replaced by the horseshoe $\supset$, so that a statement "if $A$ then $B$" is written "$A \supset B$". $\vee$, $\wedge$ and $\supset$ are called (*sentential* or *propositional*) *connectives*, because they are used to form new statements by connecting other statements. A statement of the form $A \supset B$ is called a *conditional* statement or an *implication*, $A \wedge B$ is a *conjunction*, and $A \vee B$ is a *disjunction*.

Using these (standard) logical symbols, the statement above becomes

$$\forall n\,\exists p(p > n \wedge p > 1 \wedge \forall m(\exists k(p = mk) \supset (m = 1 \vee m = p)))$$

In this symbolic version, parentheses have been used to indicate what is called the *scope* of quantifiers, that is, what statement about $k$ a phrase "for every $k$" refers to. Parentheses are also used to disambiguate potentially ambiguous expressions such as

$$a = b \wedge a = c \vee e = f \supset a = f$$

which might be read as a conjunction, a disjunction, or a conditional statement. We will follow some standard conventions which allow us to leave out some parentheses. Chief of these is that an expression is read as an implication, when possible, rather than as a conjunction or disjunction. Thus the symbolic formulation of "there are infinitely many primes" will ordinarily be written

$$\forall n\,\exists p(p > n \wedge p > 1 \wedge \forall m(\exists k(p = mk) \supset m = 1 \vee m = p))$$

Note that the statement

$$p > n \wedge p > 1 \wedge \forall m(\exists k(p = mk) \supset m = 1 \vee m = p)$$

can be read as a conjunction in two ways. More generally, a statement of the form $A \wedge B \wedge C$ can be read as $(A \wedge B) \wedge C$ or as $A \wedge (B \wedge C)$. Since the meaning of the statement is the same however it's read, we don't usually need to consider the matter, but in some formal contexts it makes a difference, and it will then be assumed that $A$ and $B$ are grouped together, so that the statement is to be read $(A \wedge B) \wedge C$. Similar remarks apply to disjunction.

Two more connectives are common in arithmetical statements: *negation* and *biconditional* or *equivalence*. A negated statement is one which says that something is not the case, for example "5 is not a prime", "there is no $k$ such that $k$ is a prime greater than 5", "no primes are greater than 3". In the formal language, such negated statements are all expressed using the construction "it is not the case that" applied to a statement: "it is not the case that 5 is a prime", "it is not the case that there is a $k$ such that $k$ is a prime greater than 5", "it is not the case that there is a prime greater than 3". The symbol for "it is not the case that" is $\neg$. A biconditional or equivalence, finally, is a statement of the form "$A$ if and only if $B$"; this is written "$A \equiv B$".

Besides the logical apparatus of quantifiers and connectives, we also use in arithmetical statements symbols for any number of number-theoretic functions, relations, and properties. These are functions and relations that we either accept at the outset as arithmetical in the informal sense, such as the operations of addition and multiplication and the relations "equal to" and "smaller than", or else functions and relations that we define in arithmetical terms, such as "$k$ is a prime", "$k$ divides $n$". The question how to characterize the arithmetical functions and relations will be dealt with below. But first there is another matter to consider. We also speak in arithmetic about finite sequences and sets of natural numbers, as in the fundamental theorem of arithmetic, and the question is how we should deal with such constructions in our formal language.

## 4.3. Finite sequences of numbers

It would be perfectly reasonable to extend the arithmetical language described above so as to allow us to talk about finite sequences of natural numbers and whatever other finite structures based on the natural numbers we wish to make use of in arithmetic. Thus one might add a special class of variables $S_1, S_2, \ldots$ for finite sequences, with corresponding quantifiers "for every finite sequence $S$", "for at least one finite sequence $S$". One would also add symbols for operations on sequences, in particular a function yielding the length of a sequence, and one picking out the $i$-th element of a sequence. We write $\mathit{length}(S)$ for the length of $S$, and $(S)_i$ for the $i$-th element of $S$, where we count the first element as the 0-th. Thus if $\mathit{length}(S) = n$, $S$ is the sequence $\langle (S)_0, \ldots, (S)_{n-1} \rangle$. A special case is the empty sequence, with length 0. Since we want $(S)_i$ to be defined for every sequence and every index $i$, we define $(S)_i = 0$ if there is no $i$-th element in $S$.

Let's consider how to express the fundamental theorem of arithmetic using this language. Let $prod(S)$ be the product of the numbers in the sequence $S$. We assume for the moment that $prod(S)$ has been defined in some way in arithmetical terms—how this can be done will emerge at a later point in the chapter. The fundamental theorem of arithmetic then states that for every $k > 1$ there is a sequence $S$ of primes such that $k = prod(S)$, and for any sequences of primes $S$ and $S'$ such that $prod(S) = prod(S')$, $S$ is a rearrangement of $S'$. That $S$ is a rearrangement of $S'$ means that $S$ and $S'$ have the same length $n$, and there is a *permutation* $P$ of the numbers $0, 1, \ldots, n-1$, which we can take to be a sequence $P$ of length $n$ in which each of the numbers $0, 1, \ldots, n-1$ occurs exactly once, such that for every $i$ from 0 to $n-1$, $(S)_i = (S')_j$ where $j = (P)_i$. In symbols we express "$S$ is a sequence of primes" as

$$\forall i\,(i < length(S) \supset prime\,((S)_i))$$

where $prime((S)_i)$ is defined as indicated earlier. We can then express the fundamental theorem of arithmetic in symbols, writing $pseq(S)$ for "$S$ is a sequence of primes", as the statement

$$\begin{aligned}
&\forall k(k > 1 \supset \exists S(pseq(S) \wedge k = prod(S))) \wedge \\
&\forall S\, \forall S'(prod(S') = prod(S) \wedge pseq(S') \wedge pseq(S) \\
&\qquad \supset \exists P\, \exists n(length(P) = n \wedge length(S) = n \wedge length(S') = n \wedge \\
&\qquad\qquad\qquad \forall m(m < n \supset \exists j(j < n \wedge (m = (P)_j) \wedge \\
&\qquad\qquad\qquad \forall m(m < n \supset \exists j(j = (P)_m \wedge (S)_m = (S')_j)))))).
\end{aligned}$$

Note that in this formulation "$P$ is a permutation of $0, 1, \ldots, n-1$" is expressed not as "$P$ is a sequence of length $n$ in which each of $0, 1, \ldots, n-1$ occurs exactly once", but rather as "$P$ is a sequence of length $n$ in which each of $0, 1, \ldots, n-1$ occurs at least once". This would ordinarily be taken as an acceptable formulation of "$P$ is a permutation of $0, 1, \ldots, n-1$" because the two conditions are equivalent: if $P$ has length $n$, it contains each of $0, 1, \ldots, n-1$ exactly once if and only if it contains each of $0, 1, \ldots, n-1$ at least once. But of course this equivalence is itself established only by mathematical reasoning. If the mathematical reasoning required had been as lengthy as the proof of the fundamental theorem itself, it would have been unacceptable to present the above as a formulation of the fundamental theorem. "Minor" variants of an arithmetical statement, which are "easily" seen to be equivalent are, however, normally accepted as equivalent formulations. Formulating a particular mathematical statement in a particular formal language is not always a matter for mechanical translation, but requires some degree of choice and judgment, and depending on the context, differences in formulation may be more or less significant.

## 4.4. Reducing sequences to numbers

Although a language of arithmetic which has special variables for finite sequences or sets of natural numbers, with corresponding operations, is very natural and gives us a handy language in which to formalize arithmetical theorems, in this book a different (and traditional) route will be taken: finite sequences of natural numbers will be reduced to natural numbers, following Gödel. The chief reason for this is that it allows us to show that the incompleteness theorems apply to a basic form of arithmetic in which we quantify only over the natural numbers.

To achieve this reduction of finite sequences of natural numbers to natural numbers we use Gödel's $\beta$-function. Recall the $\beta$-lemma from §2.6: for every finite sequence $\langle a_0, \ldots, a_n \rangle$ of natural numbers there are natural numbers $k$ and $m$ such that $\beta(k, m, i) = a_i$ for $i = 0, \ldots, n$. Thus any sequence of natural numbers can be represented through two natural numbers. But note that $\beta(k, m, i)$ is defined for all $k$, $m$, $i$, so there is nothing in $k$ and $m$ to indicate that they represent the sequence $\langle \beta(k, m, 0), \ldots, \beta(k, m, n) \rangle$ rather than the sequence $\langle \beta(k, m, 0), \ldots, \beta(k, m, n + 1) \rangle$. We therefore introduce the convention that $\beta(k, m, 0)$ gives the length of the sequence represented by $k, m$. Thus the sequence $\langle a_0, \ldots, a_n \rangle$ is represented by $k, m$ such that $\beta(k, m, 0) = n + 1$ and $\beta(k, m, i + 1) = a_i$ for $i = 0, \ldots, n$.

This representation of finite sequences of natural numbers by pairs of natural numbers is quite sufficient to allow us to formulate such statements as the fundamental theorem of arithmetic by quantifying only over natural numbers. The quantifier "for every sequence $S$" becomes "for all natural numbers $k, m$", with a suitable translation of statements about $S$ into statements about $k$ and $m$. However, in arithmetic we also consider functions with finite sequences as values, and sets of finite sequences. We therefore need to carry the reduction a step further, and replace the pair of natural numbers representing $S$ by a single number.

To achieve this, we need a correspondence between pairs of natural numbers and single natural numbers. We will use an enumeration obtained by grouping pairs on the basis of the sum of the two numbers:

$$\langle 0, 0 \rangle, \langle 0, 1 \rangle, \langle 1, 0 \rangle, \langle 0, 2 \rangle, \langle 1, 1 \rangle, \langle 2, 0 \rangle, \langle 0, 3 \rangle, \langle 1, 2 \rangle, \langle 2, 1 \rangle, \langle 3, 0 \rangle, \ldots$$

Here we get to the pair $\langle n, k \rangle$ by first going through the pairs that add up to numbers smaller than $n + k$, and then count the pairs $\langle m, n + k - m \rangle$ for $m = 0, \ldots, n$. There are $m + 1$ pairs that add up to $m$, so if we assign $\langle 0, 0 \rangle$ to 0, any later pair $\langle n, k \rangle$ is assigned the number

$$(1 + 2 + \cdots + n + k - 1) + n + 1 = ((n + k)(n + k + 1)/2) + n.$$

Thus if we define $OP(0,0) = 0$, and for other pairs

$$OP(n,k) = \frac{(n+k)(n+k+1)}{2} + n$$

we find that $OP$ is a one-one mapping between pairs of numbers and numbers (that is, different pairs correspond to different numbers, and every number corresponds to some pair), so there are functions $P_1$ and $P_2$ such that $P_1(OP(n,k)) = n$ and $P_2(OP(n,k)) = k$. From the definition of $OP$, we see that $P_1(m) < m$ and $P_2(m) < m$ for every $m > 1$.

Using this pairing function, we define the sequence number encoding the sequence $\langle a_0, \ldots, a_n \rangle$ as the smallest number $a > 1$ such that if $k = P_1(a)$ and $m = P_2(a)$, we have $\beta(k,m,0) = n+1$ and $\beta(k,m,i+1) = a_i$ for $i = 0, \ldots, n$. We denote this sequence number by $\langle a_0, \ldots, a_n \rangle$, that is, by the same expression used to denote the sequence itself. (Context will decide what is intended.) A number $a$ is a sequence number if $a = \langle a_0, \ldots, a_n \rangle$ for some $a_0, \ldots, a_n$. As a special case, the empty sequence number, denoted $\langle\ \rangle$, is the smallest $a > 1$ such that $\beta(P_1(a), P_2(a), 0) = 0$. More generally, we define $length(a) = \beta(P_1(a), P_2(a), 0)$ and $(a)_i = \beta(P_1(a), P_2(a), i+1)$. Since $\beta(k,m,i) \leq k$ for all $k$, $m$, $i$, it follows that $length(a) < a$ and $(a)_i < a$ for every $a, i$.

It may occur to the reader to wonder why we don't use the pairing function to define sequence numbers, since a sequence $\langle a_0, a_1, a_2 \rangle$ can be represented as $\langle\langle a_0, a_1 \rangle, a_2 \rangle$, and similarly for longer sequences. The answer is again (as in §2.6) that we want to be able to define the function giving the $i$-th element of a sequence number using only addition and multiplication.

By representing sequences through sequence numbers, we can formulate arithmetical statements as statements that refer only to natural numbers. But this does not necessarily mean that such reformulations are in every respect faithful to the original meaning of the reformulated statement. As an extreme case, consider the statement in §2.6 of the $\beta$-lemma: For every finite sequence $\langle a_0, \ldots, a_n \rangle$ of natural numbers there are natural numbers $k$ and $m$ such that $\beta(k,m,i) = a_i$ for $i = 0, \ldots, n$. Translating this using the representation of sequences by sequence numbers results in essentially the statement that for all $k$ and $m$ there are $k'$ and $m'$ such that $\beta(k',m',i) = \beta(k,m,i+1)$ for $i < \beta(k,m,0)$, which does not convey the same information. The $\beta$-lemma cannot be translated using our encoding of sequences in a way that preserves the information in the lemma, for it expresses the mathematical fact underlying the translation scheme itself. (Similarly, if we choose to represent sequences by exponents in prime factorizations as explained in §2.6, it won't be possible to state the fundamental theorem of arithmetic in any reasonable way.) However, for each choice of $n$, the information in the $\beta$-lemma applied to sequences of length $n$ can be faithfully expressed, by formalizing "for all numbers $a_0, \ldots, a_n$ there are numbers $k$ and $m$ such that $\beta(k,m,i) = a_i$ for $i = 0, \ldots, n$."

## 4.5. Arithmetical statements defined

We can now define an arithmetical statement as one built up from numerals and symbols for suitable number-theoretic functions and relations using the propositional connectives and quantifiers $\forall$ and $\exists$. The variables used with the quantifiers range over the natural numbers—that is, $\forall k$ is read "for every natural number $k$", and similarly for the existential quantifier.

But what are these suitable number-theoretic functions and relations? We'll begin with an answer that is useful only in a theoretical context: the basic relations are the equality relation $=$ and the ordering relation $<$ between natural numbers, and the basic functions are addition, multiplication, and the function mapping $n$ to $n+1$ (the *successor function*), which we denote by $s$. Of the numerals we keep only 0. This will be called a *primitive arithmetical statement*: one formulated using only $s, +, *, 0, <, =$ and the logical apparatus of the connectives and quantifiers.

An arithmetical statement which is not primitive is allowed to use, in addition to the basic relations and functions, any function and relation that can be defined by means of arithmetical statements. Such functions and relations will also be called arithmetical. Thus if we define a relation symbol $R$ by the equivalence

$$(1) \qquad R(k_1,\ldots,k_n) \equiv A(k_1,\ldots,k_n)$$

where $A$ is an arithmetical statement expressing a condition on $k_1,\ldots,k_n$, $R$ can also be used in arithmetical statements. If we define a function symbol $F$ by the equivalence

$$(2) \qquad F(k_1,\ldots,k_n) = m \equiv A(k_1,\ldots,k_n,m)$$

where $A$ is arithmetical, the function symbol $F$ can also be used in arithmetical statements. Definitions of the form (1) and (2) will be called *explicit* definitions of $R$ and $F$ respectively. They are explicit in the sense that it is assumed that the statement A does not itself contain $R$ or $F$, so that $R$ and $F$ are fully defined in terms of other functions and relations already introduced, together with some logical apparatus.

Note that (2) defines a function only if for all natural numbers $k_1,\ldots,k_n$ there is one and only one number $m$ such that $A(k_1,\ldots,k_n,m)$. So before saying "We define $F$ by (2)", we need to have proved that this condition is satisfied. In defining $R$ by (1), there is no corresponding condition.

Thus, beginning with the primitive arithmetical statements, we can use these to define function and relation symbols to be used in other arithmetical statements, and these new statements can in turn be used in defining further function and relation symbols. By using the definitions, any arithmetical statement can be translated into a primitive arithmetical statement, or can at least be assumed to be thus translatable for theoretical purposes. For this reason, explicit definitions are also sometimes called *eliminative* definitions. The translation consists in eliminating, given the definition (1), all statements of the form $R(t_1, \ldots, t_n)$ and replacing them with the corresponding statement $A(t_1, \ldots, t_n)$, and similarly (2) is used to replace a statement of the form $B(F(t_1, \ldots, t_n))$ with "there is an $m$ such that $A(t_1, \ldots, t_n, m)$ and $B(m)$".

When we are talking about arithmetical statements, and about other such matters having to do with formally defined languages, we need from time to time to clarify whether we are talking about symbols for functions and relations (called function symbols and relation or predicate symbols) or about functions and relations denoted by those symbols. Inevitably there will be a certain looseness of usage. Thus, above it was said that arithmetical statements are allowed to use "any function and relation that can be defined by means of arithmetical statements", or in other words any arithmetical function and relation. It may be thought that arithmetical statements more properly should be said to use *symbols* for those functions and relations, rather than the functions or relations themselves. Also, (1) and (2) were said to define a function symbol $F$ and a relation symbol $R$, but at the same time (1) and (2) were spoken of as defining a function and a relation, respectively. Both of these descriptions of (1) and (2) make good sense, from different perspectives. Defined function symbols and relation symbols always have, in any context where they have been defined, a unique definition, given by a statement such as (1) or (2). This definition determines, as a matter of convention, the meaning of the function symbol or relation symbol in question, and it is the definition we use when we eliminate defined symbols from statements, reducing them to primitive notation. The definitions imply that the function symbol is interpreted as a particular function, and the relation symbol as a particular relation. This function or relation is defined by (1) or (2) in the sense that it is the unique relation $R$ or function $F$ for which (1) or (2) is true. But any definable function can be defined in many ways. Given a definable function $F$, which is to say a function for which there is a statement $A$ such that (2) is true for all $k_1, \ldots, k_n, m$, there will be infinitely many different statements $A$ for which (2) holds. For example, the squaring function *square* can be defined by

$$\mathit{square}(a) = b \equiv b = a * a$$

but also by

$$\mathit{square}(a) = b \equiv (a + 1) * (a + 1) = b + 1 + 2a.$$

Thus defined symbols have, in any given context, a unique definition, but the relations and functions defined by these definitions can be defined in different ways (and of course be given different names).

A reader with a natural bent for logical pedantry may find it objectionable that in the above comments, there are references both to "the relation $R$

defined by (1)" and "the relation symbol $R$ defined by (1)". Is $R$ a relation or a relation symbol? The answer to this question is that we will sometimes use the letter "$R$" as the name of a relation, sometimes as a relation symbol associated with a particular definition, and sometimes as a name for itself. This last usage is illustrated by references to "the relation symbol $R$ defined by (1)". From experience in logic and philosophy, we know that it is sometimes important to be aware of these distinctions and to use a notation that brings them out, but we have also learned that there is such a thing as trying to be too precise.

The $\beta$-function and the other functions associated with sequence numbers can easily be shown to be arithmetical. For example, "$a$ is a sequence number" can be defined as "there is no $b < a$ such that $length(b) = length(a)$ and for every $i$ from 0 to $length(a) - 1$, $(b)_i = (a)_i$". In this definition, $(b)_i$ can in turn be defined using the $\beta$-function and the functions $P_1$ and $P_2$, and these can be defined using logic, addition, and multiplication.

What can not easily be seen is that for example the exponential function is definable in the language of arithmetic. $k^n$ is $k$ multiplied by itself $n$ times, but how is such a definition to be formulated using only addition, multiplication, and the logical apparatus of connectives and quantifiers? It is thus not clear that arithmetical statements as defined above can express even such central arithmetical results as the fundamental theorem of arithmetic. In fact, however, all the functions and relations commonly used in arithmetic can be defined by arithmetical statements, as we shall see. To convince ourselves of this, we first shift perspective a bit, to consider not the language of arithmetic, but number-theoretic functions and relations.

## 4.6. Primitive recursion

The *primitive recursive functions* are generated from certain basic functions by operations called composition and primitive recursion, in a way now to be explained.

The *basic functions* are the zero function $Z$, defined by $Z(k) = 0$ for all $k$, the successor function $s$, defined by $s(k) = k+1$, and the projection functions $P^i_k$, where $P^i_k$ is a $k$-ary function defined by $P^i_k(a_1, \ldots, a_k) = a_i$.

The function $f$ obtained by *composition* from an $n$-ary function $h$ and $k$-ary functions $g_1, \ldots, g_n$ is defined by

$$f(a_1, \ldots, a_k) = h\left(g_1(a_1, \ldots, a_k), \ldots, g_n(a_1, \ldots, a_k)\right)$$

The function $f$ obtained by *primitive recursion* from functions $h$ and $g$, finally, is defined by

$$f(0, a_1, \ldots, a_k) = h(a_1, \ldots, a_k)$$
$$f(b+1, a_1, \ldots, a_k) = g(b, a_1, \ldots, a_k, f(b, a_1, \ldots, a_k))$$

The primitive recursive functions are those that can be obtained by starting with the basic functions and repeatedly applying composition and primitive recursion to functions already obtained.

The arithmetical operations of addition, multiplication, and exponentiation are easily shown to be primitive recursive. For example, addition is primitive recursive since

$$a + 0 = 0$$
$$a + (b + 1) = s(a + b)$$

and similar equations define multiplication (using addition) and exponentiation (using multiplication). The use of $Z$ and the projection functions has been left implicit in the above equations, as will consistently be done in presenting definitions of primitive recursive functions.

A relation $R$ is said to be primitive recursive if its characteristic function is primitive recursive. The *characteristic function* of an $n$-ary relation $R$ is the $n$-ary function $C_R$ defined by

$$C_R(a_1, \ldots, a_n) = 1 \text{ if } R(a_1, \ldots, a_n),\ C_R(a_1, \ldots, a_n) = 0 \text{ otherwise.}$$

To speak of "the function obtained by primitive recursion from $h$ and $g$" is to presuppose that there is one and only function $f$ that satisfies the above two equations, which we will refer to as the *primitive recursion equations* determined by $h$ and $g$. But the equations do not give an explicit definition of $f$, unlike the definition of the function obtained by composition from other functions, so we need to be able to justify this presupposition.

Informally, it should be clear enough that the primitive recursion equations define a unique function $f$. In particular, if the functions $h$ and $g$ are computable (in the sense explained in connection with Euclid's algorithm in §2.3), the equations define a unique computable function $f$, for we can compute $f(b, a_1, \ldots, a_k)$ by successively computing

$$f(0, a_1, \ldots, a_k), f(1, a_1, \ldots, a_k), \ldots, f(b - 1, a_1, \ldots, a_k), f(b, a_1, \ldots, a_k),$$

using the first equation for the case 0, and after that the second equation together with the result of the previous computation. So, since the basic functions are clearly computable and both composition and primitive recursion give new computable functions from old, we find that all primitive recursive functions are computable.

We want, however, to see that every primitive recursive function can also be defined by an explicit arithmetical definition, and therefore take a more formal approach.

To prove that there can be *at most* one function $f$ satisfying the primitive recursion equations for given $h$ and $g$, we use mathematical induction. If $f_1$ and $f_2$ both satisfy the primitive recursion equations, we easily prove by ordinary mathematical induction on $n$ that $f_1(n) = f_2(n)$ for all $n$.

To prove that there is at least one function $f$ satisfying the equations, we give an explicit definition of $f$:

$f(b, a_1, \ldots, a_k) = c$ if and only if there is a sequence $c_0, c_1, \ldots, c_b$, where $c_b = c$, $c_0 = h(a_1, \ldots, a_k)$, and for every $i < b$, $c_{i+1} = g(i, a_1, \ldots, a_k, c_i)$.

That this explicit definition defines a function satisfying the recursion equations is again proved by mathematical induction on $b$. This definition also shows that all primitive recursive functions are arithmetical, since we know that finite sequences can be handled in the language of arithmetic, and there is no problem with expressing composition or the basic functions in that language.

In particular we see that the exponential function is arithmetical, since it is primitive recursive. So the fundamental theorem of arithmetic and other such statements involving exponentiation can be formulated in the language of arithmetic.

Primitive recursive functions will figure prominently in later chapters, for the following reasons. The primitive recursive functions are a subclass of the general recursive functions, which are just the computable functions formally characterized. When we introduce axiomatizations of arithmetic, an important property of general recursive functions will be that any computation involving such functions has a corresponding formal derivation in the axiomatic theory. This indeed will be the basis for the formal definition of the computable functions to be given in this book. Most of the computable functions actually introduced in the following will be primitive recursive functions, and the simplest way of showing them to be computable is often to show that they are primitive recursive. Further, the definitions of primitive recursive functions have the property that it is provable in all the axiomatic theories of arithmetic that we will consider that they do define a function, which is not always the case for definitions of general recursive functions. This property too will play an important role.

Next we'll introduce some general principles for showing that predicates and functions are primitive recursive. These principles will be used frequently in the following, although often tacitly.

## Connectives, bounded quantifiers, and the bounded $\mu$-operator

The bounded universal quantifier $\forall a < b$ is read "for every $a$ smaller than $b$", so that "$\forall a < b(a \mid b \supset a = 1 \vee a = b)$" means "for every $a$ smaller than $b$, if $a$ divides $b$ then $a = 1$ or $a = b$". Similarly the bounded existential quantifier $\exists a < b$ is read "for at least one $a$ smaller than $b$". Since we can define the bounded quantifiers using the ordinary quantifiers and propositional connectives, they can be used freely in arithmetical statements.

Less obviously, they can also be used in defining new primitive recursive functions and relations: we can apply both these bounded quantifiers and the usual propositional connectives to relations and functions already shown to be primitive recursive to obtain new primitive recursive functions and relations.

To see this, first consider the connectives. If the relation $R$ is defined by

$$R(a) \equiv S(a) \land Q(a)$$

where $S$ and $Q$ are primitive recursive, $R$ is also primitive recursive, since

$$C_R(a) = max\,(C_S(a), C_Q(a))$$

Here *max* is the function yielding the maximum of two numbers, and it is left to the reader to show that *max* is primitive recursive. Similar definitions show that the other propositional connectives applied to primitive recursive predicates yield new primitive recursive predicates. A similar argument also applies to definitions by cases: if $g_1$, $g_2$ are primitive recursive functions and $R$ is a primitive recursive relation, then the function $f$ defined by

$$f(a_1, \ldots, a_n) = \text{if } R(a_1, \ldots, a_n) \text{ then } g_1(a_1, \ldots, a_n) \text{ else } g_2(a_1, \ldots, a_n)$$

is also primitive recursive.

For the bounded universal quantifier, we note that if the relation $R$ is defined by

$$R(a, b) \equiv \forall x < aS(x, a, b)$$

where $S$ is primitive recursive, $R$ is primitive recursive since

$$C_R(0, b) = 1$$
$$C_R(n_1, b) = C_R(n, b) * C_S(n, a, b)$$

A similar argument applies to the bounded existential quantifier.

The bounded $\mu$-operator $\mu x < a$ means "the smallest $x$ smaller than $a$ such that". Thus $\mu x < a\ (x * x > a)$ is the smallest number smaller than $a$ with a square greater than $a$. Actually, when we use the bounded $\mu$-operator, there is not always a guarantee that there is any $x < a$ at all satisfying the indicated condition. We want the expression $\mu x < aA(x)$ to have a value anyway, and therefore take it to have the value 0 if there is no $x$ smaller than $a$ such that $A(x)$.

The bounded $\mu$-operator can also be used to define primitive recursive relations. For if $f$ is defined by

$$f(a, b) = \mu x < bR(x, a)$$

we have that

$$f(a, 0) = 0$$
$$f(a, n+1) = \text{if } \exists x < nR(x, a) \text{ then } f(a, n) \text{ else if } R(a, n) \text{ then } n \text{ else } 0$$

Finally we will establish that the arithmetical functions and predicates associated with sequence numbers are primitive recursive. First we note that the function $rm(a, b)$ = the remainder when dividing $a$ by $b$, or 0 if $b$ is 0, is primitive recursive. This is shown by the definition

$$rm(a, b) = \mu x < a \exists y < b\ (b = ky + x)$$

It follows that the $\beta$-function is primitive recursive. The quotient function, $quot(a, b)$ = the quotient when dividing $a$ by $b$ is similarly shown to be primitive recursive. It follows that the ordered pair function $OP$ is primitive recursive, and so are the functions $P_1$ and $P_2$. $P_1$, for example, is defined by

$$P_1(a) = \mu x < a \exists y < a\ (a = OP(x, y))$$

This tells us that the length function is primitive recursive, and also $(a)_i$ is a primitive recursive function of $a$ and $i$. Further, for every $n$, the function giving the sequence number of its $n$ arguments is primitive recursive:

$$\begin{aligned}\langle a_0, \ldots, a_n \rangle &= \mu a < expr \\ &\Big(\beta\big(P_1(a), P_2(a), 0\big) = n + 1 \wedge \beta\big(P_1(a), P_2(a), 1\big) \\ &\qquad = a_0 \wedge \cdots \wedge \beta\big(P_1(a), P_2(a), n\big) = a_n\Big)\end{aligned}$$

Here $expr$ stands for a suitable expression giving an upper bound for $k$ and $m$ in the $\beta$-lemma, established in §2.6. This expression is easily shown to define a primitive recursive function of $a_1, \ldots a_n$, so the function giving the sequence number of $a_1, \ldots a_n$ is also primitive recursive, by the above equation.

The property of being a sequence number is also primitive recursive:

$$\begin{aligned}seq(x) \equiv x > 1 \wedge \\ \neg\exists y < x\Big(length(y) = length(x) \wedge \forall i < length(x) (x)_i = (y)_i\Big)\end{aligned}$$

An operation on sequence numbers that will be used below is that of *concatenation*, which we denote by $*$. $\langle a_1, \ldots, a_n \rangle * \langle b_1, \ldots, b_m \rangle$ is the sequence number $\langle a_1, \ldots, a_n, b_1, \ldots, b_m \rangle$, while $\langle\rangle * s = s$ and $s * \langle\rangle = s$. To see that $*$ is a primitive recursive operation, we first note that for sequence numbers $s$ and $s'$, $s * s'$ is the smallest $s''$ such that $length(s'') = length(s) + length(s')$ and for every $i < length(s'')$, $(s'')_i = (s)_i$ if $i < length(s)$ and $(s'')_i = (s')_{i-length(s)}$ if $length(s) \leq i < length(s'')$. So it follows that $*$ is primitive recursive, provided there is an upper bound for $s * s'$ which is a primitive recursive function of $s$ and $s'$. Such an upper bound is again obtained from the upper bound for $k$ and $m$ in the $\beta$-lemma.

## Course-of-values recursion and relative recursiveness

For any $n+1$-ary number-theoretic function $g$, recursive or not, we let $\overline{g}$ be the function defined by $\overline{g}(b, a_1, \ldots, a_n) = \langle g(0, a_1, \ldots, a_n), \ldots, g(b-1, a_1, \ldots, a_n) \rangle$ for $b > 0$, and $\overline{g}(0, a_1, \ldots, a_n) = \langle\rangle$. Thus $\overline{g}(b, a_1, \ldots, a_n)$ collects into a sequence number the sequence of values of $g(x, a_1, \ldots, a_n)$ for $x$ from 0 to $b-1$. If $g$ is primitive recursive, so is $\overline{g}$, as can be seen by formulating suitable recursion equations using the concatenation operator. A more informative statement is that $\overline{g}$ is *primitive recursive in* $g$. What this means is that the function $\overline{g}$ is obtainable using composition and primitive recursion, starting from the basic primitive recursive functions together with $g$. Generally, to say that a function $g$ is primitive recursive in a set $M$ of functions is to say that it is obtainable through repeated application of composition and primitive recursion, beginning with the basic primitive recursive functions and the functions in $M$. Conversely, $g$ is primitive recursive in $\overline{g}$, since $g(b, a_1, \ldots, a_n) = (\overline{g}(b+1, a_1, \ldots, a_n))_b$.

Now suppose we define a function $g$ by

(3) $$g(b, a_1, \ldots, a_k) = h(b, a_1, \ldots, a_k, \overline{g}(b, a_1, \ldots, a_k))$$

This equation defines $g(b, a_1, \ldots, a_k)$ in terms of the values of $g(x, a_1, \ldots, a_k)$ for $x$ smaller than $b$. To see that the function $g$ thus defined is primitive recursive in $h$, we note that it implies the following equations for $\overline{g}$:

$$\overline{g}(0, a_1, \ldots, a_k) = \langle\rangle$$

$$\overline{g}(b+1, a_1, \ldots, a_k) = \overline{g}(b, a_1, \ldots, a_k) * \langle h(b, a_1, \ldots, a_k, \overline{g}(b, a_1, \ldots, a_k)) \rangle$$

Thus $\overline{g}$ is primitive recursive in $h$, so the same holds for $g$.

A definition of the type (3) is called a definition by *course-of-values recursion.* The fact that we can use course-of-values recursion in defining primitive recursive functions and predicates will often be tacitly used in later chapters.

## 4.7. Summary

A formal language has been introduced, the language of arithmetic. Everything that can be expressed in this language can in theory be expressed using only the logical apparatus of connectives and quantifiers (over the natural numbers) together with symbols for identity, multiplication, addition, zero, and the successor function. In practice we also use other symbols, defined using the primitive notation of the language. A number-theoretic function or relation is *arithmetical* if there is a statement in the language of arithmetic defining (in the sense of uniquely characterizing) that function or relation. Among the arithmetical functions and relations are those that are primitive recursive. All primitive recursive functions are computable, and among them are found all the computable operations commonly used in arithmetic, such as the greatest common divisor, the factorial, exponentiation. It will be shown in later chapters that not every arithmetical function or relation is computable, and that not every computable function or relation is primitive recursive. Because it is possible to code finite sequences of numbers as sequence numbers, which can be defined in the language of arithmetic, it is also possible to express in the language of arithmetic all statements that have been informally described as arithmetical. In later chapters we will also introduce formal axiomatic theories for proving statements in the language of arithmetic.

CHAPTER 5

# THE LANGUAGE OF ANALYSIS

## 5.1. The natural numbers revisited

The natural numbers can be described as the numbers obtainable from 0 by repeated application of the operation "add one". Here the second "numbers" is redundant, since everything obtained from a number by adding one is a number, so that an equivalent formulation is that the natural numbers are the (mathematical) objects obtainable from 0 by repeatedly adding one.

We can take a more abstract view of the operation of adding one to a natural number. Forgetting about the concept of addition, we just call this operation the *successor operation* and denote it by $s$. This operation applied to a natural number $n$ yields the next greater number $s(n)$, more usually written $n+1$. For the series $0, s(0), s(s(0)), \ldots$ we normally use decimal notation $0, 1, 2, \ldots$.

Just what is the operation $s$? If we are only interested in the *structure* of the number series, $s$ is characterized by being an injective function—meaning that $s(x) = s(y)$ only if $x = y$—for which $s(x)$ is never identical with 0. These two properties guarantee that repeated applications of $s$ starting from 0 will yield different numbers $0, s(0), s(s(0)), \ldots$. This in turn implies that any two such sequences, using different starting objects 0 and $0'$ and different successor functions $s$ and $s'$ are *isomorphic*, that is, have identical mathematical structure. In formal terms: there is one-one correspondence $F$ between the two sequences such that $F(0) = 0'$ and $F(s(a)) = s'(F(a))$ for every $a$ in the first sequence. That $F$ is a one-one correspondence, or *bijection*, means that every element in the first sequence is mapped to an element in the second sequence, and every element in the second sequence is the image of one and only one element in the first sequence.

In computations, we use various concrete representations of some finite subset of the natural numbers. We may be manipulating numbers written down in decimal notation or numbers stored as bit sequences in computer memory. There is then some particular manipulation of these representations that leads from the representation of a number $n$ to the representation of $s(n)$. There will necessarily be a limit on the physical size of representations and on how many times the concrete successor operation can be applied, depending on our computational resources. In mathematical reasoning, we disregard all such limitations, as well as all uncertainties surrounding the physical properties of actual representations of natural numbers, and simply either stipulate or prove mathematically that $s$ has the properties required of it: for any number $x$, $s(x)$ is also a number; if $s(x) = s(y)$ then $x = y$; $s(x)$ is never 0. We then go on to prove the existence, in a purely mathematical sense, of large natural numbers, and to use notations for them, such as $2^{1000}$, which we manipulate in our calculations.

The statement that the natural numbers are the objects obtainable from 0 by repeated application of the successor operation can be expressed in different but equivalent terms. One such variant formulation states, redundantly, that the natural numbers are the numbers obtainable from 0 by repeatedly applying the successor operation any finite number of times. Here "any finite number of times" is redundant because it has no obvious meaning to speak of the object or objects obtained by applying the successor operation an infinite number of times. As long as we haven't presented any explanation giving meaning to the phrase "applying the successor operation an infinite number of times" there is no point in excluding such infinite application.

Another equivalent formulation is in terms of *rules*:

Rule 1: 0 is a natural number.

Rule 2: If $a$ is a natural number, $s(a)$ is a natural number.

The above two rules are often referred to as rules for *generating* the natural numbers. This terminology shouldn't be taken too literally, since there isn't usually any actual generating going on when we're talking about the natural numbers. Rather, the two rules generate the natural numbers only in the sense that they describe the natural numbers as the objects obtainable (in a mathematical or idealized sense) from the number 0 by iterating the successor operation.

As they stand, rules 1 and 2 are compatible with other objects than $0, s(0), s(s(0)), \ldots$ being natural numbers, so when the natural numbers are characterized by these rules, there is an implicit or explicit third rule, sometimes formulated

Rule 3: Nothing is a natural number unless its being so follows by rules 1 and 2.

This last rule is sometimes called the *extremal clause*, with Rule 1 being the *basic clause* and Rule 2 the *inductive clause*. The "follows by rules 1 and 2" in Rule 3 is to be understood to mean that a natural number can be shown to be such only by repeated application of the two rules, which is again to say that the natural numbers are *those and only those* objects obtainable by repeatedly applying the successor operation, starting with 0.

Yet another way of putting this is to say that the natural numbers are "$0, s(0), s(s(0))$, and so on." It might be argued that this "and so on" is vague and indefinite, but in fact—that is, in actual usage—it is no more vague and indefinite than any of the other formulations. Unintended or deviant interpretations of "and so on" can indeed be imagined, but this is equally true of the other formulations. However, the "and so on" formulation is less explicit: it doesn't immediately suggest any formal rules for drawing conclusions about all natural numbers, such as we need to be able to formulate in a mathematical context.

In a set-theoretical version of the characterization of the natural numbers in terms of rules, $n$ is a natural number if and only if it belongs to every set $A$ which contains 0 and is closed under the successor operation, that is, for which $s(a)$ is in $A$ whenever $a$ is in $A$. This formulation presupposes that there *are* sets that contain 0 and are closed under the successor operation, and thus presupposes the existence of infinite sets. There are variants of the set-theoretical characterization which do not make this presupposition. One such (see Feferman and Hellman [1999], Quine [1969]) states that $n$ is a natural number if and only if every set that contains $n$, and contains $a$ whenever it contains $s(a)$, contains 0. This expresses that $n$ is a natural number if $n$ is identical with 0 or 0 is obtainable from $n$ by repeatedly applying the predecessor operation (going from $s(n)$ to $n$).

The various equivalent characterizations of the structure of the natural number series given above haven't been referred to as *definitions* of the natural numbers. This is because it has not been assumed that either 0 or the successor operation has been given a definition, but only that 0 and $s$ have the properties stipulated. The descriptions uniquely characterize the structure of the natural numbers (in the sense explained), but they don't uniquely specify any particular instance of that structure. In some contexts we wish to define a specific such instance, and this can be done in different ways. For example, if we are talking about the real numbers, we can single out the natural numbers as a subset of the reals by identifying 0 with the real number 0, and $s(n)$ with $n+1$, where $+$ is the addition operation for real numbers. The set-theoretical characterization then defines the natural numbers as a subset of the reals. In pure set theory, Zermelo [1908] defined the natural numbers by taking 0 to be the empty set, and $s(n)$ to be the set $\{n\}$. Today, it is more common in a set theoretical context to define the natural numbers as the so-called finite von Neumann ordinals, taking 0 to be the empty set and $s(n)$ to be the set $n \cup \{n\}$. For the purposes of the philosophical discussions in this book, the natural numbers can be identified with the objects in any such sequence, since the questions considered turn only on the structure of the natural number series, not on the individual identity of natural numbers.

The conception, view, or explanation of the natural numbers as generated by indefinitely repeating a simple operation is no doubt fundamental to our understanding of them. This kind of definition or characterization of a structure or set as generated by repeated application of one or more operations starting with some basic objects, which is called an *inductive* definition or characterization, is one that is used all the time in logic, and we have in fact already used it in several cases without being explicit about it. Inductive definitions will be given closer consideration in the next chapter. In this chapter, the main topic is a category of mathematical objects that is *not* explained or understood inductively, namely that of *sets* of natural numbers.

## 5.2. Sets and functions

Just as rational numbers and integers can be reduced, in the sense explained in earlier chapters, to natural numbers, so real numbers can be reduced in various ways to *subsets* of the set **N** of natural numbers, or to *functions* from **N** to **N**. For example, a real number $r$ can be identified, for foundational purposes, with the set of rational numbers smaller than $r$, which can in turn be identified with a set of natural numbers, via some representation of rational numbers in terms of natural numbers. An explicit definition of the real numbers in such terms was first given by Dedekind in 1872. No such reduction will be used in this book, because we will not be concerned with the formalization of statements about real numbers or of proofs that use real numbers. But in addition to the language of arithmetic, described in the preceding chapter, we will also use what will be called the *language of analysis* (or the language of *second order arithmetic*), which in addition to variables ranging over the natural numbers has variables ranging over *sets* of natural numbers. The name "language of analysis" derives from the fact that a large part of analysis (mathematics involving real numbers, sets and sequences of real numbers, limits, integrals, logarithms, and so on) can be formulated in this language. In the context of this book, we will instead be concerned with the possibility of proving *arithmetical* statements using this language (and suitable axioms).

Sets and functions (by which, in this chapter, will be meant sets of natural numbers and number-theoretic functions, unless otherwise noted) can be finite or infinite. If we restrict ourselves to *finite* sets, and to functions defined on finite sets, they can be described as generated by rules, just like the natural numbers. For example, the finite sets are generated by the rules

The empty set $\varnothing$ is a finite set with no members.

If $k$ is a natural number and $A$ a finite set, $A \cup \{k\}$ is a finite set, the members of which are $k$ and the members of $A$.

Combining these with the two rules for generating natural numbers, we have four rules for generating the finite sets of natural numbers from 0 and $\varnothing$, using the successor operation and the operation of adding a number to a set. This means that they form a structure generated from a finite number of initial objects by a finite number of rules, just like the natural numbers. And indeed we know that for theoretical purposes we can replace all reasoning involving finite sets with reasoning about numbers, by representing sets as numbers (using sequence numbers or in some other way).

*Infinite* sets and functions are a different matter. They do not in any obvious way form a totality that can be described or defined as generated by rules. In the foundations and philosophy of mathematics, infinite sets have been and continue to be a controversial topic, and in logic there are distinctions between different categories and uses of infinite sets.

### Sets as convenient fictions

What is a set of natural numbers? One possible answer is that sets are only convenient *fictions*, in the sense that statements or arguments apparently referring to sets can be reformulated so as to avoid such references. We know that this is in fact always the case when we restrict ourselves to finite sets, but sometimes it can also be done for infinite sets. Consider for example the proof in §2.3 involving the set $I$ of integers of the form $kx + my$. The role of $I$ is only to make it easier to formulate the argument, and every reference to $I$ could be eliminated, at the cost of making formulations somewhat longer and clumsier, by using instead the condition defining $I$. Following Quine [1969] we can call sets thus conceived *virtual* sets. Their use is mainly guided by two principles, the principles of *abstraction* and *extensionality*. The principle of abstraction tells us that for some range of conditions $F(x)$, expressed in some suitable language, there is a set $\{x \mid F(x)\}$ (the set of natural numbers $x$ such that $F(x)$), and it holds that

$$\forall k(k \in \{x \mid F(x)\} \equiv F(k))$$

Here $\in$ is read as "is a member of", and the set is said to be *defined by abstraction* using the condition $F(x)$. We say that a set is *definable* in a language $L$ if it is identical (in the sense of the principle of extensionality) with a set defined by abstraction using some condition in the language $L$.

The principle of extensionality tells us that for any sets $A$ and $B$,

$$A = B \equiv \forall k(k \in A \equiv k \in B)$$

That is, the identity of a set is only a matter of what members it has.

The principle of abstraction and the principle of extensionality do not define the term "set", but give *contextual definitions* which allow us to eliminate or rewrite references to sets in specific contexts. Other such contextual definitions can be introduced, for example

$$A \subseteq B \equiv \forall k(k \in A \supset k \in B)$$
$$k \notin A \equiv \neg k \in A$$
$$A \text{ is infinite} \equiv \forall k \exists n(n > k \land n \in A)$$

These definitions do not need to be taken as implying the existence of any such objects as sets. Instead we use them to eliminate references to sets in specific contexts. For example, that the set of prime numbers is an infinite subset of the natural numbers is the same as saying that every prime number is a natural number, and for every natural number there is a greater prime number. If our use of sets is such that contextual definitions can be used to *eliminate* all references to sets in a proof of an arithmetical statement, we are justified in regarding such references as a mere manner of speaking in that proof. This can be done in the case of the argument using the set $I$. Instead of saying "$n$ is a member of $I$" we can say, in accordance with the principle of abstraction, "$n$ is a number of the form $kx + my$", and the argument still goes through.

## Definable sets and diagonalization

In many contexts, however, references to sets cannot be eliminated in this way. Instead we must think of sets as mathematical objects in their own right. When we do, we always think of them as purely extensional collections, which is to say as satisfying the principle of extensionality, but we may or may not think of them as definable in some particular language using the principle of abstraction. We speak, for example, of *arithmetical* sets when we wish to restrict attention to sets definable by abstraction in the language of arithmetic. Thus the set of primes is arithmetical, and the set of sentences in a formal language (once such sentences have been defined as numbers, as will be done in Chapter 7).

By enumerating the conditions $F_1(x), F_2(x), \ldots$ which define sets in some particular language $L$, we can define a new set $D$ by *diagonalization*, as the set of $k$ such that $F_k(k)$ is not true. If $M_k$ is the set $\{x \mid F_k(x)\}$ we see that $D$ is different from all the $M_k$, since for every $k$, $k \in M_k$ if and only if $k \notin D$. This is the diagonal construction first introduced by Cantor and later applied in many different ways in logic, in particular in the theory of computable functions, and also in the proof of Gödel's theorem.

Since $D$ is different from all the $M_k$, and therefore not itself definable in $L$, the definition of $D$ must use some concept not available in $L$. If we can talk in $L$ about enumerations of conditions and similar things, this can only be the concept of truth for sentences in $L$. And indeed, as we shall see in Chapter 7, we can use just this argument to show that "true arithmetical sentence" cannot be defined in the language of arithmetic.

Given a sufficiently liberal notion of set, the diagonal construction yields a straightforward proof that not every set is definable in $L$, but the construction does not compel us to conclude that there are sets not definable in $L$. It is perfectly possible, for example, to stipulate that by "set" is meant "arithmetical set", and thus simply disallow the use of "true arithmetical sentence" in defining sets. This, however, is not very satisfactory if we are exploring a general concept of "definable set". Diagonalization seems to show that there is an inexhaustibility phenomenon for definability similar to that for provability: whatever language we introduce for defining sets, we can extend that language so as to be able to define more sets. In particular, if we start with the language of arithmetic and arithmetically definable sets, we can go on to consider sets definable in terms of natural numbers and arithmetical sets, and so on. To refer to a supposed totality of definable sets, containing all sets definable in *any* language, is problematic in several ways. What totality of languages and associated definitions are we talking about, and how are the interpretations of those languages determined? Does it make good sense to speak even of the totality of sets definable in the English language, and if so, what are we to make of the diagonalization of an enumeration of those sets? These are not questions that will be considered in this book, but they are posed here only to emphasize the problematic character of the general concept of "definable set". However, the suggested hierarchy of definable sets beginning with the arithmetical sets has a connection with questions about inexhaustibility in connection with Gödel's theorem, as will emerge in later chapters.

### Arbitrary sets

Instead of assuming sets to be definable, we may in other contexts emphasize that we are talking about what are called *arbitrary* sets of natural numbers, for which there is no assumption that they are definable by abstraction in any particular language, or indeed that they are in any sense definable at all.

It should be noted that this notion of arbitrary set does not entail any notion of "absolutely undefinable set". An arbitrary set is not assumed to be in any way definable, but we cannot conclude from this that it may be in some absolute sense undefinable, or that it makes any sense to speak of "absolutely undefinable sets". Rather, definability simply doesn't enter into it when we talk about arbitrary sets. Instead we think of these sets as "extensional totalities", where infinite sets differ from finite sets only in being infinite, and reason freely about these totalities as though they could be inspected and manipulated, just as we do when reasoning about finite sets. Paul Bernays (see Bernays [1935]) coined the expression "quasi-combinatorial" to describe this way of thinking about infinite sets.

The diagonal construction applied to arbitrary sets shows that there is no enumeration $M_0, M_1, \ldots$ of all subsets of the natural numbers. If we introduce in our mathematical reasoning the set of all subsets of **N**—the *power set* of **N**—this will be an *uncountably infinite* set, which is to say an infinite set $A$ for which there is no (arbitrary) function $f$ from **N** to the whole of $A$, or in other words no sequence $a_0, a_1, \ldots$ containing every element of $A$.

Unlike the case of definable sets, there is nothing in the diagonal construction applied to arbitrary sets that suggests any inexhaustible series of extensions of the notion of arbitrary set. The reason is that the ("quasi-combinatorial") idea of an arbitrary set of natural numbers is not connected with any particular enumeration of such sets, unlike the idea of arithmetical or otherwise definable sets of natural numbers. In this respect the notion of an arbitrary set is less problematic than that of a definable set. The diagonalization of a sequence of arbitrary sets is simply itself a set in the same sense as the sets in the sequence.

Yet the notion of an arbitrary subset of the natural numbers is often also considered problematic. There is a difference between the natural numbers, generated from 0 by iterating the successor operation, and arbitrary sets of natural numbers, which are not on the face of it generated in any way at all. If we do speak of them as generated, it can only be as generated by "infinite sequences of choices". Thus any infinite binary sequence 01010101... determines a set of natural numbers, with each 1 or 0 determining whether the corresponding natural number 0, 1, 2,... belongs to the set or not, and conversely every set of natural numbers determines such a sequence. However, "make an infinite sequence of binary choices" is not a rule in the sense that "add 1" is a rule—it is not a mathematical idealization of an operation that can be carried out concretely when this is computationally feasible—and so those who feel ill at ease with talk of arbitrary sets will not regard "making an infinite sequence of binary choices" as an operation that generates the sets of natural numbers.

One common and natural way of expressing this unease about the notion of arbitrary set is to ask what it means for an arbitrary set to *exist*, if it is not assumed to be definable. Sometimes this question is based on misgivings which are just as applicable to definable sets, or to the natural numbers themselves, which after all have no apparent physical existence. We don't need to establish the existence of natural numbers in any physical or metaphysical sense in order to work with them in mathematics, and there is no a priori reason why we should need to do so in the case of sets. But the question more often springs from a perceived difference between "theoretically possible finite sequences" and "theoretically possible infinite sequences", where the former are perceived as more definite and unproblematic. Even though "arbitrary natural number" is a purely theoretical construct, it is widely seen as a simpler and more readily comprehensible construct than "arbitrary set of natural numbers". In practical terms, this is seen in the fact that there is no distinction in mathematics between different notions of "natural number", whereas different subsets of the power set of **N** are singled out and used in different mathematical contexts.[1]

---

[1] Distinctions between different notions of "natural number" are made in "ultrafinitist" philosophy of mathematics, which does not however correspond to anything in ordinary mathematical practice (unlike "finitist" philosophy of mathematics).

## Speaking of arbitrary sets

In spite of such misgivings, the notion of an arbitrary set of natural numbers is in many contexts the clearest and most adequate concept to use. Consider the induction principle for the natural numbers, formulated in terms of sets: for any set $X$, if 0 is in $X$ and $s(a)$ is in $X$ whenever $a$ is, then every natural number is in $X$. Here "for any set $X$" cannot be replaced by "for any set $X$ definable by abstraction in the language $L$" for any language $L$. The induction principle does indeed hold, for example, for arithmetically definable sets as a special case, but it is not limited to sets definable in any particular language. We might choose to formulate it as a principle for *definable* sets in general, but this would be to introduce the problems surrounding the notion of "definable set" in a context where they are irrelevant.

Let's compare the formulation of the induction principle in terms of arbitrary sets with another formulation given earlier: whatever is true of 0, and is true of $s(a)$ whenever it is true of $a$, is true of all natural numbers. This formulation conveys essentially the same information as the set-theoretical formulation. What the set-theoretical formulation adds is not a precise specification of just what "whatever" refers to (which would require a precise specification of just how sets can be defined), but a restriction of the "whatever" to conditions that are mathematically precise, and in particular definite, in the sense of not being vague. For the classical paradox of the heap gives an apparent counterexample to the induction principle (since we cannot from "1 dollar is definitely cheap for a house", and "if $n$ dollars is definitely cheap for a house, then $n + 1$ dollars is definitely cheap for a house" conclude "for every $n$, $n$ dollars is definitely cheap for a house"). When the principle of mathematical induction is presented and seen as compelling, we are always thinking of or invoking "definite properties" or "well-defined sets". Just what is the range of this notion of "definite property" or "well-defined set" may be unclear, but the induction principle is no more evident formulated in terms of properties or sets definable in some particular language than it is when formulated in terms of arbitrary sets. On the contrary, the induction principle restricted to sets definable in a particular language is evident only because we recognize it as a particular instance of the general principle. That this is the case emerges when we set out to justify the induction principle. If we wish to make the principle applied to arithmetical sets—sets definable in the language of arithmetic—convincing or evident to ourselves or to somebody else, our argument will not make any use of the structure of arithmetical sets. Instead we will give arguments in pictures or words that apply to arbitrary sets.

There is however a significant distinction between the use of arbitrary sets in such contexts as the formulation of the induction principle, which we might call *schematic* uses, and their use in contexts where we quantify over arbitrary sets in defining other sets. If we accept the notion of an arbitrary set, we can go on to consider sets definable in terms of this notion, as is done systematically in descriptive set theory. Suppose for example we define a set $A$ of numbers as the set of all $k$ for which some condition $P(k, B)$ holds for every set $B$. Here we will in general define different sets $A$ depending on whether "every set $B$" refers to arbitrary sets, to arithmetical sets, or to some other special category of sets. Also, any uncertainty we may feel about the notion of an arbitrary set will in this case imply an uncertainty about the set $A$, which we may or may not regard as well-defined. This is in contrast to the schematic use, as exemplified by the induction principle or by the proof that $X \cup Y = Y \cup X$, where we recognize a principle as valid for arbitrary sets and can thereafter freely use it for any sets that we may introduce. There is in this case nothing that will be interpreted differently or have a different truth value depending on just what sets we are talking about, but only a general schema which we accept as valid whatever sets we may later be talking about.

## 5.3. The language of analysis defined

The language of arithmetic, as defined in Chapter 4, is what is called a *first order* language. What this means is that there are only variables ranging over natural numbers, and none ranging over sets of natural numbers, or number-theoretic relations or functions. The language of *second order arithmetic*, also called the language of analysis, extends the first order language by adding variables $X, Y, \ldots$ for sets of natural numbers, with corresponding quantifiers. We also add equalities between set variables and the symbol $\in$ for the membership relation between natural numbers and sets. Thus in the second order language, we can say such things as "$k$ belongs to every set which contains 0 and is closed under $F$", where $F$ is some arithmetically definable function.

A statement in the language of analysis will be called *arithmetical* if it contains no *bound* set variables. As will be seen in Chapter 10, it is only when we start using quantifiers on set variables that sets cannot be explained away as a theoretically inessential convenience.

There is an important difference in semantics between the language of arithmetic and the language of analysis. There is never any question of how the first order logical apparatus is to be understood in an arithmetical statement. In particular, the variables always range over the natural numbers. In the case of a second order statement, as has been commented on above, it is perfectly possible to have in mind different interpretations of the set variables in different contexts. In one context we may wish to interpret them as ranging over arithmetical sets, in another as ranging over arbitrary sets. Thus when we speak of the truth or falsity of a second order statement, we need to stipulate a range for the set variables. The general rule is that unless there is a statement to the contrary, the variables are taken to range over arbitrary sets.

There is no necessity about introducing a second order language, with two styles of variables, in talking about numbers and sets. We could introduce instead a first order language in which we talk about both numbers and sets. The statements in this first order language would quantify over "objects", where an object can be either a number or a set of numbers, and we could then make the same remarks about the interpretations of "$x$ is a number" and "$x$ is a set" in this language as were made above for the second order language. However, there are logical distinctions between first and second order languages that are more clearly brought out by using two styles of variables.

Sometimes it is more convenient to speak of *functions* than of sets. As noted above, there is a natural one-one correspondence between sets and infinite binary sequences, or functions from $\mathbf{N}$ to $\{0, 1\}$. We can translate statements about sets in our language into statements about functions, and conversely. For the translation from sets $A$ to binary sequences $a$, we just replace $k \in A$ by $a(k) = 1$, and a statement "for every set $A, \ldots$" becomes "for every function $a$ from $\mathbf{N}$ to $\{0, 1\}, \ldots$". For the converse, we identify a function $a$ from $\mathbf{N}$ to $\mathbf{N}$ with its *graph*, the set of $\langle k, a(k)\rangle$ for natural numbers $k$, and then interpret the pairs $\langle k, a(k)\rangle$ as sequence numbers. So "for every function $f, \ldots$" is translated "for every set $A$, if for every $n$ there is a unique $k$ such that $\langle n, k\rangle$ is in $A, \ldots$" and $f(k) = m$ is translated $\langle k, m\rangle \in A$.

Thus it makes no difference from a theoretical point of view whether we have function variables or set variables or both in our second order language. We will assume that we have set variables only, but could as well include function variables.

CHAPTER 6

# ORDINALS AND INDUCTIVE DEFINITIONS

## 6.1. Two related concepts

### Inductive definitions

The inductive definition or characterization (depending on whether zero and the successor function have been defined or not) of the natural numbers was commented on in Chapter 5. Other inductively defined sets have been introduced in earlier chapters, but with less emphasis on their inductive structure. Take for example the statements of arithmetic. These are generated from atomic statements of the form $P(t_1, \ldots, t_n)$ by rules for forming compound statements using connectives and quantifiers. The $t_1, \ldots, t_n$ are *terms*, expressions formed using function symbols, variables, and numerals. The terms are also inductively generated, starting with variables and numerals. Similarly the primitive recursive functions are generated from the basic primitive recursive functions using composition and primitive recursion.

In logic we very often use such inductive definitions, defining sets and relations as the sets and relations generated from some basic objects using some kind of rules or operations. Several more such definitions will be introduced in later chapters. Most of them are similar to the ones that have been given so far, and can be understood informally in analogy to the basic example of the natural numbers. We will however also have occasion to use inductive definitions that differ from the ones mentioned in an essential respect. The natural numbers, the formulas of a language, the theorems of a theory, are generated by rules that refer only to a *finite* number of objects, and there are corresponding finite constructions showing that something is a natural number, a formula, a theorem, and so on. In inductively defining (in later chapters) arithmetical truth or ordinal notations, we use rules with *infinitely* many premises, and therefore an explanation of such definitions in set-theoretical terms may be helpful.

The property of sets of being *countable* will be prominent in this chapter. A set is countable if it is either finite or denumerably (or countably) infinite, that is, there is a one-one correspondence between the members of the set and the natural numbers. In other words, a set is countable if it is not uncountably infinite in the sense defined in §5.2.

## Ordinals

It is not difficult to appreciate how we arrive at the idea of counting beyond the natural numbers in connection with Gödel's theorem and the inexhaustibility phenomenon. Given a theory $T$ to which Gödel's theorem applies and which we accept as valid, we will also accept as valid the theory $T'$ obtained by adding to $T$ the axiom "$T$ is consistent". By Gödel's second incompleteness theorem, $T'$ is logically stronger than $T$. But Gödel's theorem applies to $T'$ as well, and we get an even stronger theory $T''$ by adding to $T'$ the axiom "$T'$ is consistent". Continuing in this way we get an infinite sequence

$$T_0, T_1, T_2, \ldots$$

of theories, where $T_0$ is $T$ and $T_{i+1}$ is obtained from $T_i$ by adding a new axiom "$T_i$ is consistent". We call this an *$\omega$-sequence* of theories, which means that it is a sequence of elements indexed by the natural numbers $0, 1, \ldots$. Inevitably we reflect that this process can be taken further. For if we let the theory $T_\omega$ have as axioms all the axioms added in the above series, this is also a valid theory, which can be strengthened by adding the new axiom "$T_\omega$ is consistent".

The question how far this process can be continued, and where it takes us, will be prominent in later chapters. Here we will introduce in abstract terms the numbers used to count such infinite sequences of theories. We call these numbers *countable ordinal numbers*, or just *ordinals* for short, since no other ordinal numbers will be considered in this book. The theory $T_\omega$ is indexed by the number $\omega$, which is the first *infinite* ordinal, that is, the first ordinal which comes after all of the natural numbers $0, 1, \ldots$. Once we have $\omega$, we can collect the theories thus far defined into a new theory, and then iterate our procedure (in this case: the procedure of forming a theory by adding a consistency statement as a new axiom) once more, and then again, getting a new infinite sequence of theories $T_\omega, T_{\omega+1}, \ldots$. The theories in this sequence are indexed by infinite ordinals denoted $\omega, \omega+1, \omega+2, \ldots$, where the next ordinal in the series is obtained by adding 1 to its predecessor. It is characteristic of the countable ordinals that as soon as we have managed to define such a countable increasing sequence of ordinals, we can count beyond that sequence, to a new ordinal. Thus after all of the ordinals $\omega, \omega+1, \omega+2, \ldots$ comes an ordinal denoted $\omega+\omega$. If we look at all the ordinals smaller than $\omega+\omega$, they form two infinite $\omega$-sequences, one following the other:

$$0, 1, 2, \ldots, \omega, \omega+1, \omega+2, \ldots, \omega+\omega$$

The ordinal $\omega+\omega$ is also denoted $\omega * 2$, and if we consider ordinals $\omega * 3$, $\omega * 4, \ldots$ obtained in a similar way, we get a countable sequence of countable sequences of increasing ordinals:

$$0, 1, 2, \ldots, \omega, \omega + 1, \omega + 2, \ldots,$$
$$\omega * 2, \omega * 2 + 1, \ldots, \omega * 3, \omega * 3 + 1, \ldots, \omega * 4, \ldots$$

and after all these ordinals comes an ordinal denoted $\omega * \omega$ or $\omega^2$. And so it continues.

The above description illustrates two intertwined aspects of the countable ordinals. They can be defined and studied in the style of abstract set theory, without regard to questions of definability and computability, but as soon as we are concerned with either the visualization of ordinals or the formalization of statements involving particular ordinals—such as statements about what can be proved in some infinite sequence of theories—we will also be concerned with how we can denote ordinals. In this book we will need both a basic grasp of the abstract set-theoretical aspects of ordinals, presented in this chapter, and some acquaintance with ways of denoting ordinals, as explained in Chapter 11.

## 6.2. Rules and their associated inductive definitions

From a set-theoretical perspective, we define a *rule* to be any way of associating with a set $X$ of *premises* a set of *conclusions* of $X$. Here the "premises" and "conclusions" can be any objects—the terminology is just intended to be suggestive. The pair $\langle X, x \rangle$ is called an *instance* of the rule $R$ if $x$ belongs to the set of conclusions of $X$ under $R$. If $X$ is empty, $x$ is also called an *axiom*.

A set $A$ is *closed* under the rule $R$ if $x \in A$ whenever $\langle X, x \rangle$ is an instance of $R$ and $X \subseteq A$. That is, $A$ is closed under $R$ if every conclusion that can be obtained using the rule $R$ from premises in $A$ also belongs to $A$. As a special case of this, every axiom of $R$ must belong to $A$.

If all sets $A$ in some collection $C$ of sets are closed under $R$, then the intersection $\cap C$ of all the $A$ in $C$, which contains those elements that belong to every $A$ in $C$, is also closed under $R$. For any set of premises included in the intersection will be included in each of the $A$, so any consequence of those premises will also belong to each of the $A$.

The set $M_R$ *generated* by $R$ is the smallest set closed under $R$. "Smallest" refers to set inclusion: $M_R \subseteq A$ for every set $A$ closed under $R$. $M_R$ exists as long as there is any set $M$ closed under $R$, for we can then define $M_R$ as the intersection of all $M$ closed under $R$, which is also closed under $R$, as noted above.

From this point of view, the set of natural numbers is $M_R$ where $R$ is the rule giving $s(x)$ as the sole conclusion of the set $\{x\}$ of premises, with 0 as its only axiom.

Note that if there are no axioms, $M_R$ is the empty set, since the empty set is then closed under $R$.

How do we know that there is, for a particular rule $R$, at least one set closed under $R$, and thereby a smallest set closed under $R$? All rules considered in this book have only countably many premises, and in axiomatic set theory the existence of a set closed under $R$ is always provable in this case. Indeed, we will only have occasion to consider rules $R$ for which we have already either defined or postulated a set containing *every* conclusion of $R$—the set of natural numbers or the set of countable ordinals—and this set is trivially closed under $R$.

For every $M_R$ there is a corresponding method of *proof by R-induction*: to prove that an assertion $F(a)$ is true of every $a$ in $M_R$, we prove that the set of $a$ in $M_R$ such that $F(a)$ is closed under $R$. The justification for this method of proof is immediate: since $M_R$ is defined as the smallest set closed under $R$, if the set of $a$ in $M_R$ such that $F(a)$ is closed under $R$, every $a$ in $M_R$ satisfies $F(a)$. Ordinary mathematical induction is a special case of this: that a set $A$ is closed under the rule defining the natural numbers means that 0 is in $A$ and that $s(x)$ is in $A$ whenever $\{x\}$ is a subset of $A$, or in other words, whenever $x$ is in $A$.

$M_R$ can be looked at from another point of view. For any set $A$, let $F_R(A)$ be the set of $x$ such that there is an instance $\langle X, x\rangle$ of $R$ with $X \subseteq A$. Thus $F_R(A)$ is the set of conclusions obtainable by one application of $R$ with premises taken from $A$. That $A$ is closed under $R$ is the same as saying that $F_R(A) \subseteq A$, so $M_R$ is the smallest $A$ for which $F_R(A) \subseteq A$. But $M_R$ also has the property that $M_R \subseteq F_R(M_R)$, which means that $M_R$ is the smallest *fixed point* of $F_R$, that is, the smallest $A$ such that $F_R(A) = A$. To see that $M_R \subseteq F_R(M_R)$, we need only observe that since $F_R$ is *monotonic*, meaning that $F_R(A) \subseteq F_R(A')$ whenever $A \subseteq A'$, the fact that $F_R(M_R) \subseteq M_R$ implies that $F_R(F_R(M_R)) \subseteq F_R(M_R)$, so $F_R(M_R)$ is closed under $R$, so $M_R \subseteq F_R(M_R)$ follows.

The above reduction of inductive definitions to explicit set-theoretical definitions is based on the "quasi-combinatorial" notion of "arbitrary set", as explained in Chapter 5 in connection with subsets of **N**. Or rather, although it may well be possible to justify that reduction from some other point of view, only the notion of arbitrary set obviously supports it. To see this, let us consider whether the definition of $M_R$ can be based on a more restricted notion of set. A moment's reflection tells us that in a sense it can, for as long as $M_R$ itself belongs to a class $C$ of sets, the intersection of all $R$-closed sets that belong to $C$ is identical with $M_R$. Thus an equivalent description of $M_R$ is obtained by restricting the $R$-closed sets considered to those in $C$. (In fact restricting the $R$-closed sets considered to those in some class $C$ which does *not* contain $M_R$ may in some cases also yield the same $M_R$.) But then we must take into account what we can *prove* about $M_R$ thus defined. In particular, if we want to use proof by $R$-induction to show that every member $x$ of $M_R$ satisfies some condition $F(x)$, we need to know that the set $\{x \mid x \in M_R \wedge F(x)\}$ belongs to $C$. Also, the proof above that $M_R$ is a fixed point of $F_R$ depended on $F_R(M_R)$ being in $C$. Thus we assumed closure conditions on $C$ that are automatically satisfied only if we are talking about arbitrary sets. But also the very definition of $M_R$ uses a set existence principle that is highly problematic if we do not accept the notion of arbitrary set. This is the so-called *impredicative comprehension principle*, which will be formulated and discussed in Chapter 10.

### Inductive definitions: one rule or many?

For theoretical purposes, we take $M_R$ to be generated by a single rule $R$. In practice, when using inductive definitions, it is often more convenient or natural to say that several rules are involved. For example, in defining the statements of arithmetic inductively, we refer to the rule

If $A$ and $B$ are arithmetical statements, then so is $A \wedge B$

for forming conjunctions, and the rule

If $A$ and $B$ are arithmetical statements, then so is $A \vee B$

for forming disjunctions. In squeezing this into the mold of the general formulation, we introduce instead a single rule whereby both $A \wedge B$ and $A \vee B$ are consequences of the set of premises $\{A, B\}$. In order to avoid confusion with logical rules of inference we may prefer instead to introduce a rule whereby "$A \wedge B$ is an arithmetical statement" and "$A \vee B$ is an arithmetical statement" both follow from $\{A$ is an arithmetical statement, $B$ is an arithmetical statement$\}$.

The locution "define by simultaneous induction", which will be used in later chapters, is to be understood similarly. For example, we may say that we define (assuming 0 and $s$ to be given) "$n$ is a natural number" and "$n < m$" by simultaneous induction through the clauses

0 is a natural number, and $0 < s(0)$

If $n$ is a natural number, then $s(n)$ is a natural number

If $n < m$ then $s(n) < s(m)$

If $n < m$ and $m < k$ then $n < k$

Considered as an instance of the general procedure, this definition uses a rule whereby $0 < s(0)$ and "0 is a natural number" are axioms, while "$s(n)$ is a natural number" follows from {"$n$ is a natural number"}, $s(n) < s(m)$ follows from $\{n < m\}$, and $n < k$ follows from $\{n < m, m < k\}$. In inductive proofs using this rule $R$, we may focus on a particular kind of conclusion, and prove for example that $m$ and $n$ are natural numbers whenever $m < n$, where the natural numbers are the $n$ for which "$n$ is a natural number" belongs to $M_R$. This means that we prove by $R$-induction that every member $x$ of $M_R$ satisfies "if $x$ is $m < n$, then $m$ and $n$ are natural numbers".

Induction can also be used to define functions on inductively defined sets. This means that we define the graph of the function inductively. For an example (which will not be used later, but has just been chosen as a non-trivial illustration), suppose we are first given the following inductive definition:

The empty set is hereditarily countable.

If $A$ is a countable set and every member of $A$ is hereditarily countable, then $A$ is hereditarily countable.

Some reflection is needed to see how to cast this definition in terms of the general scheme described. What is the relevant rule? The closure condition in the inductive clause of the definition of the class HC of hereditarily countable sets says that if $A$ is a countable subset of HC, then $A$ is a member of HC. This means that HC is closed under the rule with instances $\langle X, X\rangle$, where $X$ is any countable set, and HC is the smallest set closed under this rule. Using this definition, we can show by induction that the membership relation is well-founded when restricted to HC, that is, every sequence $A_0, A_1, \ldots$ where all the $A_i$ are hereditarily countable and $A_{i+1} \in A_i$ must end with the empty set. (Since $A$ is hereditarily countable, all sets in such a sequence must also be countable.)

We can now define a function $F$ on HC inductively by

$F(\varnothing) = \varnothing$.

$F(A) = \{x \mid x \in F(a) \text{ or } x = F(a) \text{ for some } a \in A\}$ if $A$ is hereditarily countable and non-empty.

This can be justified as follows. We define the relation $R$ (= the graph of $F$) inductively by

$R(\varnothing, \varnothing)$.

If $g$ is a function defined on a non-empty hereditarily countable set $A$, and $R(a, g(a))$ for every $a$ in $A$, then $R(A, G)$, where $G = \{x \mid x \in g(a) \text{ or } x = g(a) \text{ for some } a \in A\}$.

We can now prove by HC-induction that $R$ is itself a function, that is, if $R(A, C)$ and $R(A, C')$, then $C = C'$. By the fixed point property of inductive definitions, if $R(A, X)$, then either $A$ and $X$ are both $\varnothing$, or there is a function $g$ defined on $A$ such that $R(a, g(a))$ holds for every $a$ in $A$, and $X = \{x \mid x \in g(a) \text{ or } x = g(a) \text{ for some } a \in A\}$. By the induction hypothesis, the function $g$ is uniquely determined by $A$, and therefore so is $X$. HC-induction also shows that for every $A$ in HC there is a $C$ such that $R(A, C)$. Thus we are justified in concluding that $R$ is the graph of a function $F$ as above. Here again the definition of $F$ and $R$ can be given by a simultaneous induction:

$\varnothing$ is hereditarily countable, and $F(\varnothing) = \varnothing$.

If $A$ is countable and every member of $A$ is hereditarily countable, then $A$ is hereditarily countable, and $F(A) = \{x \mid x \in F(a) \text{ or } x = F(a) \text{ for some } a \in A\}$.

Inductive definitions will in the following be introduced in semi-formal terms, through base clauses and inductive clauses, and it will usually be left to the reader to squeeze them into the general mold. Also, we will usually say that a conclusion $x$ follows from premises $x_0, \ldots, x_n$, meaning that $x$ follows from the set $\{x_0, \ldots, x_n\}$ of premises.

## The case of finite rules

If $X$ is finite in every instance of $R$ we call $R$ a *finite rule*. For finite rules, we can describe the members of $M_R$ in terms of finite sequences: a belongs to $M_R$ if and only if there is a sequence $x_0, \ldots, x_n$, called a *derivation* of $a$ using $R$, such that $x_n$ is $a$, and every $x_i$ is the conclusion of a rule instance all of whose premises are found among the $x_j$ for $j < i$. (Thus in particular $x_0$ must be an axiom.) Most of the inductive definitions that will be introduced in this book use finite rules.

If $R$ is a finite rule, we can also give a more perspicuous description of $M_R$ in terms of $F_R$. The axioms of $R$ are the members of $F_R(\varnothing)$, while $F_R(F_R(\varnothing))$ contains the axioms together with the consequences of axioms, and so on. Now take the union

$$M = F_R(\varnothing) \cup F_R(F_R(\varnothing)) \cup \cdots = M_0 \cup M_1 \cup \cdots$$

If $R$ is finite, $M$ is closed under $R$, since any finite subset $A$ of $M$ will be a subset of some $M_i$, and therefore every consequence of $A$ is in $M_{i+1}$. By induction on $i$, using the fact that $F$ is monotonic, we can also conclude that for any set $N$ closed under $R$, $M_i \subseteq N$ and so $M \subseteq N$. Thus $M$ is the smallest set closed under $R$.

If the premises and conclusions are natural numbers, and $R$ is not only finite, but there is also an algorithm for deciding whether a pair $\langle X, x\rangle$ is an instance of the rule, the above description can be used to show that the elements of $M_R$ can be generated by an algorithmic procedure, which we express by saying that $M_R$ is *effectively enumerable*. The class of effectively enumerable sets will be formally defined and some basic properties of such sets proved in Chapter 9.

A finite rule doesn't necessarily have any finite upper bound on the number of members of $X$ in an instance $\langle X, x\rangle$. We may for example take $R$ to be the rule whereby $\{x_0, \ldots, x_n\}$ follows from the premises $x_0, \ldots, x_n$, for any finite set $\{x_0, \ldots, x_n\}$. In other words, the instances of $R$ are the pairs $\langle X, X\rangle$, where $X$ is any finite set. Then $M_R$ is the smallest set $M$ such that $\varnothing \in M$ and $\{x_0, \ldots, x_n\} \in M$ whenever all of $x_0, \ldots, x_n$ are in $M$. This $M_R$ is known as the set of *hereditarily finite sets*, or more long-windedly as the set of hereditarily finite pure well-founded sets. They are the sets obtainable by repeatedly forming finite sets starting from the empty set. Equivalently, a set $A$ is hereditarily finite if $A$ is finite, and every member of $A$ is a finite set, and every member of a member of $A$, and so on, and furthermore every sequence $x_0, x_1, \ldots$ where $x_0$ is a member of $A$ and $x_{i+1}$ is a member of $x_i$ for every $x_{i+1}$ in the sequence ends at some index $i$ with the empty set.

## 6.3. Ordinals

### An axiomatic characterization of the ordinals

We can characterize the ordinals by a set of axioms for the set $\Omega$ of countable ordinals and the ordering relation $<$ between the ordinals. The axioms are the following:

1. $\Omega$ is strictly and totally ordered by $<$. Furthermore, there is a smallest element in this ordering, denoted 0.

2. Every $\alpha$ in $\Omega$ has an *immediate successor* $\alpha+$, defined by the conditions that $\alpha < \alpha+$ and there is no $\beta$ such that $\alpha < \beta$ and $\beta < \alpha+$.

3. Every strictly increasing $\omega$-sequence $\alpha_0 < \alpha_1 < \alpha_2 \ldots$ of ordinals (called a *fundamental sequence*) has a least upper bound or *supremum*, which is denoted $sup\, \alpha_i$. That is, $\alpha_i < sup\, \alpha_i$ for every $i$, and there is no $\beta < sup\, \alpha_i$ such that $\alpha_i < \beta$ for every $i$. Thus $sup\, \alpha_i$ is the first number greater than all of $\alpha_0, \alpha_1, \alpha_2, \ldots$.

   As a consequence of 1–3, for two fundamental sequences $\alpha_0 < \alpha_1 < \alpha_2 \ldots$ and $\beta_0 < \beta_1 < \beta_2 \ldots, sup\, \alpha_i < sup\, \beta_i$ if and only if there is some $j$ such that for all $i$, $\alpha_i < \beta_j$. For given such a $j$, $sup\, \alpha_i \leq \beta_j < sup\, \beta_i$, and conversely, if $sup\, \alpha_i < sup\, \beta_i$ there is some $j$ for which $sup\, \alpha_i \leq \beta_j$, so that $\alpha_i < \beta_j$ for every $i$.

4. The countable ordinals are precisely the objects generated from 0 using the successor operation and the supremum operation. In other words, any set of ordinals $A$ for which $0 \in A$, and $\alpha+ \in A$ whenever $\alpha \in A$, and $sup\, \alpha_i \in A$ whenever $\alpha_i \in A$ for all $i$ contains all ordinals in $\Omega$.

The finite ordinals are $0, 0+, 0++, \ldots$. We will identify the natural numbers with the finite ordinals whenever there is any point in doing so. An ordinal $\alpha$ is a *successor ordinal* if $\alpha$ is $\beta+$ for some $\beta$, and a *limit ordinal* if it is $sup\, \alpha_i$ for some fundamental sequence. This fundamental sequence is never unique—the first limit ordinal, $\omega$, is the supremum of each of the uncountably many fundamental sequences of natural numbers. (Two such fundamental sequences are the sequence of odd numbers and the sequence of even numbers.) A successor ordinal is never a limit ordinal, since there is, for a limit ordinal $\alpha$, no largest ordinal smaller than $\alpha$. Using the induction principle for ordinals postulated in the fourth axiom above, we can prove that any ordinal is either 0, or a successor ordinal, or a limit ordinal.

A sequence $\alpha_0 \leq \alpha_1 \leq \alpha_2 \ldots$ of ordinals which is increasing but not necessarily strictly increasing also has a least upper bound. If after a certain point in the sequence there is no greater ordinal, the least upper bound is the largest ordinal in the sequence, and otherwise it is the supremum of any strictly increasing infinite *subsequence* of $\alpha_0, \alpha_1, \alpha_2 \ldots$, which is to say a sequence $\alpha_{i_0}, \alpha_{i_1}, \ldots$ where $i_0 < i_1 < i_2 \ldots$ is a strictly increasing sequence of natural numbers.

A subset $A$ of $\Omega$ is *bounded* if there is a $\beta$ in $\Omega$ (called an upper bound of $A$) such that $\alpha < \beta$ for every $\alpha$ in $A$, and otherwise *unbounded*. The axioms imply that every bounded set of ordinals has a least upper bound. For this, we prove by induction on $\beta$ that if $\alpha < \beta$ for every $\alpha$ in $A$, then $A$ has a least upper bound. For $\beta = 0$, the assumption implies that $A$ is empty, so 0 is the least upper bound of $A$. If $\alpha < \beta+$ for every $\alpha$ in $A$, $\alpha < \beta$ for every $\alpha$ in $A \setminus \{\beta\}$, and if $\gamma$ is the least upper bound for $A \setminus \{\beta\}$, either $\gamma$ or $\beta$ is the least upper bound for $A$. Finally, if $\alpha < sup\, \alpha_i$ for every $\alpha$ in $A$, the induction hypothesis yields that $A_i = \{\alpha \in A \mid \alpha < \alpha_i\}$ has a least upper bound $\gamma_i$ for every $i$, and since $A$ is the union of the $A_i$ and $\gamma_0 \leq \gamma_1 \leq \gamma_2 \ldots$, the supremum of the $\gamma_i$ is also the least upper bound of $A$.

Every countable set of ordinals is bounded. For if there is an enumeration $\alpha_0, \alpha_1, \alpha_2 \ldots$ of the ordinals in the infinite set $A$ and $A$ has no largest element, we can define a strictly increasing subsequence by taking first $\alpha_0$, then the first ordinal in the sequence greater than $\alpha_0$, and so on. Every ordinal in $A$ is smaller than some ordinal in this strictly increasing subsequence, so every ordinal in $A$ is smaller than the supremum of this subsequence. It follows that $\Omega$ itself is not countable.

The converse also holds, that the set of ordinals smaller than a given ordinal $\alpha$, and thereby any bounded set, is countable. This is proved by induction. The case $\alpha = 0$ is trivial, and given an enumeration of the ordinals smaller than $\alpha$ we get an enumeration of the ordinals smaller than $\alpha+$ by putting $\alpha$ first, followed by the given enumeration. Finally, if for every $\alpha_i$ in a fundamental sequence for $\alpha$ we have an enumeration $\alpha_{i,0}, \alpha_{i,1}, \ldots$ of the ordinals smaller than $\alpha_i$, we can define an enumeration $\beta_0, \beta_1, \ldots$ of the ordinals smaller than $\alpha$ by letting $\beta_k = \alpha_{m,n}$ where $k = OP(m,n)$, using the one-one correspondence between natural numbers and pairs of natural numbers defined in §4.4. (An observant reader will note in this proof the use of an axiomatic principle similar to the principle of dependent choice, since we need to choose an enumeration for each $\alpha_i$.)

The ordinals actually used in the book will be the constructive ordinals—a concept defined in Chapter 11—which form a *proper initial segment* of $\Omega$: they are the ordinals smaller than $\alpha$ for a particular $\alpha$ known as the Church-Kleene ordinal.

Looking downwards instead of upwards, the axioms also imply that every non-empty set $A$ of ordinals has a smallest element (so that $<$ is a well-ordering in the sense defined in §2.3). For let $\alpha$ be the supremum of the set of $\beta$ such that $\beta$ is smaller than every ordinal in $A$. If $\alpha$ is itself smaller than every ordinal in $A$, $\alpha+$ is the smallest ordinal in $A$, and otherwise $\alpha$ is the smallest ordinal in $A$.

## Defining the ordinals

The above axioms for the countable ordinals tell us all we need to know about them to use them in counting. There is a smallest ordinal 0, after every ordinal there is a next ordinal, and after every infinite strictly increasing $\omega$-sequence of ordinals there is also a next ordinal. The axioms do not amount to an inductive *definition* of the ordinals, for we haven't defined the ordering relation $<$ or proved the existence of successors and upper bounds. But as with the inductive characterization of the natural numbers, we can show by an inductive argument that the axioms uniquely characterize the mathematical structure of the ordinals, in the sense that any two sets $\Omega$ and $\Omega'$ with corresponding orderings $<$ and $<'$ that both satisfy the above axioms are isomorphic.

In set theory, it is possible to define in various ways a set $\Omega$ with an associated ordering $<$ satisfying the axioms characterizing the countable ordinals. One of many possible ways of approaching the definition of $\Omega$ in set theory will be sketched here, because of its connection with the later treatment of ordinal notations.

Every ordinal is 0, a successor ordinal, or a limit ordinal, which suggests that we might start by introducing an abstract notion of *names* for ordinals. We define by induction the set of names and the relation $u < u'$, with the intended meaning "$u$ and $u'$ are names of ordinals, and the ordinal named by $u$ is smaller than the ordinal named by $u'$". Our names will be 0 together with $s(u) = \langle 1, u\rangle$ for any name $u$, and $sup(u) = \langle 2, u\rangle$ for certain $\omega$-sequences $u$ of names of increasing ordinals. Here the brackets denote an ordered pair, not a sequence number. These pairs are not names in any concrete sense, since arbitrary $\omega$-sequences are infinite objects, and there are uncountably many of them.

The clauses in the inductive definition are the following:

0 is a name.

If $u$ is a name, then $s(u)$ is a name, and $u < s(u)$ and $0 < s(u)$.

If $u = u_0, u_1, \ldots$ is a sequence of names satisfying $u_i < u_{i+1}$ for every $i$ then $sup(u)$ is a name, and $u_i < sup(u)$ for every $i$.

If $u < u'$ and $u' < u''$ then $u < u''$.

This definition makes good sense as a way of assigning names (in an abstract set-theoretical sense) to ordinals, and we could have included in the inductive definition a definition of "the ordinal named by $u$", assuming the ordinals as given. But the idea now is to start instead from the names and use them to define the ordinals. The simplest procedure would be to identify ordinals and names. This can't be done because every limit ordinal is the supremum of infinitely many different fundamental sequences, so every limit ordinal will have infinitely many names $sup(u)$. As a consequence, the relation $<$ is necessarily *stronger* than the intended relation "the ordinal named by $u$ is smaller than the ordinal named by $u'$", for if $u$ and $u'$ are different names for the same limit ordinal, they are *incomparable* with respect to $<$, that is, neither $u < u'$, nor $u' < u$ holds. In order to obtain ordinals from names, we must identify names that denote the same ordinal. One way of doing this is to define inductively a relation "$u$ and $u'$ name the same ordinal" (without actually using or presupposing the existence of the ordinals) and then define ordinals as equivalence classes of names under this relation, with $\alpha < \beta$ defined as "for some names $u$ in $\alpha$ and $u'$ in $\beta$, $u < u'$". Another way, which is less transparent but technically smoother, is to define directly by induction the ordinal named by $u$ as follows:

0 names the empty set.

If $u$ names $x$, $s(u)$ names $x \cup \{x\}$.

If $u_i$ names $x_i$ for $i = 0, 1, \ldots$, $sup(u)$ names the union of the sets $x_0, x_1, \ldots$.

The sets thus named are the von Neumann ordinals, for which $\alpha < \beta$ is defined simply as $\alpha \in \beta$. The verification that they satisfy the axioms for $\Omega$ is left to the set-theoretically inclined reader. In Chapter 11, we will return to the idea of naming ordinals, but then in a more concrete sense, and introduce a subset of the names defined above as names for the ordinals in a subset of $\Omega$.

In preparation for the later treatment of ordinal notations, we next need to formulate mathematically some ways of counting up to large ordinals.

## Defining functions on $\Omega$ by recursion

Using primitive recursion, we can define a function $F$ on **N** by defining $F(0)$ and then defining $F(n+1)$ in terms of $n$ and $F(n)$. This is not quite sufficient in the case of functions on $\Omega$, since in defining $F(\alpha)$ we also need to consider the case when $\alpha$ is a limit ordinal. A natural procedure in order to generalize definitions by recursion on **N** to $\Omega$ is to add to the definition of $F(\alpha+)$ in terms of $F(\alpha)$ a definition of $F(sup\, \alpha_i)$ in terms of the values $F(\alpha_i)$ for $i = 0, 1, \ldots$. But then we must take into account that while the immediate predecessor $\alpha$ of a successor ordinal $\alpha+$ is uniquely determined, there are always infinitely many fundamental sequences with a given limit ordinal $\alpha$ as supremum. For example, we can't define a function on $\Omega$ by the stipulation that $F(sup\, \alpha_i) = F(\alpha_0)$, since the value of $F(\alpha_0)$ depends on the ordinals in the sequence $\alpha_0, \alpha_1, \alpha_2, \ldots$ and not only on their supremum.

We can however directly generalize definition by course-of-values recursion (as described in §4.6) from number-theoretic functions to functions on $\Omega$. For any function $F$ defined on ordinals, we define $\overline{F}$ by

$$\overline{F}(\alpha, \beta_0, \beta_1, \ldots, \beta_k) = \text{the set of all } \langle \beta, F(\beta, \beta_0, \beta_1, \ldots, \beta_k) \rangle \text{ for } \beta < \alpha.$$

We can then prove, for any given $G$, that there is a unique function $F$ such that

$$(1) \quad F(\alpha, \beta_0, \beta_1, \ldots, \beta_k) = G(\alpha, \beta_0, \beta_1, \ldots, \beta_k, \overline{F}(\alpha, \beta_0, \beta_1, \ldots, \beta_k))$$

for all ordinals $\alpha, \beta_0, \beta_1, \ldots, \beta_k$. Thus in defining $F(\alpha, \beta_0, \beta_1, \ldots, \beta_k)$ we can freely use $F(\beta, \beta_0, \beta_1, \ldots, \beta_k)$ for $\beta < \alpha$.

As in the case of functions on **N**, we can justify this style of definition by showing that it can be replaced by an explicit definition of $F$. Leaving out the parameters $\beta_0, \ldots, \beta_k$ to shorten the notation, the explicit definition of $F$ is in this case:

> $F(\alpha) = \gamma$ if and only if there is a sequence $\gamma_\beta$, indexed by the ordinals smaller than $\alpha+$, such that $\gamma_\beta = G(\beta, \{\langle \delta, \gamma_\delta \rangle \mid \delta < \beta\})$ for every $\beta < \alpha+$, and $\gamma_\alpha = \gamma$.

That this explicit definition does in fact define a function which satisfies (1) is proved by induction.

As a special case of this scheme, we can define functions $F$ from $\Omega$ to $\Omega$ by defining $F(0)$, defining $F(\alpha+)$ in terms of $F(a)$, and defining $F(\mathit{sup}\,\alpha_i)$ in terms of $F(\alpha_i)$ for $i = 0, 1, \ldots$, whenever we can show that the value of $F(\mathit{sup}\,\alpha_i)$ depends only on the supremum $\mathit{sup}\,\alpha_i$, and not on the particular fundamental sequence chosen. We will be interested in doing this for so-called normal functions on $\Omega$, which are used to count up to large ordinals.

## Normal functions on $\Omega$

A function $F$ from $\Omega$ to $\Omega$ is *normal* if it is both strictly monotonic—which here means that if $\alpha < \beta$ then $F(\alpha) < F(\beta)$ for all ordinals $\alpha, \beta$—and *continuous*, which means that $F(\mathit{sup}\,\alpha_i) = \mathit{sup}\,F(\alpha_i)$ for every strictly increasing sequence. For any strictly monotonic $G$ and initial value $\alpha$, we can recursively define a normal function $F$ by $F(0) = \alpha$, $F(\alpha+) = G(F(\alpha))$, and $F(\mathit{sup}\,\alpha_i) = \mathit{sup}\,F(\alpha_i)$. This last clause is justified by the observation that if $\mathit{sup}\,\alpha_i = \mathit{sup}\,\beta_i$ and $F$ is monotonic, it follows that $\mathit{sup}\,F(\alpha_i) = \mathit{sup}\,F(\beta_i)$, so $F$ is in fact well-defined by course-of-values recursion. Below, in defining various normal functions we will leave the condition $F(\mathit{sup}\,\alpha_i) = \mathit{sup}\,F(\alpha_i)$ implicit.

For strictly monotonic $F$, an inductive argument shows that $\alpha \leq F(\alpha)$ for every $\alpha$. It follows that $\alpha, F(\alpha), F(F(\alpha)), \ldots$ is an increasing (but not necessarily strictly increasing) sequence for any $\alpha$.

If $F$ has more than one argument, we say that $F$ is normal in the first argument if for every $\beta_1, \ldots, \beta_k$ the function mapping $\alpha$ to $F(\alpha, \beta_1, \ldots, \beta_k)$ is normal, and similarly for other arguments. We will use the notation $F_\beta$ for the function mapping $\alpha$ to $F(\beta, \alpha)$.

For a first example of a normal function, *addition* of ordinals is defined by $\alpha + 0 = 0$ and $\alpha + \beta+ = (\alpha + \beta)+$. In particular, $\alpha + 1 = \alpha + 0+ = \alpha+$, and we will often write $\alpha + 1$ instead of $\alpha+$. The ordinal $\alpha + \beta$ is the ordinal we get to by first counting to $\alpha$ and then counting to $\beta$, starting with $\alpha$. Thus $\omega + 1$ is the first ordinal greater than $\omega$, while $1 + \omega$ is $\omega$.

Ordinal multiplication is similarly defined by $\alpha * 0 = 0$, $\alpha * \beta+ = \alpha * \beta + \alpha$, and exponentiation by $\alpha^0 = 1$, $\alpha^{\beta}+ = \alpha^{\beta} * \alpha$.

Using ordinal addition, multiplication, and exponentiation, and starting with 0 and 1, we can name quite a lot of ordinals. More precisely, as we shall see in Chapter 11, we can name in this way just those ordinals that are smaller than an ordinal famous in logic, known as $\varepsilon_0$. This is the first ordinal $\alpha$ for which $\omega^\alpha = \alpha$. More generally, the fixed points of the function mapping $\alpha$ to $\omega^\alpha$ are known as the *epsilon numbers*. Normal functions have many fixed points, and in order to name larger ordinals, the traditional first step is to invoke those fixed points, as will now be described.

The set $Fix_F$ of fixed points of a normal function $F$ is a *closed unbounded set*. A subset $A$ of $\Omega$ is *closed* if $sup\ \alpha_i$ belongs to $A$ for every strictly increasing sequence $\alpha_0 < \alpha_1 < \alpha_2 \ldots$ of ordinals in $A$. The property of being both closed and unbounded is a richness property of subsets of $\Omega$. The set $\Omega$ itself is closed and unbounded, and any subset of $\Omega$ that is closed and unbounded contains a "large portion" of the ordinals. That $Fix_F$ is closed follows from the continuity of $F$, and it is also unbounded, since for any $\beta$, the supremum of the sequence $\beta, F(\beta), F(F(\beta)), \ldots$ (which is increasing since $F$ is monotonic) is a fixed point, again by the continuity of $F$.

Conversely, with any closed unbounded set $A$ of ordinals is associated a normal function $F_A$ which enumerates the members of $A$. That is, $F_A(\alpha) =$ the smallest member of $A$ which is greater than $F_A(\beta)$ for every $\beta < \alpha$. $F_A$ is defined (by course-of-values recursion) for every $\alpha$ since $A$ is unbounded, it is strictly increasing by definition, and it is continuous since $A$ is closed. We call $F_A(\alpha)$ the $\alpha$-th member of $A$.

If $A_0, A_1, \ldots$ is an $\omega$-sequence of closed unbounded subsets of $\Omega$, their intersection $A$ is also closed and unbounded. That $A$ is closed follows immediately from the assumption that every $A_i$ is closed. To show that $A$ is unbounded, we define a strictly increasing sequence $\alpha_0 < \alpha_1 < \alpha_2 \ldots$ of ordinals in $A$ as follows. $\alpha_0$ is the smallest ordinal in $A_0$ which is greater than $\beta$. For every $k$, $\alpha_{k+1}$ is the smallest ordinal greater than $\alpha_k$ which is in $A_m$, where $m$ is $(k+1)_0$.

Because there are, for every $m$, infinitely many $k$ such that $m$ is $(k+1)_0$, this implies that the sequence has, for every $m$, an infinite subsequence of elements in $A_m$. Thus $sup\, \alpha_i$ is in $A_m$ for every $m$, so $sup\, \alpha_i$ is in $A$.

We can now define another famous ordinal, the ordinal called $\Gamma_0$. Let $F(0, \alpha)$ be $\omega^\alpha$. For $\beta > 0$, we define $F(\beta, \alpha)$ to be the $\alpha$-th common fixed point of all the functions $F_\gamma$ for $\gamma < \beta$. Here we use the fact that the set of fixed points of $F_\gamma$ is closed and unbounded for every $\gamma$ and that the set of $\gamma < \beta$ is countable, so that the set of common fixpoints of all the $F_\gamma$ for $\gamma < \beta$ is also closed and unbounded, by the preceding argument.

Now let $H(\beta) = F(\beta, 0)$. In other words, $H(\beta)$ is the first ordinal $\alpha$ such that $F_\gamma(\alpha) = \alpha$ for every $\gamma < \beta$. $H$ is itself a normal function. To verify that $H$ is strictly monotonic, we need only observe that if $\beta < \beta'$, $F_\beta(0) < F_\beta(F_{\beta'}(0)) = F_{\beta'}(0)$. For continuity, we need to show that if $\beta = sup\, \beta_i$, $H(\beta) = sup\, H(\beta_i)$. Let $\gamma = sup\, H(\beta_i)$. $\gamma$ is a common fixed point of every $F_\delta$ for $\delta < \beta$, since for such $\delta$, $F_\delta(\gamma) = sup\, F_\delta(H(\beta_i)) = sup\, H(\beta_j)$ for $j$ large enough so that $\delta < \beta_j$, so $F_\delta(\gamma) = \gamma$. $\gamma$ is the smallest such fixed point, for if $F_\delta(\beta') = \beta'$ for every $\delta < \beta$, $H(\beta_i) = F(\beta_i, 0) < F(\beta_i, \beta') = \beta'$ for every $i$, so $\gamma \leq \beta'$.

Having established that $H$ is normal, we can now define $\Gamma_0$ as the smallest ordinal $\beta$ such that $F(\beta, 0) = \beta$. Thus $F_\gamma(\Gamma_0) = \Gamma_0$ for every $\gamma < \Gamma_0$, which means that $\Gamma_0$ has strong closure properties. We will take a closer look at $\varepsilon_0$ and $\Gamma_0$ in Chapter 11.

Ordinals play a large role in the part of logic known as proof theory, where there are many highly refined studies and results relating what can be proved in different theories to specific ordinals associated with those theories. In this book, no attempt will be made to state or prove any such results. Mostly, difficult and quantitative results will be set aside in favor of simple and qualitative ones, intended to bring out, clarify, and resolve the philosophical conundrums connected with inexhaustibility.

CHAPTER 7

# FORMAL LANGUAGES AND THE DEFINITION OF TRUTH

## 7.1. Formal syntax for first and second order languages

In earlier chapters, the languages of arithmetic and analysis have been introduced in semi-formal terms. In later chapters we will need to have available a highly formal description of these languages, chiefly because we will need to convince ourselves that various assertions about statements in these languages can be formulated and proved in formal theories of arithmetic.

We need to define *term* and *formula* for first and second order languages. For these languages we will need an unlimited supply of function symbols and predicate symbols, of all arities. (The *arity* of a function or predicate symbol is the number of arguments it takes.) We will use, when referring to the formal syntax, the symbol $f^i_j$ for the $j$-th function symbol of arity $i$, and similarly we use $p^i_j$ for the $j$-th predicate symbol of arity $i$. The first order variables are $v_0, v_1, \ldots$ and the second order variables $V_0, V_1, \ldots$.

0-ary function symbols are admitted, and are called *individual constants*. Similarly we could admit 0-ary predicate symbols, *propositional constants*, although these will not in fact be used in this book.

There is a certain indefiniteness in the above formulations. Just what is $f^i_j$? From a mathematical point of view, it doesn't matter. What is required is only that we have a countably infinite set $F$ of objects, which we call function symbols, and an arity function from $F$ to $N$, such that for every $n$ in $N$ there are infinitely many $f$ in $F$ with arity $n$.

Since all the specific information about $f^i_j$ is given by the indices $i$ and $j$, we can, when we want to be explicit about just what $f^i_j$ is, define it as identical with the pair $\langle i, j\rangle$. Or rather, since we want to have similarly infinite supplies of other syntactic objects, we define it as the triple $\langle 0, i, j\rangle$ where the 0 indicates that we are dealing with a function symbol. Similarly we can define $p^i_j$ as the triple $\langle 1, i, j\rangle$ (the 1 indicating that this is a predicate symbol), $v_i$ as $\langle 2, i\rangle$ and $V_i$ as $\langle 3, i\rangle$.

A *term* is either a variable, a constant, or (in the first-order case) a compound term of the form $f(t_1, \ldots, t_n)$ where $f$ is a function symbol of arity $n$ and $t_1, \ldots, t_n$ are terms. (This, it will be noted, is shorthand for an inductive definition.) Mathematically, we identify this compound term with the $n+2$-tuple $\langle 4, f, t_1, \ldots, t_n \rangle$, where the 4 indicates that we are dealing with a compound term.

But now let us note that we have used an ambiguous notation above. For natural numbers $m$ and $n$, $\langle m, n \rangle$ may be either the ordered pair containing $m, n$ or the sequence number encoding that ordered pair, as defined in §4.4. So which is it that we identify e.g., the variable $v_0$ with? Is it the ordered pair $\langle 2, 0 \rangle$ or the sequence number $\langle 2, 0 \rangle$?

In this book the rule is that we identify the variable $v_0$ with the sequence number $\langle 2, 0 \rangle$, and similarly for all other syntactical objects. Thus variables, terms, formulas, and other finite syntactic objects to be introduced, like proofs in theories, are all *numbers*.

An alternative procedure would be not to identify these syntactical objects with natural numbers, but instead use natural numbers to *encode* the syntactical objects. Thus we might say that the sequence number $\langle 2, 0 \rangle$ encodes or represents the variable $v_0$ rather than say that it is identical with the variable $v_0$. The difference between these alternatives isn't really significant from a mathematical point of view. The second alternative is probably the more widely known, under the designation *Gödel numbering*. The reason why the first alternative is chosen here (following Feferman [1961]) is that it will simplify the notation at some points where a simplified notation is much to be desired.

So having defined the terms, let's continue with the formulas. An atomic formula is in the first-order case a sequence number $\langle 5, p, t_1, \ldots, t_n \rangle$, where $p$ is a predicate symbol of arity $n$ and $t_1, \ldots, t_n$ are terms. In the second order case we also have atomic formulas $t \in V_i$, which we identify with the sequence number $\langle 6, t, V_i \rangle$. For compound formulas we need to introduce the quantifiers and connectives. We identify $\forall$ with 5, $\exists$ with 6, $\neg$ with 7, $\vee$ with 8, $\wedge$ with 9, $\supset$ with 10, $\equiv$ with 11. A universally quantified first-order formula is a sequence number $\langle 7, 5, x, \phi \rangle$ where $\phi$ is a formula and $x$ a variable, a conjunction is a sequence number $\langle 7, 9, \phi_1, \phi_2 \rangle$ where $\phi_1$ and $\phi_2$ are formulas, and similarly in the remaining cases.

Thus function symbols, predicate symbols, variables, compound terms, atomic formulas and compound formulas are all sequence numbers. Function symbols begin with 0—that is, $(n)_0 = 0$ if $n$ is a function symbol—predicate symbols with 1, variables with 2 or 3, compound terms with 4, atomic formulas with 5 or 6, and compound formulas with 7.

Among the predicate symbols, a special role will be played by $p_0^2$, which we write as $=$, and $p_1^2$, which we write as $<$. The function symbol $f_0^0$ we will write as 0, $f_1^1$ as $s$, $f_0^2$ as $+$ and $f_1^2$ as $*$. These, it will be noted, are the primitive predicate symbols and function symbols of the languages of arithmetic and analysis.

A first or second order *language* is defined by its set of function symbols and predicate symbols. In this book, it will be assumed that every language includes $=$, $<$, $0$, $s$, $+$, and $*$ among its predicate and function symbols. The *primitive* language of arithmetic or analysis has only these predicate and function symbols, while an *extended* language may have further such symbols. A formula in the primitive language is called a primitive formula. Whenever this makes an argument shorter or more convenient, we will also assume that the logical apparatus has been trimmed down to negation, disjunction and existential quantification. The basis for this trimming-down will be presented in Chapter 8.

## 7.2. Syntax and primitive recursion

The set $term_1$ of first-order terms is primitive recursive, since

$$term_1(n) \equiv var_1(n) \vee (seq(n) \wedge (n)_0 = 4 \wedge func((n)_1) \wedge \\ (length(n) = arity((n)_1) + 2) \wedge \forall i < arity((n)_1)\, term_1((n)_{i+2})))$$

Here $var_1$ is the set of first order variables, defined by

$$var_1(n) \equiv seq(n) \wedge length(n) = 2 \wedge (n)_0 = 2$$

$func(a)$, meaning "$a$ is a function symbol" is similarly defined, and $arity(a)$ is defined as $(a)_1$. That $term_1$ is primitive recursive follows by the results in Chapter 4, since it is defined using primitive recursive functions, bounded quantifiers, connectives, and course-of-values recursion.

Similar arguments establish that the set $term_2$ of second-order terms and the sets $form_1$ and $form_2$ of first and second order formulas are primitive recursive. Also the substitution functions defined in this chapter are primitive recursive, and in fact all of the syntactic sets and operations which are clearly computable are also primitive recursive. This can always be shown by routine applications of the methods invoked above to show that $term_1$ is primitive recursive (since formulas and terms have been defined so that, for example, $\phi$ is always a smaller number than $\phi \vee \psi$), so the observation that such a set or operation is primitive recursive will mostly just be taken for granted on the basis of an informally worded definition.

## 7.3. Talking about terms and formulas

Terms and formulas in the formal languages of arithmetic and analysis have been identified with numbers, but of course we don't want to talk about "the formula 198918237918237912" and so on. Nor do we refer to formulas by giving sequences of numbers. Instead, when we want to talk about a particular formula we use ordinary semi-formal notation. We may for example speak of "the formula $\forall x \exists y(y > x \wedge prime(y))$", where *prime* is some numerically unspecified predicate symbol.

We will also need to speak in general terms about formulas and various operations on formulas, and unfortunately there is ample opportunity for unclarity and ambiguity when such statements about terms and formulas are themselves formalized. We will use the following basic conventions. The letters $\phi$, $\psi$, and $\theta$, sometimes indexed, will be used as variables for formulas. Boldface lower case letters $\boldsymbol{x}$, $\boldsymbol{y}$, $\boldsymbol{z}$ will be used to vary over first order variables, and boldface upper case $\boldsymbol{X}$, $\boldsymbol{Y}$, $\boldsymbol{Z}$ will vary over second order variables. $\boldsymbol{s}$, $\boldsymbol{t}$ are similarly used as variables for (first order) terms. For $n$ a natural number, $\overline{n}$ is a particular term in the primitive language of arithmetic, namely the term $s(s(\dots(0)\dots))$, with $n$ occurrences of $s$. We will call this the (formal) *numeral* for $n$, and it will be said to have *value* $n$. (This is a special case of the general definition of the value of a variable-free term in the formal language given below in §7.6.) It is easy to verify that the function *num* defined by $num(n) = \overline{n}$ is a primitive recursive function.

There is a special reason for having two notations for the numeral for $n$, $num(n)$ and $\overline{n}$. In an arithmetical statement we may use various function and predicate symbols, but often we will assume, for theoretical purposes, that all such defined symbols have been eliminated through a translation of the statement into primitive notation. In particular, when we consider a formula written using symbols for primitive recursive functions it will often be assumed to have been translated into a primitive formula using the method given in §4.6. For readability we will also use the same names for functions and predicates in formulas and in the running text. Since the formulas referred to will often be formalizations of statements about formulas, an ambiguity arises when we speak of for example the formula $p(\overline{v_0})$. Does this refer to a formalization of "the numeral for $v_0$ has property $p$", or is it instead the formula consisting of $p$ applied to the numeral for the variable $v_0$? It is this ambiguity that is resolved by the stipulation that the overline notation always indicates the occurrence in a formula of a numeral. *num* is used in formulas that formalize statements about numerals. Thus $p(\overline{v_0})$ does not contain any variable at all, but only the numeral whose value is the variable $v_0$ ($v_0$, it will be recalled, is the natural number $\langle 2, 0\rangle$), while $p(num(v_0))$ is a formula containing the variable $v_0$. And in general, whenever there is a reference to a formula $\phi(\overline{e})$, where $e$ is any expression, not necessarily an expression in the formal language, this is to be understood as a synonymous with "$\phi(\overline{n})$, where $n$ is $e$". For example, the formula $\phi(\overline{\text{the number of planets}})$ is the formula $\phi(\overline{n})$, where $n$ is the number of planets, and the formula $\phi(\overline{1+2})$ is the formula $\phi(\overline{n})$ where $n$ is $1+2$, which is to say the formula $\phi(\overline{3})$, or $\phi(s(s(s(0))))$.

These conventions for talking about formulas will be further illustrated and explained in later chapters. They are intended to make it possible to put things precisely but readably, with a minimum of special notation. The problem of combining precision with readability in this context is an unwelcome but unavoidable consequence of the fact that a formal treatment of Gödel's theorem involves the formalization of statements which are themselves about formalized languages.

## 7.4. Quantifiers, scope, and substitution

The syntactical complications that come with formulas mostly have to do with variables and substitution. The *scope* of a quantifier $\forall v_i$ or $\exists v_i$ in a formula is the quantifier itself together with the formula immediately following the quantifier. Any occurrence of a variable $x$ in the scope of a quantifier $\forall x$ or $\exists x$ is *bound*, while other occurrences are *free*. Similarly for second order quantification. A *closed* formula, or *sentence*, is one that has no free occurrences of variables. Semantically, a sentence is a formula that we can interpret as a complete statement. A closed *term* is the same as a variable-free term, one that does not contain any variables.

Given a universally quantified formula $\forall x\phi$ and a term $t$, we get a new formula $\phi_x(t)$ by substituting $t$ for every free occurrence of $x$ in the formula $\phi$. The intention is that $\phi_x(t)$ should specialize the general statement made by $\forall x\phi$ to $t$, and so it does, unless some variable in $t$ is captured by a quantifier in $\phi$ (that is, falls within the scope of the quantifier) when the substitution is made. If this does not happen, we say that *$t$ is free for $x$* in $\phi$. Whenever the notation $\phi_x(t)$ is used, it is presupposed that $t$ is free for $x$ in $\phi$. The variable $x$ will sometimes be suppressed when it is irrelevant or clear from the context, and we will write only $\phi(t)$.

In putting these definitions on a more formal basis—which we need to do in order to be able to verify that various functions and relations are primitive recursive—we don't need to formalize "occurrence" or "scope", but only the substitution operation. Since primitive recursive functions must be defined for all natural numbers as arguments, we use ordinary functional notation. $sub(n_1, n_2, n_3)$ is defined as follows. If $n_1$ is not a term, $n_2$ is not a variable, or $n_3$ is not a term or formula, $sub(n_1, n_2, n_3)$ is 0. For any terms $t_1$ and $t_2$ and variable $x$, $sub(t_1, x, t_2)$ is the term that results from substituting $t_1$ for every occurrence of $x$ in $t_1$. Similarly for every formula $\phi$, $sub(t, x, \phi)$ is the result of substituting $t$ for every free occurrence of $x$ in $\phi$, unless this results in some variable in $t$ being captured by a quantifier in $\phi$, in which case the value of $sub(t, x, \phi)$ is 0.

To see that this function *sub* is primitive recursive, we first need to verify that the predicate *occurs* is primitive recursive, where $occurs(n_1, n_2)$ is true if and only if $n_1$ is a variable and $n_2$ a term or formula and $n_1$ has a free occurrence in $n_2$ (where "free" makes a difference only if $n_2$ is a formula). The function *sub* can then be seen to be primitive recursive from the following recursive definition (where the clause stating that $sub(n_1, n_2, n_3)$ is 0 if the arguments are inappropriate is left implicit):

$sub(t, x, y)$ is $t$ if $y$ and $x$ are the same variable, otherwise $y$.

$sub(t, x, f(s_1, \ldots, s_n))$ is $f(sub(t, x, s_1), \ldots, sub(t, x, s_n))$.

$sub(t, x, p(s_1, \ldots, s_n))$ is $p(sub(t, x, s_1), \ldots, sub(t, x, s_n))$.

$sub(t, x, \phi_1 \vee \phi_2)$ is 0 if $sub(t, x, \phi_1)$ or $sub(t, x, \phi_2)$ is 0, and otherwise $sub(t, x, \phi_1) \vee sub(t, x, \phi_2)$. Similarly for the other connectives.

$sub(t, x, \forall y\phi)$ is $\forall y\phi$ if $x$ and $y$ are the same variable. Otherwise, if $occurs(y, t)$ and $occurs(x, \phi)$, $sub(t, x, \forall y\phi)$ is 0, and otherwise it is $\forall y\, sub(t, x, \phi)$. Similarly for existential quantification.

Note that we define $sub(t, x, n)$ in terms of $sub(t, x, m)$ for $m < n$, which means that we can use course-of-values recursion to establish that $sub$ is primitive recursive.

We will need a further substitution operation, $subseq$. If $s$ is a sequence number $\langle t_1, \ldots, t_n \rangle$ and $s'$ a sequence number $\langle x_1, \ldots, x_n \rangle$, where the $x_i$ are different variables, $subseq(s, s', \phi)$ is the result of simultaneously substituting the terms in $s$ for the variables in $s'$ in the formula $\phi$ (provided each $t_i$ is free for the corresponding $x_i$ in $\phi$). "Simultaneously" means that e.g., $subseq(\langle v_0, v_1 \rangle, \langle v_1, v_0 \rangle, p(v_1, v_0))$ is $p(v_0, v_1)$, so that the occurrences of $x_i$ substituted for are only those which are in $\phi$ originally. $sub(v_0, v_1)$, $sub(v_1, v_0, p(v_1, v_0))$, by contrast, is $p(v_0, v_0)$. $subseq(s, s', t)$ is defined similarly, and again the value is 0 if the arguments are inappropriate. $subseq$ is seen to be primitive recursive on the basis of a recursive definition similar to that of $sub$.

Above, we have only treated substitution for first order variables. There are corresponding operations for making substitutions for second order variables, except that there are no function symbols in second order terms, making substitution for second order variables a simpler operation.

## 7.5. The semantics of first and second order languages

When we talk about a sentence being true or false, we must implicitly or explicitly invoke some interpretation of the language in which the sentence is formulated. Formal languages are used in this book only for the purpose of studying axiomatizations of arithmetic, systems of axioms and rules for deriving statements about the natural numbers. When we speak of a first order sentence in the primitive language of arithmetic as true or false, we always mean that it is true or false when we read the quantifiers as ranging over the natural numbers, take $<$ to mean "is strictly smaller than", take $s$ to be the successor operation and read $+$ and $*$ as standing for the operations of multiplication and addition. This is usually referred to in logic as "truth in the standard model". When we speak of sentences in an extended first order language as true or false, we always have in mind some interpretation of the extra function symbols and predicate symbols as number-theoretic functions and relations.

In the case of second order languages, the interpretation of the quantifiers over sets may vary from one context to another. Sometimes they are taken to range over some restricted class of subsets of **N**, and sometimes they range over arbitrary subsets. The intended reading of the set quantifiers will be clear from the context.

In some arguments, however, and in particular in the proof of the completeness theorem in the next chapter and in later applications of that theorem, the concept of a *general* interpretation of a formal language will also be needed. In a general interpretation, the quantifiers don't necessarily range over the natural numbers, and the function and predicate symbols can have any arbitrary meaning, except that $=$ is always interpreted as identity.

In formal terms, an *interpretation* of a first order language $L$ is provided by giving (1) a non-empty domain $D$ for the first order variables to range over, (2) for every function symbol $f_j^i$ in the language a function $F_{ij}$ from $D^i$ to $D$, and (3) for every predicate symbol $p_j^i$ in the language an $i$-ary relation $R_{ij}$ defined on $D$, or in other words a subset of the set $D^i$ of sequences of length $i$ of elements of $D$. Given an interpretation $I$ of $L$ and a sentence $\phi$ of $L$ we say that $\phi$ is *true in* $I$ if $\phi$ is true when we read the quantifiers as ranging over the domain of $I$ and also interpret the function symbols and predicate symbols that occur in $\phi$ as stipulated in the interpretation $I$. The quantifiers and connectives are of course interpreted the way they usually are in mathematics, as will be commented on in the next chapter in connection with the rules of logic. The identity symbol is interpreted as identity in the domain $D$.

If $L$ is a second order language, there is a further ingredient in an interpretation of $L$: a non-empty set $M_D$ of subsets of the domain $D$. The epsilon symbol is always interpreted as the membership relation, and the identity symbol applied to sets is interpreted as identity in $M_D$.

General interpretations are studied in the part of logic known as model theory, but will only be touched on occasionally in this book. No formal definition of "true in $I$" was given above, since the informal explanation will be quite sufficient for the uses of model theory in the book. Truth in the standard model is a different matter. This will be of central importance in later chapters, were we will need to consider formalizations of statements about arithmetical truth, and therefore need a formal definition of "true arithmetical sentence".

## 7.6. Defining arithmetical truth

In considering truth for statements in the language of arithmetic, we may restrict ourselves to terms and formulas in the primitive arithmetical language, since statements in an extended arithmetical language can be assumed translated into primitive notation. Thus unless otherwise noted, "sentence", "formula", and "term" in this section refer to the primitive first-order language.

## True atomic sentences

If $t$ is a variable-free term, it has a certain *value*, which is a natural number obtained by computation using addition and multiplication. The value of $t$ is given by the primitive recursive function *val*, with the defining equations $val(0) = 0$, $val(s(t)) = val(t) + 1$, $val(s + t) = val(s) + val(t)$, $val(s * t) = val(s) * val(t)$. If $n$ is not a variable-free term, $val(n)$ is defined to be 0. *val* is primitive recursive, since for example the last equation means that if $n$ is a sequence number of length 4 where $(n)_0 = 4$, $(n)_1 = f_1^2$ and $(n)_2$ and $(n)_3$ are variable-free terms, $val(n) = val((n)_2) * val((n)_3)$, so *val* is defined by course-of-values recursion.

Using *val*, we see that the set $True_{Vf}$ of *variable-free* true sentences is primitive recursive, and thus in particular arithmetically definable. For $\phi$ is in $True_{Vf}$ if and only if $\phi$ is $s = t$ for closed terms $s$, $t$ with $val(s) = val(t)$, or $s < t$, where $val(s) < val(t)$, or $\neg\psi$ where $\psi$ is a variable-free formula not in $True_{Vf}$, or $\psi_1 \vee \psi_2$, where $\psi_1$ and $\psi_2$ are variable-free formulas and at least one of them is in $True_{Vf}$.

This definition of $True_{Vf}$, it will be noted, is in agreement with ordinary usage. For example, most likely everybody will agree in any ordinary mathematical context that an equation $s = t$ is true if and only if $s$ and $t$ have the same value. Similarly, the negation of a sentence is true (and the sentence itself false) if and only if the sentence itself is not true, and the disjunction of two sentences is true if and only if at least one of the sentences is true. It is of course perfectly possible to adopt some usage on which this no longer holds, but the justification for presenting the above definition of $True_{Vf}$ without any particular preamble or introductory explanation is only that this usage is assumed to be familiar and acceptable.

In the context of a philosophical discussion, it may well happen that a distinction is made between saying e.g., that "3 + 5" and "8" are terms with the same value, and saying that "3 + 5 = 8" is a true statement. In particular, it may be thought that referring to "3 + 5 = 8" as a true statement carries some philosophical implication or presupposition concerning the existence of numbers or the nature of mathematics in general. As was emphasized in §1.3, there are no such implications in the use of "true" applied to mathematical statements in this book.

## Tarski's theorem and Gödel diagonalization

Moving on to the set *True* of all true arithmetical sentences (to be formally defined below), it would be surprising indeed if it were primitive recursive, since there would then be an algorithm for deciding the truth or falsity of an arithmetical sentence. In fact *True* is not even arithmetically definable. This result is known as *Tarski's theorem*, but it was also used by Gödel in arriving at his incompleteness theorem. (About this, see Feferman [1984].)

We can prove Tarski's theorem by applying the diagonal argument outlined in §5.2 Suppose *True* is definable in the language of arithmetic by the formula $\phi$ with the free variable $v_0$. Then the formula

$$form_1(v_0) \wedge free(v_0, \overline{v}_0) \wedge \neg\phi_x(sub(num(v_0), \overline{v}_0, v_0))$$

(where $free(n, k)$ means that $n$ is a formula with $k$ as its only free variable) also has $v_0$ as its sole free variable, but since it cannot express the same condition as any of the formulas with free variable $v_0$, a contradiction follows.

Let's consider somewhat more carefully a variant of this argument. This variant consists in showing that if arithmetical truth were arithmetically definable, a version of the statement put forward in the ancient paradox of the Liar ("This sentence is false") could be formulated as an arithmetical statement, which would then be true if and only if it is false. For this, a construction is used which we will later return to in connection with the proof of Gödel's theorem.

Let $\psi$ be any formula with a single free variable $x$. We will define a sentence $\phi$, called the *diagonalization* of $\psi$, which is true if and only if it satisfies $\psi$, that is, if and only if $\psi_x(\overline{\phi})$ is true.

$\phi$ will be defined, using a method invented by Gödel, as a sentence which "says of itself that it satisfies $\psi$". For this we use substitution, expressing in the language of arithmetic a sentence corresponding to

> The result of substituting the quotation of "The result of substituting the quotation of $x$ for '$x$' in $x$ satisfies $\psi$." for '$x$' in "The result of substituting the quotation of $x$ for '$x$' in $x$ satisfies $\psi$." satisfies $\psi$.

The point of this strange-looking sentence is that it says of a certain sentence, defined as the result of a substitution operation, that it satisfies $\psi$. If we carry out the indicated substitution operation, we find that it yields as its result precisely the sentence just formulated. The strange-looking sentence is in this sense *self-referential.*

For the formalized version, we just let $\theta$ be (a translation into primitive notation of) the formula $\psi_x(sub(num(x), \overline{x}, x))$, and then let $\phi$ be the formula $\theta_x(\overline{\theta})$, which is to say the formula $\psi_x(sub(num(\overline{\theta}), \overline{x}, \overline{\theta}))$. The value of $sub(num(\overline{\theta}), \overline{x}, \overline{\theta})$ is the formula obtained by substituting the numeral of $\theta$ for the variable $x$ in $\theta$, which is the formula $\phi$, so $\phi$ is true if and only if $\psi_x(\overline{\phi})$ is true.

Tarski's theorem is now an easy consequence. For if the formula $\psi$ defines the class of true arithmetical sentences, the diagonalization of the formula $\neg\psi$ is true if and only if it is not true. So there is no such definition.

## Inductive definitions of truth

So how is the set *True* of true arithmetical sentences to be defined? We can define *True* and the set *False* of false arithmetical sentences through a simultaneous inductive definition, as follows:

For atomic $\phi$, if $\phi$ is in $True_{Vf}$ it is true, otherwise $\phi$ is false.

If $\phi_1$ or $\phi_2$ is true, $\phi_1 \vee \phi_2$ is true, and if $\phi_1$ and $\phi_2$ are both false, $\phi_1 \vee \phi_2$ is false.

If $\phi$ is true, $\neg\phi$ is false, and if $\phi$ is false, $\neg\phi$ is true.

If $\phi_x(\overline{n})$ is true for some $n$, $\exists x\phi$ is true, and if $\phi_x(\overline{n})$ is false for every $n$, $\exists x\phi$ is false.

When this inductive definition is expressed in set-theoretical terms, as in §6.2, it can be formulated in the language of analysis. As an exercise in squeezing inductive definitions into the general set-theoretical mold, let's spell out what this definition will look like. Representing "$\phi$ is true" by the pair $\langle\phi, 1\rangle$ and "$\phi$ is false" by the pair $\langle\phi, 0\rangle$, a sentence $\phi$ is true if and only if $\langle\phi, 1\rangle$ belongs to every set $A$ such that

$\langle\phi, 1\rangle$ belongs to $A$ for every atomic $\phi$ in $True_{Vf}$,

$\langle\phi, 0\rangle$ belongs to $A$ for every atomic sentence $\phi$ not in $True_{Vf}$,

$\langle\phi_1 \vee \phi_2, 1\rangle$ belongs to $A$ whenever $\langle\phi_1, 1\rangle$ or $\langle\phi_2, 1\rangle$ belongs to $A$,

$\langle\phi_1 \vee \phi_2, 0\rangle$ belongs to $A$ whenever $\langle\phi_1, 0\rangle$ and $\langle\phi_2, 0\rangle$ belong to $A$,

$\langle\neg\phi, 1\rangle$ belongs to $A$ whenever $\langle\phi, 0\rangle$ belongs to $A$,

$\langle\neg\phi, 0\rangle$ belongs to $A$ whenever $\langle\phi, 1\rangle$ belongs to $A$,

$\langle\exists x\phi, 1\rangle$ belongs to $A$ whenever $\langle\phi_x(\overline{n}), 1\rangle$ belongs to $A$ for some $n$,

$\langle\exists x\phi, 0\rangle$ belongs to $A$ whenever $\langle\phi_x(\overline{n}), 0\rangle$ belongs to $A$ for every $n$.

Note that this is an example of an inductive definition using a rule that is not finite. This is because of the last clause above, where infinitely many premises are needed to obtain $\langle\exists x\phi, 0\rangle$. If we leave out the last clause in the definition, what remains is an inductive definition of truth for sentences that make no universal assertions. The corresponding rule is finite, indicating that there is an algorithmic procedure for generating all the true sentences of this restricted type, whereas there is no such procedure for sentences in general. In Chapter 10 it will emerge that arithmetical truth can be defined using quantification over *arithmetical* sets, and that reasoning in terms of arithmetical truth and reasoning in terms of arithmetical sets are in fact interchangeable.

## Characterizing truth

There are other possible mathematical definitions of the true arithmetical sentences. No particular definition occupies a privileged position. In particular, in taking truth to be inductively defined as shown above, we need not necessarily presuppose any reduction of inductive definitions to explicit set-theoretical definitions, but may accept inductive definitions as justified in other terms, or as a primitive idea in mathematics, that is, one that is understood and justified on a basic level. What is essential to the argument of this book is only that we understand "true" as a mathematical property of sentences, and that we define or understand it so as to be able to conclude that for all arithmetical sentences $\phi$, $\phi_1$, $\phi_2$,

if $\phi$ is atomic, $\phi$ is true if and only if $\phi$ is in $True_{Vf}$,

$\phi_1 \lor \phi_2$ is true if and only if $\phi_1$ is true or $\phi_2$ is true,

$\neg\phi$ is true if and only if $\phi$ is not true, and

$\exists x\phi$ is true if and only if $\phi_x(\overline{n})$ is true for some $n$.

To prove these characterizing properties of truth using the inductive definition given, first note that by the fixed point characterization of inductively defined sets in §6.2, we know from this definition that

for atomic $\phi$, $\phi$ is true if and only if $\phi$ is in $True_{Vf}$,

for atomic $\phi$, $\phi$ is false if and only if $\phi$ is not in $True_{Vf}$,

$\phi_1 \lor \phi_2$ is true if and only if $\phi_1$ or $\phi_2$ is true,

$\phi_1 \lor \phi_2$ is false if and only if $\phi_1$ and $\phi_2$ are false,

$\exists x\phi$ is true if and only if $\phi_x(\overline{n})$ is true for some $n$,

$\exists x\phi$ is false if and only if $\phi_x(\overline{n})$ is false for every $n$,

$\neg\phi$ is true if and only if $\phi$ is false,

$\neg\phi$ is false if and only if $\phi$ is true.

Therefore, to derive the characterizing properties of truth, it is necessary and sufficient to show by induction on $\phi$ that $\phi$ is false if and only if $\phi$ is not true. Here "induction on $\phi$" does not refer to induction on natural numbers, but induction on formulas, that is, induction on the basis of the inductive definition of the set of formulas. We can however also use induction on the natural numbers, by casting the argument as a proof by induction on the number of occurrences of quantifiers and connectives in $\phi$. The proof is straightforward, using the above equivalences.

# CHAPTER 8

# LOGIC AND THEORIES

## 8.1. Logical reasoning

Mathematical proofs, as usually represented in logic, consist in logical deductions from axioms stated in a formalized language. We have a set of general logical rules (and in some formulations axioms), applicable in all mathematical reasoning, and a set of particular mathematical axioms, depending on what mathematics is being represented, from which we draw conclusions using the logical rules. In using such a representation of mathematical proofs, we are not supposing that mathematicians do or could prove theorems in this way, even though the axiomatic method on a more informal level has been very useful in actual mathematics. (For an elaboration of this point, see the comments on formal and actual provability in §1.3.)

In this book, the mathematical axioms will be formulas in a first or second order language as defined in Chapter 7. The basic mathematical axioms will be presented in Chapters 9 and 10. Exactly what logical rules are used doesn't matter for the results and questions to be considered here, but we need to give a precise definition of just what "logically deducible" means for our formal languages, both in order to establish that various relations having to do with logical deductions can be defined arithmetically, and to prove the completeness theorem for predicate logic, which will be used at some points in later chapters.

The basic relation to be defined in this chapter is $\Gamma \Rightarrow \phi$, which we read as "$\phi$ can be logically deduced from the formulas in $\Gamma$". Here $\Gamma$ is a set, possibly infinite, of formulas in some first or second order language, and $\phi$ a formula in that language. The formulas in $\Gamma$ may be axioms in a mathematical theory, but they can also be assumptions used in hypothetical reasoning. The relation $\Rightarrow$ is defined inductively, and since all the rules defining the relation are finite rules, we will automatically get a concept of *logical deduction* in the sense of a finite sequence establishing that the relation holds, as described in Chapter 6 for finite rules in general. Such a definition of "logical deduction" is quite adequate for the purposes of this book. There are many different ways of defining logical deductions, and in other logical investigations, in particular those belonging to proof theory, the structure of deductions plays an important role.

$\Rightarrow$ is a purely syntactic relation, that is, the definition refers only to the forms of formulas, and not to any meaning given to them. But of course the justification for the rules lies in the meaning of the connectives and quantifiers, as they are in fact used in mathematics.

The rules to be given define the deducibility relation for what is called *classical predicate logic*. A few of the many variants of this logic will be touched upon below. The rules given include rules for the identity relation, which express properties of identity that one expects to hold in all ordinary mathematical contexts. They are not always included in formulations of predicate logic, so a distinction is often made between predicate logic with and without identity.

There are no rules for the connective $\equiv$. This is common in formulations of predicate logic, and the reason is that the biconditional is fully explained by saying that $\phi \equiv \psi$ is an abbreviation of $(\phi \supset \psi) \wedge (\psi \supset \phi)$. In the context of any result or discussion in this book, $\phi \equiv \psi$ can be replaced by this longer formula without loss, and will be regarded as purely an abbreviation of the longer formula. The other connectives cannot be eliminated in the same way. Even though for example $\phi \vee \psi$ has logically exactly the same content as $\neg(\neg\phi \wedge \neg\psi)$, and the two formulas are logically equivalent in the sense that each is logically deducible from the other, there is a distinction between the two in that there is a "direct" way of deducing a disjunction $\phi \vee \psi$ and a different "direct" way of deducing the logically equivalent formula $\neg(\neg\phi \wedge \neg\psi)$, as will be explained below. In what is called *constructive* or *intuitionistic* logic, also defined below, this difference becomes a difference in logical content, and $\phi \vee \psi$ and $\neg(\neg\phi \wedge \neg\psi)$ are no longer logically equivalent. In contrast, the direct way of proving an equivalence $\phi \equiv \psi$ is just the same as the direct way of proving $(\phi \supset \psi) \wedge (\psi \supset \phi)$, both in classical and intuitionistic logic. It should be added that in other contexts where the propositional connectives are used, such as the logical analysis of digital circuits, there are indeed reasons for formulating rules specifically for the connective $\equiv$.

## 8.2. The deducibility relation

The relation $\Rightarrow$ will be inductively defined through rules [L1]–[L16]. The accompanying comments draw heavily on the intended role of these rules as capturing forms of reasoning which are in fact used in mathematics and which we are inclined to regard as purely logical. Below, $\Gamma, \Delta$ stands for $\Gamma \cup \Delta$, and $\Gamma, \phi$ for $\Gamma \cup \{\phi\}$. We will also write $\phi \Rightarrow \psi$ for $\{\phi\} \Rightarrow \psi$.

### Axioms and assumptions

[L1] $\Gamma \Rightarrow \phi$ if $\phi \Rightarrow \Gamma$.

This is the basis in the inductive definition, together with the identity principle [L14]. It stipulates that every formula that belongs to $\Gamma$ can be logically deduced from $\Gamma$. This is to be understood only in the sense that from the assumption that something is true, it follows that it is true. The deduction consists essentially in an annotation: "is an assumption" or "is an axiom".

In mathematics one doesn't normally speak of axioms as "provable from themselves", even though one may well say that immediate consequences of axioms have ("trivial" or "obvious") proofs. In logic, this fine distinction is not made, and we don't hesitate to say that for example the induction principle is provable in a formal theory where it is taken as an axiom.

## Conjunction and disjunction

[L2] $\Gamma \Rightarrow \phi \wedge \psi$ if and only if $\Gamma \Rightarrow \phi$ and $\Gamma \Rightarrow \psi$.

An "if and only if" statement such as [L2] when used to formulate rules in an inductive definition is to be understood as two "if" statements, which in turn are to be read as rules. In the present case, [L2] tells us that if $\Gamma \Rightarrow \phi$ and $\Gamma \Rightarrow \psi$ then $\Gamma \Rightarrow \phi \wedge \psi$, if $\Gamma \Rightarrow \phi \wedge \psi$ then $\Gamma \Rightarrow \phi$, and if $\Gamma \Rightarrow \phi \wedge \psi$ then $\Gamma \Rightarrow \psi$.

Note that the corresponding principle is not true for disjunction. We do indeed have that

[L3] If $\Gamma \Rightarrow \phi$ or $\Gamma \Rightarrow \psi$ then $\Gamma \Rightarrow \phi \vee \psi$.

but the converse does not hold, since e.g., "$k$ is even or a prime" is deducible from "$k$ is even or a prime", although neither "$k$ is even" nor "$k$ is a prime" can be deduced. But there is another principle involving disjunction that does hold, namely that of "reasoning by cases" or "constructive dilemma":

[L4] If $\Gamma, \phi \Rightarrow \theta$ and $\Gamma, \psi \Rightarrow \theta$ then $\Gamma, \phi \vee \psi \Rightarrow \theta$.

An example of reasoning by constructive dilemma is Euclid's proof of the inexhaustibility of the primes. There is in mathematics also reasoning by "non-constructive dilemma", in which we conclude that a statement $A$ holds after verifying that both the assumption $B$ and the assumption not-$B$ lead to the conclusion $A$. A famous instance of this was the proof by J. E. Littlewood of a theorem $A$ in number theory with $B =$ the Riemann hypothesis, an outstanding open problem in mathematics. Proofs by non-constructive dilemma can be represented using the rules for negation and implication to be given below.

## Implication

[L5] If $\Gamma, \phi \Rightarrow \psi$ then $\Gamma \Rightarrow \phi \supset \psi$.

This encapsulates the method of "hypothetical reasoning": to deduce "if $A$ then $B$" from a set of axioms or assumptions $M$, assume for the moment that $A$ is true (together with the premises in $M$) and deduce $B$ from these assumptions. This is used in mathematics all the time, except that the final step (concluding that "if $A$ then $B$" follows from $M$, since $B$ follows from $M$ together with $A$) is usually left implicit.

A particular aspect of the rule should be noted: it is not required that $\phi$ is at all used in deducing $\psi$. A special case of the rule is

If $\Gamma \Rightarrow \psi$ then $\Gamma \Rightarrow \phi \supset \psi$.

(We don't need to include this as a special case, for it follows from [L5] and the monotonicity property of $\Rightarrow$ to be established below.) This strikes some as odd or undesirable when formulated as a general principle. It is felt that "if $A$ then $B$" should follow only if we have used the assumption $A$ in some meaningful way in arriving at the conclusion $B$. The subject of *relevant logic* deals with rules of reasoning that respect this idea. Such a system of rules will be a great deal more complicated and difficult than the rules of classical logic, and will not be adequate for the formalization of ordinary mathematical reasoning, where one does not require that all assumptions stated are used. This of course doesn't mean that it wouldn't be possible to reorganize mathematics on the basis of relevant logic, but it does imply that only the general rule [L5] will allow us to formalize mathematical reasoning as it now exists.

The second principle for reasoning with conditionals is the rule traditionally known as *modus ponens* (or more formally, *modus ponendo ponens*):

[L6] If $\Gamma \Rightarrow \phi$ and $\Gamma \Rightarrow \phi \supset \psi$ then $\Gamma \Rightarrow \psi$.

## Negation

Now for negation. We have three principles.

[L7] If $\Gamma, \phi \Rightarrow \psi$ and $\Gamma, \phi \Rightarrow \neg\psi$ then $\Gamma \Rightarrow \neg\phi$.

This is the most basic principle for negation: if contradictory conclusions can be drawn from $A$ (perhaps together with some further assumptions), then we can conclude that $A$ is false (given those further assumptions).

[L7], which is one form of the principle of *reductio ad absurdum*, is sometimes confused with a different principle:

[L8] If $\Gamma, \neg\phi \Rightarrow \psi$ and $\Gamma, \neg\phi \Rightarrow \neg\psi$ then $\Gamma \Rightarrow \phi$.

This second principle is the principle of *indirect proof*, also known as *classical reductio*: if a contradiction can be derived from the assumption that $A$ is false, we conclude that $A$ is true. This is "indirect" because there is also a "direct" way of proving a statement, depending on the logical form of the statement. Thus the "direct" way of proving a disjunction "$A$ or $B$" is to prove either $A$ or $B$, and the "direct" way of proving an existential statement "for some $x$, $A(x)$" is to prove $A(t)$ for some particular $t$. In classical mathematics, it is not always possible to give a direct proof of a statement $A$, but instead one must use an indirect proof, showing that the assumption that $A$ is false leads to a contradiction.

For example, in proving the "fundamental theorem of algebra", that every polynomial over the complex numbers (other than constant non-zero polynomials) has at least one zero, one often argues indirectly. Assuming that $p(x)$ is different from 0 for every complex $x$, some complex calculus yields a result contradicting known theorems. A direct proof would in this case be a proof that $p(t)$ is 0, where $t$ is some term defined using the coefficients of the polynomial $p$. Such a direct proof can be given for second degree polynomials $p$, where $t$ can be chosen in accordance with the standard formula for solving second degree equations, as a term involving square roots. (In the general case, where there is no such root expression $t$, the theorem can still be proved "constructively", rather than as a "pure existence proof", but these distinctions between more or less constructive proofs will not play any role in this book.)

In contrast, Euclid's proof that the square root of two is irrational, or more properly, that there is no rational square root of two, is not an indirect proof, although it is often said to be such. The proof does indeed proceed by assuming that there is a rational square root of two and deriving a contradiction, but this is not an indirect proof since "irrational" is defined as "not rational". To assume that $A$ is true and derive a contradiction is the *direct* way of proving the negative statement "it is not the case that $A$".

Similarly, Euclid's proof that there are infinitely many primes is not an indirect proof. To prove that there is, for any sequence of primes $q_1, \ldots, q_n$, a prime not in the sequence, Euclid argues that $q_1 \ldots q_{n+1}$ is either itself a prime or is divisible by a prime not in the sequence. In both cases there is a prime not in the sequence. This is an argument by constructive dilemma, not an indirect argument. The argument is sometimes formulated as an indirect proof by prefacing it with the assumption "Suppose $q_1, \ldots, q_n$ are all the primes". However, since this assumption isn't even used in the proof, the reformulation is pointless.

On the other hand, Euler's proof that there are infinitely many primes, presented in Chapter 3, *is* indirect. The argument is that if there are only finitely many primes, $\sum p^{-1}$, where the sum is taken over all primes $p$, will converge, but since it doesn't, there are infinitely many primes. Thus formulated, the proof gives no upper bound below which a prime greater than a given number $k$ must be found. On the other hand, since the smallest prime greater than $k$ *can* always be computed, quite independently of how we have proved that there is such a prime, it is not immediately clear in this case just what separates a direct from an indirect proof. There are in fact some subtleties involved, which are studied in proof theory, where it is also shown how Euler's proof can be transformed into an argument giving such an upper bound.

Indirect proofs have traditionally been somewhat controversial in mathematics. A direct (or, more generally, constructive) proof of a disjunction or an existential statement usually gives more information than an indirect proof, and many mathematicians have also regarded indirect proofs as philosophically suspect. Mathematics teachers also tend to take a somewhat dim view of indirect proofs, because in an indirect proof one argues from logically inconsistent premises, which means that it is easy to get confused and to arrive at the desired contradiction by various illicit steps. This is true also of proofs that apply the principle [L7], but unlike indirect proofs, such proofs have never been controversial among those who at all accept the use of negation in mathematical statements.

The third rule for negation has again given rise to controversy:

[L9] If $\Gamma \Rightarrow \psi$ and $\Gamma \Rightarrow \neg\psi$ then $\Gamma \Rightarrow \phi$.

This is traditionally known as the rule of *ex falso quodlibet*—"anything from a falsehood"—although it should more properly be called "anything from a logically inconsistent set of premises". If a set $M$ of premises is inconsistent, in the sense that both $B$ and the negation of $B$ can be logically derived from it, then any statement can be derived from $M$. This has struck many as odd or undesirable. And indeed a system of rules containing [L9] is not adequate for formalizing the ordinary informal practice whereby we "contain" contradictions. We may have and even be aware of contradictions in our theory or system of beliefs, but still be able to "work around" those contradictions, without ever concluding that anything goes since our premises are inconsistent, and without drawing arbitrary or irrelevant conclusions on the basis of the contradiction. However, again a formal logic that does not include [L9] must be a great deal more complicated than classical logic and does not accord well with ordinary mathematical reasoning. Note in particular that [L9] is a consequence of the following two rules (the second of which is known as the principle of disjunctive syllogism):

(1) If $\Gamma \Rightarrow \psi$ then $\Gamma \Rightarrow \phi \vee \psi$.
(2) If $\Gamma \Rightarrow \phi \vee \psi$ and $\Gamma \Rightarrow \neg\psi$ then $\Gamma \Rightarrow \phi$.

To avoid [L9], we would have to give up one of these. (1) and (2) are used all the time in mathematics (which is to say that arguments are used which have the form expressed in the general principles (1) and (2)), and it is far from clear what more restrictive principles could be used instead.

Classical logic, then, incorporates all three principles [L7]–[L9] for negation. What is called *minimal logic* uses only [L7], and *intuitionistic logic* uses [L7] and [L9]. All of the other rules for $\Rightarrow$ are valid also in minimal and intuitionistic logic.

If we are only interested in classical logic, there is a redundancy in the rules for negation, since [L7] and [L9] both follow from [L8] in combination with the other rules. However, negation occupies a rather special place in logic, being associated with several traditional philosophical disputes and conundrums, and so it is often helpful to be aware of the two weaker logical systems minimal and intuitionistic logic.

### The universal quantifier

[L10] If $\Gamma \Rightarrow \phi$ and $x$ does not occur free in any formula in $\Gamma$, then $\Gamma, \Delta \Rightarrow \forall x\phi$.

This again is in agreement with ordinary mathematical reasoning. If we want to deduce from a set of assumptions $M$ that $A(x)$ holds for every real number $x$, we usually do so by arguing about an unspecified and arbitrary real number $r$, and showing that $A(r)$ holds. The final step expressed in the rule above, "since $r$ was an arbitrary real number, it follows that for every real number $x$, $A(x)$", is usually left implicit. Of course for $r$ really to be arbitrary, we must not have made any special assumptions about $r$, which in the formal rule becomes the requirement that $x$ does not occur free in any formula in $\Gamma$.

The reason why $\Delta$ appears in the conclusion of [L10] is to ensure that the monotonicity property explained in §8.3 below holds. Informally, what the introduction of $\Delta$ amounts to is a relaxation of the condition that $r$ does not occur in any premise: it's allowed to occur in a premise, as long as we don't use that premise in arriving at the conclusion $A(r)$.

If we are dealing with a second order language, the rule [L10] also covers the case of set quantifiers:

If $\Gamma \Rightarrow \phi$ and $X$ does not occur free in any formula in $\Gamma$,
then $\Gamma, \Delta \Rightarrow \forall X\phi$

The second rule for the universal quantifier expresses the principle that what is true of everything is true of any particular individual:

[L11] If $\Gamma \Rightarrow \forall x\phi$ then $\Gamma \Rightarrow \phi_x(s)$ for any term $s$ free for $x$ in $\phi$.

Again this also holds for set variables, given a second order language.

### The existential quantifier

The first principle is that used in direct proofs of existential statements: whatever is true of some particular individual is true of something:

[L12] If $\Gamma \Rightarrow \phi_x(s)$ then $\Gamma \Rightarrow \exists x\phi$, for any term $s$ free for $x$ in $\phi$.

Note the order of substitution here: $\phi_x(s)$ is obtained by substituting $s$ for all free occurrences of $x$ in $\phi$. Thus the rule also covers such cases as "$5 \leq 5$, so there is an $x$ such that $x \leq 5$".

The last quantifier rule formalizes a common form of argument in which one uses an existential premise. Assuming that there is an $x$ such that $A(x)$, we let $y$ stand for an arbitrary such $x$. We reason from the assumption $A(y)$ to some conclusion $B$ which is not about $y$, and then conclude that $B$ follows from the assumption that there is an $x$ such that $A(x)$. Again this last step is usually left implicit in ordinary mathematical reasoning. The formal rule includes conditions on the variable $y$ to ensure that we haven't assumed anything about $y$ except that it is some arbitrary individual satisfying $A(y)$:

[L13] If $\Gamma, \phi_x(y) \Rightarrow \psi$ and $y$ is not free in any formula in $\Gamma$ or in the formula $\psi$, then $\Gamma, \Delta, \exists x\phi \Rightarrow \psi$.

As in the case of the universal quantifier, these rules also apply to set quantifiers.

Recall the discussion in §2.3 of the axiom of dependent choice: if for every $x$ in $M$ there is a $y$ in $M$ such that $S(x, y)$, then there is an infinite sequence $x_0, x_1, \dots$ of elements in $M$ such that $S(x_i, x_{i+1})$ for every $i$. Let us call a sequence in which consecutive elements are related by $S$ an *$S$-sequence.* If we only want to prove that every finite $S$-sequence $x_0, x_1, \dots, x_n$ can be extended by the addition of a further element, we don't need the axiom of dependent choice, but only principle [L13]. In informal terms, the argument is "since there is an $x$ such that $S(x_n, x)$, just let $x_{n+1}$ be any such, yielding a longer $S$-sequence $x_0, x_1, \dots, x_n, x_{n+1}$." Thus using logic alone we can prove for every $n$ that there is an $S$-sequence of length $n$, and using logic and mathematical induction we can prove "for every $n$, there is an $S$-sequence of length $n$". However, to prove the existence of an infinite $S$-sequence, we cannot say "just let $x_{n+1}$ be any such" infinitely many times, so we must either explicitly define the $S$-sequence or else invoke a special existence principle, the axiom of dependent choice.

## Identity

The logical rules for identity express the axiom that everything is identical with itself, and the principle of substitution of equals for equals: if $a = b$, whatever is true of $a$ is true of $b$.

[L14] $\Gamma \Rightarrow \forall \boldsymbol{x}(\boldsymbol{x} = \boldsymbol{x})$

[L15] If $\Gamma \Rightarrow \boldsymbol{s} = \boldsymbol{t}$ and $\Gamma \Rightarrow \phi_{\boldsymbol{x}}(\boldsymbol{s})$ then $\Gamma \Rightarrow \phi_{\boldsymbol{x}}(\boldsymbol{t})$.

The rules also apply in the second order case. Note the following consequences of these rules:

If $\Gamma \Rightarrow \boldsymbol{s} = \boldsymbol{t}$ then $\Gamma \Rightarrow \boldsymbol{t} = \boldsymbol{s}$.

This follows using [L14], [L11], and [L15], with $\phi$ in [L15] chosen as $\boldsymbol{x} = \boldsymbol{s}$. Similarly,

If $\Gamma \Rightarrow \boldsymbol{s} = \boldsymbol{t}$ and $\Gamma \Rightarrow \boldsymbol{t} = \boldsymbol{u}$ then $\Gamma \Rightarrow \boldsymbol{s} = \boldsymbol{u}$.

## Extensionality

The final rule is the only one which has to do specifically with the concept of set, and it expresses the one principle satisfied by sets on any interpretation, no matter what sets we take the set quantifiers to refer to, namely the principle of extensionality:

[L16] If $\Gamma \Rightarrow \forall x(x \in X \equiv x \in Y)$ then $\Gamma \Rightarrow X = Y$

## 8.3. Some properties of the deducibility relation

The deducibility relation $\Rightarrow$ is defined as the smallest relation between sets of formulas and formulas satisfying [L1]–[L16]. Thus to prove by induction that some relation $R$ holds between $\Gamma$ and $\phi$ whenever $\Gamma \Rightarrow \phi$, we prove that [L1]–[L16] also hold with $R$ instead of $\Rightarrow$.

There is some room for confusion with regard to this inductive definition because of an overlapping of terminology. The instances of the rule defining the relation $\Rightarrow$ have premises and conclusions (using the terminology of the general presentation of inductive definitions in §6.2) of the form $\langle \Gamma, \phi \rangle$. If $\Gamma \Rightarrow \phi$ holds, it is also true, using ordinary logical terminology, that $\phi$ is a (logical) conclusion of the premises in $\Gamma$. For a given $\Gamma$, it is possible to define the set of $\phi$ such that $\Gamma \Rightarrow \phi$ by a suitable inductive definition with a finite rule. In the present exposition, however, the focus is on the finite rule defining the relation $\Rightarrow$.

An easy induction shows that $\Rightarrow$ has the *monotonicity property*: if $\Gamma \Rightarrow \phi$ and $\Gamma \subseteq \Delta$, then $\Delta \Rightarrow \phi$. There are forms of reasoning, studied in the field of *non-monotonic logic*, for which the monotonicity property does not hold, but instead new information may prompt the rejection of earlier conclusions. In mathematical deductive reasoning, monotonicity is taken for granted. We could have included the monotonicity property as a clause in the inductive definition, but this isn't necessary, since it follows by induction using the clauses given.

Another easy inductive argument (using monotonicity) shows that if $\Gamma \Rightarrow \phi$ then $\Gamma' \Rightarrow \phi$ for some finite subset $\Gamma'$ of $\Gamma$. This is because in each of the rules only finitely many formulas in $\Gamma$ are actually used. We will refer to this as the *finiteness property* of $\Rightarrow$. (In proof theory, for theoretical purposes, deducibility relations which do not have the finiteness property are also studied.)

A third property of $\Rightarrow$ is the *cut property*: if $\Gamma \Rightarrow \phi$ and $\Delta, \phi \Rightarrow \psi$ then $\Gamma, \Delta \Rightarrow \psi$. This property plays an important role in proof theory. In the context of our definition of $\Rightarrow$ it is a simple consequence of the other rules. For if $\Gamma \Rightarrow \phi$ and $\Delta, \phi \Rightarrow \psi$ then $\Gamma \Rightarrow \phi$ and $\Delta \Rightarrow \phi \supset \psi$ by [L5], so by the monotonicity property $\Gamma, \Delta \Rightarrow \phi$ and $\Gamma, \Delta \Rightarrow \phi \supset \psi$, so $\Gamma, \Delta \Rightarrow \psi$ by [L6].

$\Gamma$ is *inconsistent* if $\Gamma \Rightarrow \phi$ and $\Gamma \Rightarrow \neg\phi$ for some formula $\phi$. Equivalently, by the ex falso quodlibet rule, $\Gamma$ is inconsistent if $\Gamma \Rightarrow \phi$ for every $\phi$. It follows from the rules that if $\Gamma, \phi$ and $\Gamma, \neg\phi$ are both inconsistent, then $\Gamma$ is inconsistent. For if $\Gamma, \phi$ is inconsistent, then $\Gamma \Rightarrow \neg\phi$ by [L7], so we have that $\Gamma \Rightarrow \neg\phi$ and $\Gamma, \neg\phi \Rightarrow \psi$ for every $\psi$, so $\Gamma \Rightarrow \psi$ for every $\psi$ by the cut property.

We will also need to use the following rather more technical *constant elimination property*. If $\Gamma \Rightarrow \phi_x(c)$, where $c$ is a constant that does not occur in $\Gamma$ or $\phi$, then $\Gamma \Rightarrow \forall x\phi$. Informally speaking, what this means is that reasoning about an arbitrary individual using an individual constant is no different from reasoning about an arbitrary individual using a variable. The constant elimination property is a consequence of a more general property which can be proved by a simple induction: if $\Gamma \Rightarrow \phi$, then $\Gamma' \Rightarrow \phi'$, where $\Gamma' \Rightarrow \phi'$ is obtained from $\Gamma \Rightarrow \phi$ by replacing an individual constant occurring in $\Gamma \Rightarrow \phi$ with a variable that does not occur in $\Gamma \Rightarrow \phi$. Given this, if $\Gamma \Rightarrow \phi_x(c)$, where $c$ does not occur in $\Gamma$ or $\phi$, we get $\Gamma \Rightarrow \phi_x(y)$, where $y$ is a new variable, and then $\Gamma \Rightarrow \forall y\phi_x(y)$ by [L10]. By [L11], [L10], and the condition on $y$, we also have $\forall y\phi_x(y) \Rightarrow \forall x\phi$, so $\Gamma \Rightarrow \forall x\phi$ by the cut property. (If $x$ has no free occurrence in $\Gamma$, the reasoning is simpler.)

## 8.4. Minimal sets of connectives and quantifiers

For theoretical purposes, it is often convenient to trim the logical apparatus to a minimum. We can for example make do with disjunction, negation, and existential quantification in formulating predicate logic, by stipulating that $\phi \wedge \psi$ is an abbreviation for $\neg(\neg\phi \vee \neg\psi)$, $\phi \supset \psi$ for $\neg\phi \vee \psi$, and $\forall x\phi$ for $\neg\exists x\neg\phi$. The justification for this lies in the following observation. For any formula $\phi$, let $\phi^*$ be the result of rewriting all conjunctions, implications and universal quantifiers in $\phi$ using these stipulations, and let $\Gamma^*$ be the result of applying this operation to all formulas in $\Gamma$. Then $\Gamma \Rightarrow \phi$ can be derived from the full system of rules if and only if $\Gamma^* \Rightarrow \phi^*$ can be derived without using the rules for conjunction, implication, or the universal quantifier. To prove this, we first need to show that the rules for the eliminated connectives and quantifier hold as *derived rules* in the remainder of the system. This means, in the case of conjunction, that we need to show that $\Gamma \Rightarrow \neg(\neg\phi \vee \neg\psi)$ if and only if $\Gamma \Rightarrow \phi$ and $\Gamma \Rightarrow \psi$. This can be done using the rules for negation and disjunction. Thus any application of the rules for conjunction can be replaced, after conjunction has been eliminated, by an application of these other rules. Similarly for implication and universal quantification. To prove the observation in the other direction, that if $\Gamma^* \Rightarrow \phi^*$ can be proved using the trimmed-down system then $\Gamma \Rightarrow \phi$ can be proved using the full system, it suffices to show that for example $\neg(\neg\phi \vee \neg\psi)$ and $\phi \wedge \psi$ are *logically equivalent* in the full system, in the sense that each is deducible from the other. The direction $\phi \wedge \psi \Rightarrow \neg(\neg\phi \vee \neg\psi)$ follows using [L7], [L4], [L2]. For the converse, $\neg(\neg\phi \vee \neg\psi) \Rightarrow \phi \wedge \psi$, we need to use indirect proof, showing $\neg(\neg\phi \vee \neg\psi), \neg\phi$ and $\neg(\neg\phi \vee \neg\psi), \neg\psi$ to be inconsistent by [L3]. Similarly for implication and the universal quantifier.

Eliminating conjunction, implication, and the universal quantifier in this way obliterates, as was noted earlier, significant distinctions between proofs. We wouldn't in practice want to reason without using the eliminated constructs. However, when we are only interested in questions of derivability, that is, whether $\Gamma \Rightarrow \phi$ holds for some particular $\Gamma$ and $\phi$, or class of such, nothing is lost by restricting attention to the three constructs singled out above. They are of course not the only possible choice. We could equally restrict ourselves to the universal quantifier, implication, and negation, or to some other combination.

## 8.5. Soundness and completeness of the rules

The rules [L1]–[L16] have been presented as formal versions of rules of logical reasoning that we actually use in mathematics, and in that sense by definition correct. These logical rules are not tied specifically to reasoning about natural numbers and sets of natural numbers, but are applied in mathematical reasoning generally. It is possible to put the observation that the rules are correct on a more formal basis, by making use of the concept of a general interpretation $I$ of a first or second order language introduced in Chapter 7, and verifying by an inductive argument that if every sentence in a set $\Gamma$ of sentences is true in $I$—which we also express by saying that $I$ is a *model* of $\Gamma$—and if $\Gamma \Rightarrow \phi$, then $\phi$ is true in $I$. For this induction to go through we actually need to consider a more general semantic concept than truth in $I$, namely that of satisfaction: a sequence $a_0, \dots, a_n$ of elements in the domain of $I$ *satisfies* a formula with free variables $x_0, \dots, x_n$ if the formula is true in $I$ when $x_i$ is interpreted as referring to $a_i$, for $i = 0, \dots, n$. The reason for this is that the formulas in $\Gamma$ are not always sentences, so we need to prove more generally that whenever a sequence of elements in the domain of $I$ satisfies all the formulas in $\Gamma$ and $\Gamma \Rightarrow \phi$, the same sequence also satisfies $\phi$, This is what is meant by saying that the rules [L1]–[L16] are provably *sound*. Of course the soundness proof does not provide any foundational or philosophical justification of the rules, for the proof itself uses precisely those rules that are being proved sound. Nor should it be supposed that the rules are necessarily sound applied to reasoning in ordinary language, where (to mention but two relevant differences between mathematical and everyday reasoning) we argue using vague concepts and are not prepared to draw arbitrary conclusions from a contradiction. But if we had *not* been able to prove soundness of the rules as a mathematical theorem, we would have been in trouble. A formal soundness proof is necessary in logic in connection with questions about the formalizability of proofs in various axiomatic theories. Thus in later chapters we will need to know for different pairs $T$ and $T'$ of theories that the consistency of $T$ can be proved in $T'$ by formalizing the simple argument "every axiom of $T$ is true and the logical rules preserve truth, so every theorem of $T$ is true, which in particular means that no contradiction is a theorem of $T$". For this, we will need to know that a formal soundness proof can be carried out in $T'$. In §10.3, the soundness proof sketched above will be made more explicit in the special case of formalized arithmetic and truth in the standard model.

The completeness of the rules is a somewhat more recondite matter. Clearly one could add any number of further rules for deducibility which agree with our ordinary mathematical understanding of the connectives and quantifiers, such as the principle of disjunctive syllogism formulated above:

If $\Gamma \Rightarrow \phi \vee \psi$ and $\Gamma \Rightarrow \neg\psi$ then $\Gamma \Rightarrow \phi$.

However, this rule is a consequence of the rules stated. For we have that if $\Gamma \Rightarrow \neg\psi$ then $\Gamma, \psi \Rightarrow \neg\psi$ and $\Gamma, \psi \Rightarrow \psi$, so $\Gamma, \psi \Rightarrow \phi$. Also $\Gamma, \phi \Rightarrow \phi$, so $\Gamma, \phi \vee \psi \Rightarrow \phi$. So $\Gamma \Rightarrow \phi \vee \psi \supset \phi$ and $\Gamma \Rightarrow \phi \vee \psi$ and therefore $\Gamma \Rightarrow \phi$.

Another such possible further rule is

(3) If $\Gamma \Rightarrow \exists x \exists y \phi$ then $\Gamma \Rightarrow \exists y \exists x \phi$.

This again is a consequence of the rules [L1]–[L16]. For $\phi \Rightarrow \phi$ by [L1], and by two applications of [L12] we get $\phi \Rightarrow \exists y \exists x \phi$. Applying [L13] then yields $\exists x \exists y \phi \Rightarrow \exists y \exists x \phi$, and the cut property gives (3).

More generally, the deducibility relation defined by [L1]–[L16] is *complete*, in the sense of a basic result of logic, first proved by Gödel in his doctoral dissertation (1929), the

**Completeness theorem.** Let $\Gamma$ be any set (finite or infinite) of sentences in $L$, and $\phi$ any sentence. If $\Gamma \Rightarrow \phi$ does not hold, there is a model $I$ of $\Gamma$ in which $\phi$ is false.

The significance of the completeness theorem will here be taken to be the following: if there is *any* argument of any kind which establishes the truth of $\phi$ assuming the sentences in $\Gamma$ as premises, and which does not invoke any other knowledge about the domain of discourse and its predicates and relations than is stated in the premises (or, in the second order case, make any assumption not stated in the premises about just which sets exist), we know that $\phi$ is also deducible from $\Gamma$ in the sense of the relation $\Rightarrow$.

Thus for example, if we let $\exists_{\leq n} x\phi$ ("there are at most $n$ $x$ such that $\phi$) stand for the formula $\exists y_1 \ldots \exists y_n \forall x(\phi \supset x = y_1 \vee \cdots \vee x = y_n)$, where $y_1, \ldots, y_n$ are different variables that do not occur in $\phi$, we can convince ourselves by an arithmetical argument in combination with the completeness theorem that

$$\neg\exists_{\leq 99} x P(x) \wedge \exists_{\leq 99} x Q(x) \wedge \forall x(P(x) \supset \exists y(Q(y) \wedge R(x, y)))$$
$$\Rightarrow \exists x_1 \exists x_2 \exists y(P(x_1) \wedge P(x_2) \wedge x_1 \neq x_2 \wedge R(x_1, y) \wedge R(x_2, y))$$

For if a class $A$ has at least 100 members and a class $B$ has at most 99 members, and every member of $A$ is associated with a member of $B$, then as a matter of arithmetic ("the pigeon hole principle"), at least two elements in $A$ must be assigned the same element in $B$. By the completeness theorem, there is a purely logical argument using the rules [L1]–[L16] that establishes this.

### A proof of the completeness theorem

The reader may or may not want to work through the details (some of which have been left in a slightly blurred state) of the proof of the completeness theorem given in this section. A general appreciation of the structure of the argument is sufficient to understand later applications of and references to the theorem. The proof given here is due to Henkin, with simplifications introduced by Hasenjäger and Mates.

For the purposes of the completeness proof, we will introduce a new syntactic category, the *set constants*. Formally, we can identify the $i$-th set constant $C_i$ with the sequence number $\langle 8, i \rangle$.

Our aim is to show that for any set $\Gamma$ (possibly infinite) of sentences and any sentence $\phi$, if $\Gamma \Rightarrow \phi$ does not hold, then there is a model of $\Gamma$ in which $\phi$ is false. For this we need only show that for any consistent set $\Gamma$ of sentences there is an interpretation in which all the sentences in $\Gamma$ are true. The completeness theorem follows from this applied to the set $\Gamma, \neg\phi$, since $\Gamma, \neg\phi$ is inconsistent if and only if $\Gamma \Rightarrow \phi$. So suppose $\Gamma$ is such a consistent set of sentences in some first or second order language $L$.

To simplify the proof, we will use only negation and disjunction as propositional connectives, and only the existential quantifier. The completeness theorem for the full logical symbolism follows from the completeness of this fragment together with the fact that the rules for the remaining connectives and the universal quantifier can be reduced to those for negation, disjunction, and the existential quantifier, as indicated above in §8.4. The cut property and monotonicity property will be used implicitly in the proof.

We set aside the individual or set constants $c_i$ and $C_i$ with even index $i$ for special use, only allowing constants with odd index to be used in formulas in $L$. This is no restriction on what we can say in formulas, since one constant is just like another, and all that matters is that there are infinitely many of them. We will refer to the constants set aside as *special constants*. The language obtained by adding all the special constants will be called the *extended* language and denoted $L_e$.

We will define explicitly an interpretation that makes all sentences in $\Gamma$ true. To do this we first define an infinite sequence $\Gamma_0, \Gamma_1, \ldots$ of sets of formulas by induction on $n$. Let $\phi_0, \phi_1, \ldots$ be an enumeration of all sentences in the *extended* language. We now define the sequence $\Gamma_0, \Gamma_1, \ldots$ as follows, with the intention that a sentence in $L_e$ is to be true in our interpretation if and only if it occurs in one of the sets $\Gamma_n$ in the sequence. Since each $\Gamma_{n+1}$ is an extension of $\Gamma_n$, a formula that occurs in $\Gamma_n$ will occur in every later set in the sequence.

$\Gamma_0$ is defined to be our initial consistent set $\Gamma$. To define $\Gamma_{n+1}$, we look at the formula $\phi_n$. If this formula is inconsistent with $\Gamma_n$, that is, if $\Gamma_n \cup \{\phi_n\}$ is an inconsistent set, we let $\Gamma_{n+1}$ be the set $\Gamma_n \cup \{\neg\phi_n\}$. If $\phi_n$ is consistent with $\Gamma_n$, we instead let $\Gamma_{n+1}$ be the set $\Gamma_n \cup \{\phi_n\}$, with one further addition in the case when $\phi_n$ is an existential formula $\exists x\phi$ or $\exists X\phi$. If $\phi_n$ is $\exists x\phi$, let $c$ be the first special individual constant which does not occur in any formula in $\Gamma_n$. (There must be such a constant, since there are infinitely many special constants not in any formula in $\Gamma_0$ and each $\Gamma_{n+1}$ only introduces one or two new formulas.) Besides $\phi_n$ we then also add to $\Gamma_n$ the formula $\phi_x(c)$ to get $\Gamma_{n+1}$. Similarly in the case where $\phi_n$ is $\exists X\phi$.

Now let $\Gamma_\omega$ be the union of all the sets $\Gamma_n$. We will define an interpretation $I$ of the extended language such that any sentence $\phi$ is true in $I$ if and only if $\phi$ is a member of $\Gamma_\omega$. In particular, every sentence in $\Gamma_0 = \Gamma$ is true in the interpretation $I$.

First note that $\Gamma_\omega$ is a *saturated* set of sentences in the sense that for every sentence $\phi_n$ in $L_e$, either $\phi_n$ or its negation is in $\Gamma_\omega$. We know this since either $\phi_n$ or its negation is added to $\Gamma_n$ to get $\Gamma_{n+1}$. If there is to be any such interpretation $I$ as we seek, $\Gamma_\omega$ must also be consistent, so that we never have that both $\phi_n$ and its negation belong to $\Gamma_\omega$. To show that $\Gamma_\omega$ is indeed consistent, we prove by induction on $n$ that every $\Gamma_n$ is consistent (the finiteness property then yields that $\Gamma_\omega$ is consistent).

$\Gamma_0$ is consistent by assumption, which takes care of the base case. Now assume $\Gamma_n$ is consistent. If $\phi_n$ is not an existential formula, $\Gamma_{n+1}$ is consistent by the definition of $\Gamma_{n+1}$. So we only need to consider what happens when we add to $\Gamma_n$ both the existential formula $\exists x\phi$ (consistent with $\Gamma_n$) and the formula $\phi_x(c)$, where $c$ is a special individual constant that does not occur in $\phi$ or in any formula in $\Gamma_n$. Suppose this results in an inconsistent set. Then $\Gamma_n, \exists x\phi \Rightarrow \neg\phi(c)$ holds. By the constant elimination property and [L11], $\Gamma_n, \exists x\phi \Rightarrow \neg\phi$, so $\Gamma_n, \phi \Rightarrow \neg\exists x\phi$ by [L7] and thus $\Gamma_n, \exists x\phi \Rightarrow \neg\exists x\phi$ by [L13], implying that $\Gamma_n, \exists x\phi$ is inconsistent. The argument is similar when $\phi_n$ is a second order existential formula.

That $\Gamma_\omega$ is saturated and consistent implies that it has the following *closure property*: if $\phi$ is in $\Gamma_\omega$ and $\phi \Rightarrow \psi$, then $\psi$ is in $\Gamma_\omega$. For otherwise $\neg\psi$ is in $\Gamma_\omega$, and then $\Gamma_\omega$ is inconsistent.

The domain of the interpretation $I$ will almost be the set of closed first order terms of the extended language. Almost, not quite, because $\Gamma_\omega$ may contain equalities between such terms, and $=$ must be interpreted as identity in the domain. We therefore identify two closed terms $s$ and $t$ if $s = t$ is a member of $\Gamma_\omega$. Mathematically speaking, the individuals of the domain $D$ of $I$ are therefore equivalence classes of closed terms under the relation $s = t \in \Gamma_\omega$. We write $[s]$ for the equivalence class containing $s$. That this is indeed an equivalence relation follows from the fact that $\Gamma_\omega$ is saturated and consistent, together with the rules for identity.

We also need to define the domain $M_D$ of sets in the interpretation. These we take to be precisely the sets that are interpretations of some set constant. The interpretation of a particular set constant $C$ we take to be the set of all $[s]$ for $s$ such that $s \in C$ is in $\Gamma_\omega$. Note that this implies that a sentence $s \in C$ in the extended language is true if and only if it is a member of $\Gamma_\omega$.

Next we need to define the interpretations of the function symbols and predicate symbols. The interpretation $F_{ij}$ of a function symbol $f_j^i$ we take to be the function mapping its $i$ arguments $[s_1], \ldots, [s_i]$ to the equivalence class $[f_j^i(s_1, \ldots, s_i)]$. Here we must check that this defines a function, or in other words that $[f_j^i(s_1, \ldots, s_i)]$ depends only on $[s_1], \ldots, [s_i]$. That this is so follows from the identity rules and the fact that $\Gamma_\omega$ is saturated and consistent. The predicate symbols are interpreted similarly, so that the interpretation of $p_j^i$ is the set of all $\langle [s_1], \ldots, [s_i] \rangle$ for which $p_j^i(s_1, \ldots, s_i)$ is in $\Gamma_\omega$.

This defines the interpretation $I$. Essentially, the individuals are the closed terms of the extended language, where we identify closed terms said to be identical in $\Gamma_\omega$, while the sets are the sets of individuals said to exist in $\Gamma_\omega$, and function and predicate symbols are interpreted so as to make it hold by definition that an atomic sentence is true on the interpretation $I$ if and only if it belongs to $\Gamma_\omega$. This property has already been established for atomic sentences other than equalities between set constants, so we consider the case of a formula $\phi$ of the form $C_i = C_j$. If $\phi$ is in $\Gamma_\omega$, it is true, by the identity rules. Conversely, if $\phi$ is true, the identity rules yield that for every $s$, $s \in C_i$ is in $\Gamma_\omega$ if and only if $s \in C_j$ is in $\Gamma_\omega$. So $\neg\exists x \neg(x \in X \equiv x \in Y)$ must be in $\Gamma_\omega$, since otherwise $\neg(c \in X \equiv c \in Y)$ is in $\Gamma_\omega$ for some $c$, implying that one of $c \in C_i$ and $c \in C_j$ is in $\Gamma_\omega$ and the other not. Thus by the extensionality rule, $C_i = C_j$ is in $\Gamma_\omega$.

It remains to verify that it holds also for non-atomic formulas $\phi$ that $\phi$ is true in the interpretation $I$ if and only if $\phi$ belongs to $\Gamma_\omega$, which we do by induction.

Suppose $\phi$ is a negation $\neg\psi$. By the induction hypothesis, $\psi$ is in $\Gamma_\omega$ if and only if $\psi$ is true in $I$, and since $\Gamma_\omega$ is saturated and consistent, $\phi$ is in $\Gamma_\omega$ if and only if $\psi$ is not in $\Gamma_\omega$, thus if and only if $\phi$ is true in $I$.

Now suppose $\phi$ is $\psi_1 \vee \psi_2$. If $\phi$ is in $\Gamma_\omega$, at least one of $\psi_1$ and $\psi_2$ must be in $\Gamma_\omega$, since otherwise $\phi$, $\neg\psi_1$ and $\neg\psi_2$ are all in $\Gamma_\omega$, rendering $\Gamma_\omega$ inconsistent. Conversely, if at least one of $\psi_1$ and $\psi_2$ is in $\Gamma_\omega$, $\phi$ is in $\Gamma_\omega$ by the closure property.

Finally, suppose $\phi$ is an existential formula. If $\phi$ is a first order quantified formula $\exists x\psi$, it follows by an inductive argument that is here omitted that $\phi$ is true in $I$ if and only if there is a closed term $t$ such that $\psi_x(t)$ is true in $I$. If there is such a $\psi_x(t)$, $\phi$ is in $\Gamma_\omega$ by the induction hypothesis and the closure property. Conversely, if $\phi$ is in $\Gamma_\omega$, it has been introduced into $\Gamma_{n+1}$ for some $n$ together with an instance $\psi_x(c)$, for some special constant $c$, so $\phi$ is true in $I$. A similar argument applies if $\phi$ is a second order quantified formula $\exists X\psi$.

## 8.6. Theories and proofs

A formal theory $T$, in the sense used in this book, is determined by a *language* $L_T$ which is a first or second order language as defined in Chapter 7, and a set of *axioms* $Ax_T$, which are formulas in that language. We speak of the theory also as a first or second order theory. The set $Thm_T$ of *theorems* of $T$ consists of those formulas $\phi$ for which $Ax_T \Rightarrow \phi$. The theory is said to be inconsistent if $Ax_T$ is, and otherwise consistent. $T$ is *complete* if for every sentence $\phi$ in the language of $T$, $\phi$ or $\neg\phi$ is a theorem of $T$, and otherwise incomplete. Theories $T$ and $T'$ are *equivalent* if they have the same language and the same set of theorems.

For convenience, the definition of a theory does not require that the axioms of $T$ are sentences. However, an axiom $\phi$ containing free variables is to be understood as standing for the *universal closure* of $\phi$, which is the sentence obtained by prefixing to $\phi$ quantifiers $\forall x$ for every variable $x$ with a free occurrence in $\phi$.

By the general argument of §6.2, $Ax_T \Rightarrow \phi$ holds if and only if there is a derivation showing this, in the form of a finite sequence $\langle \Gamma_1, \phi_1 \rangle, \ldots, \langle \Gamma_n, \phi_n \rangle$, where $\langle \Gamma_n, \phi_n \rangle$ is $\langle Ax_T, \phi \rangle$ and each $\langle \Gamma_i, \phi_i \rangle$ follows by one of the rules [L1]–[L16] from earlier pairs in the sequence. By the finiteness property of $\Rightarrow$, we can assume that each $\Gamma_i$ is a *finite* set, and this in turn allows us to define derivations as numbers. For this, we replace finite sets (of formulas) by sequence numbers, and define a *derivation* as a sequence number $\langle s_0, \ldots, s_n \rangle$ where each $s_i$ is a pair $\langle t_i, \phi_i \rangle$ with $t_i$ a sequence number of formulas, and $s_i$ is related as indicated in the rules [L1]–[L16] to zero, one, or two earlier pairs in the sequence. $s$ is a derivation *of* $\phi$ *from* $\Gamma$ if $s_n$ is $\langle t_n, \phi \rangle$ where every formula in the sequence $t_n$ is a member of $\Gamma$. A *proof* in a theory $T$ is a derivation from $Ax_T$.

By the usual methods, we find that the set of derivations is primitive recursive. The set of derivations from $\Gamma$ is not in general primitive recursive, but it is primitive recursive in $\Gamma$ (in the sense defined in §4.6).

The definition of a theory admits any arbitrary set of axioms. For example, *true arithmetic* is the theory whose axioms is the set of true arithmetical sentences. However, if a theory $T$ is to be a formalization of some part of our mathematical knowledge or potential mathematical knowledge, so that the theorems of $T$ can be seen as truths implicit in what we know or assume (in the sense explained in §1.3), a minimal requirement is that we have an *algorithm* for generating the theorems of $T$, since otherwise we have no grounds for claiming that the theorems of $T$ are in fact implicit in our mathematical knowledge. In the absence of such an algorithm—a uniform method requiring no further invention or insight—more work would be required to show that our mathematical knowledge suffices to determine the truth of the consequences of the axioms of a theory. If $T$ is *effectively axiomatized*, in the sense that there is an algorithm for deciding whether or not a given sentence is an axiom of $T$, it follows that such an algorithm for generating the theorems of $T$ exists, since the set of proofs in $T$ is primitive recursive in the set of axioms. To generate the theorems of $T$, we can just go through the set of derivations, and for each derivation of a formula $\phi$ check whether it a proof in $T$, in which case $\phi$ is a theorem of $T$.

For this reason, we will assume unless otherwise stated that all theories $T$ considered are *effectively axiomatizable* in the sense that $T$ is equivalent to an effectively axiomatized theory. It will not generally be assumed that $T$ is effectively axiomatized, however. What will be always be the case is that the axioms of $T$ can be effectively generated, even if there is no algorithm for deciding whether a formula is an axiom. By an observation of Craig's, any theory satisfying this condition is in fact effectively axiomatizable. For if the axioms of $T$ can be effectively generated, then so can the theorems of $T$, and given any way of effectively generating the theorems $\phi_1, \phi_2, \phi_3, \ldots$ of $T$, the theory $T'$ with axioms $\phi_1, \phi_1 \wedge \phi_2, \phi_1 \wedge \phi_2 \wedge \phi_3, \ldots$ is effectively axiomatized and has the same theorems as $T$. The reason why $T'$ is effectively axiomatized is that given a sentence $\phi$, we can inspect it to see if it is the conjunction of $n$ formulas, and if so, generate the first $n$ theorems of $T$ to check whether $\phi$ is an axiom.

A theory with a *finite* set of axioms is a special case of an effectively axiomatized theory. A theory is *finitely axiomatizable* if it is equivalent to some finitely axiomatized theory. Although all theories to be considered in later chapters are effectively axiomatizable, only some of them are finitely axiomatizable.

## Completeness and incompleteness

The completeness theorem (in a somewhat less general form) was first proved by Gödel, and is therefore sometimes called Gödel's completeness theorem. There are in fact many completeness theorems in logic, and the basic result we are considering here is known as the completeness theorem for first order logic, or, in the case of second order languages, the weak completeness theorem for second order logic. The "weakness" of the latter theorem consists in the fact that the domain $M_D$ of sets in the model $I$ defined in the proof of the completeness theorem does not contain *every* subset of the domain. The significance of this will emerge below.

To speak of "Gödel's completeness theorem" may be to invite confusion with Gödel's incompleteness theorem, and there is indeed room for reflection on the relation between these two results. Completeness in the sense of the completeness theorem is a property of the rules of logical deduction: they suffice to deduce the logical consequences of any set of axioms. Completeness in the sense of the incompleteness theorem, on the other hand, is a property of a particular set of axioms: they suffice to prove or refute every sentence in some particular formal language. Although no complete theory will be used in this book, there are in fact effectively axiomatizable and complete theories that are far from trivial. A prominent example is the elementary theory of the real field, also known as the theory of real-closed fields. The language of this theory contains predicate symbols for equality and the ordering relation between real numbers, function symbols for addition and multiplication, and symbols for 0 and 1. By a theorem first proved by Tarski, every first order sentence true in the field of real numbers is provable in the theory. (Because of the incompleteness theorem, we see from this that there is no way of defining the natural numbers as a subset of the real numbers using only this first order language.)

Now suppose we have a theory $T$ with a set of axioms in the language of arithmetic or analysis, and that (Rosser's strengthening of) Gödel's incompleteness theorem applies to $T$. We then know that if $T$ is consistent, it is incomplete, and indeed that there is, even for a second order theory, an arithmetical sentence $\phi$ such that neither $\phi$ nor its negation is provable in $T$. Since all formulas logically deducible from the axioms of $T$ are provable in $T$, the completeness theorem tells us that even if $\phi$ is true in the standard model, there is an interpretation of the language of arithmetic in which all the axioms of $T$ are true, but $\phi$ is false. Such an interpretation will be a *non-standard* interpretation: its domain $D$ will contain elements (known as infinite elements) which are not the interpretation of $\overline{n}$ for any $n$. Even without invoking the incompleteness theorem, we can observe that it is not possible to exclude such non-standard interpretations by any set of consistent axioms, as long as general interpretations are admitted. For given any set $M$ of axioms in the language of analysis, if $M$ has an infinite model then so does the set $M'$ obtained by extending $M$ with the additional axioms $c \neq 0, c \neq s(0), \ldots$ where $c$ is an individual constant that does not occur in any formula in $M$. For every finite subset of $M'$ has a model (just interpret $c$ as any object different from all of the finitely many objects said to be different from $c$ in the finite subset) and is therefore consistent. This means that $M'$ is consistent, and therefore has a model, in which the interpretation of $c$ is an infinite element. We can rule out non-standard models by only admitting interpretations for which the domain $M_D$ of sets contains every (arbitrary) subset of the domain, since for such interpretations, the induction principle in the form of the formula

$$\forall X(0 \in X \land \forall x(x \in X \supset s(x) \in X) \supset \forall x(x \in X))$$

implies that there are no infinite elements, and we can show that any such model of the axioms of arithmetic must be isomorphic to the natural numbers. But then we learn from the incompleteness theorem that there can be no completeness theorem for these restricted interpretations corresponding to the completeness theorem for general interpretations. That is, it is not possible to formulate any set of rules of deduction such that either $\phi$ is deducible from

$\Gamma$ using these rules, or else there is an interpretation where the set variables range over arbitrary subsets of the domain and in which all the sentences in $\Gamma$ are true and $\phi$ is false. The reason is that if $\phi$ is false in such an interpretation, with the induction principle included in $\Gamma$, $\phi$ is false in the standard model as well, so we would have a way of proving every true arithmetical sentence.

## 8.7. Extensions of theories

A theory $T'$ is an *extension* of a theory $T$, and $T$ a *subtheory* of $T'$, if $L_{T'}$ extends $L_T$—every predicate or function symbol of $L_T$ is also in the language $L_{T'}$—and every axiom of $T$ is provable in $T'$. This latter condition implies that $Thm_T$ is a subset of $Thm_{T'}$. That $T$ and $T'$ are equivalent is thus the same as saying that each of $T$ and $T'$ extends the other.

One common type of extension of a theory $T$ is an extension by definitions. Suppose $\phi$ is a formula in $L_T$ with the free variables $x_1, \ldots, x_n$, and $p$ is an $n$-ary predicate symbol not in $L_T$. We say that the formula

$$\forall x_1 \ldots \forall x_n\, (p(x_1, \ldots, x_n) \equiv \phi)$$

is a *defining axiom* for $p$. Similarly, if $\phi$ is a formula in $L_T$ with the free variables $x_1, \ldots, x_n, y$ and $f$ is an $n$-ary function symbol not in $L_T$, we say that the formula

$$\forall x_1 \ldots \forall x_n \forall y\, (f(x_1, \ldots, x_n) = y \equiv \phi)$$

is a defining axiom for $f$ (relative to $T$), provided

$$\forall x_1 \ldots \forall x_n \exists z \forall y (z = y \equiv \phi)$$

is provable in $T$. (Here $z$ is a variable that does not occur in $\phi$.) The defining axioms for $p$ and $f$, it will be noted, are not axioms in $T$, since $f$ and $p$ are not symbols in the language of $T$. An *extension by definitions* of $T$ is a an extension $T'$ of $T$ such that every symbol in $L_{T'}$ which is not in $L_T$ has an associated defining axiom relative to $T$, and the only axioms of $T'$ which are not theorems of $T$ are these defining axioms. In general, infinitely many symbols with their corresponding defining axioms may be added in this way. A *finite* extension by definitions is one which adds only finitely many new function and predicate symbols.

Note that the defining axioms are just formal versions of the explicit definitions of function and predicate symbols described earlier in §4.5. As in §4.5, if $T'$ is an extension by definitions of $T$, every formula $\phi$ in the language of $T'$ can be translated into a formula $\phi^*$ in the language of $T$, by systematically eliminating the defined predicate and function symbols. An atomic formula $p(t_1, \ldots, t_n)$ is replaced by a corresponding substitution instance $\phi(t_1, \ldots, t_n)$ of the formula $\phi$ in the defining axiom for $p$, if necessary with bound variables in $\phi$ replaced by other variables so as to ensure that $t_1, \ldots, t_n$ are free for $x_1, \ldots, x_n$ in $\phi$. An occurrence in a formula $\phi$ of a term

$f(t_1, \ldots, t_n)$ can be eliminated in one of two ways. For a typical example, consider a formula $q(f(t, \ldots, t_n), u)$. Here we can eliminate the defined function symbol $f$ by rewriting the formula either as $\forall y(\phi(t_1, \ldots, t_n) \supset q(y, u))$ or as $\exists y(\phi(t_1, \ldots, t_n) \wedge q(y, u))$, since the defining axiom for $\phi$ implies that these are equivalent formulations. By repeated applications of such rewrites, $\phi$ is transformed into a formula $\phi^*$ which does not contain any of the defined predicate or function symbols. This formula $\phi^*$ is not uniquely determined, since the rewrites can be carried out in any order. However, in the theory $T'$, the formula $\phi$ and its translation $\phi^*$ are equivalent, and furthermore $\phi$ is provable in $T'$ if and only if its translation $\phi^*$ is provable in $T$. We can prove this by invoking the completeness theorem. First we note that $\phi \equiv \phi^*$ is provable in $T'$, since we have an informal argument for this equivalence (essentially just the observation that $\phi_x(t)$ says the same thing as $\forall x(x = t \supset \phi)$ and $\exists x(x = t \wedge \phi)$) which only invokes the defining axioms and some logical reasoning. So if $\phi^*$ is provable in $T$, $\phi$ is provable in $T'$. If $\phi$ is provable in $T'$, $\phi^*$ is also provable in $T'$, and we need to conclude that $\phi^*$ is therefore provable in $T$. This is so because $T'$, being an extension by definitions of $T$, is thereby a *conservative* extension of $T$. This means that every formula in the language of $T$ which is provable in $T'$ is also provable in $T$. Informally, this seems clear enough, since $T'$ does not add any new assumptions to $T$, only the convenience of the defined predicate and function symbols, and anything that can be proved using this convenience must also be provable without it. Again the completeness theorem gives us a formal proof. If a sentence $\phi$ in the language of $T$ is not provable in $T$, there is an interpretation $I$ of $T$ in which $\phi$ is false. But this interpretation $I$ can be extended to an interpretation $I'$ of $T'$ by interpreting the new predicate and function symbols so as to satisfy the corresponding defining axioms, so $\phi$ is not a theorem of $T'$ either.

The topic of conservative extensions will recur in Chapter 10.

CHAPTER 9

# PEANO ARITHMETIC AND COMPUTABILITY

## 9.1. The axioms of PA

The basic formal theory of arithmetic to be used in this book is known as Peano Arithmetic, or PA for short. This is a first order theory whose language is the primitive language of arithmetic, with the exception of $<$. The axioms are the following:

Successor axioms: $\forall x(s(x) \neq 0)$ and $\forall x \forall y(s(x) = s(y) \supset x = y)$
Axioms for addition: $\forall x(x+0 = x)$ and $\forall x \forall y(x+s(y) = s(x+y))$
Axioms for multiplication:
$\forall x(x * 0 = 0)$ and $\forall x \forall y(x * s(y) = x * y + x)$
Induction axioms: for every $\phi$, the formula

$$\phi_x(0) \wedge \forall x(\phi \supset \phi_x(s(x))) \supset \forall x \phi$$

is an axiom.

Why these particular axioms? The induction axioms express the principle of mathematical induction for every *arithmetical* property of the natural numbers. Since the language of PA does not include any variables ranging over sets, the induction principle can be included only in the form of an *axiom schema*, every instance of which is taken to be an axiom in the theory. Thus PA has infinitely many axioms, and in fact (although this will not be proved or used here) is not finitely axiomatizable. Since we know that there are properties of the natural numbers that cannot be expressed in the language of arithmetic—such as the property of being a true arithmetical sentence—we know that the induction principle for the natural numbers is not fully captured by the induction schema of PA. But of course this in itself doesn't exclude the possibility of every true *arithmetical* statement being provable using only the restricted induction principle. The incompleteness theorem shows that this is not the case.

The axioms for the successor function stipulate just those properties of the successor function that were presented as essential in Chapter 5. The axioms for multiplication and addition are also straightforward: they are simply the equations defining these operations as primitive recursive functions.

The omissions from the axioms are at first sight more puzzling. Although multiplication and addition are clearly central to arithmetic, we also use many other operations, like the power function and the factorial. There are no axioms for these operations in PA, because they can (for theoretical purposes) be assumed to be defined in terms of addition and multiplication, as has been shown in Chapter 5 and will be further commented on in this chapter. But even addition and multiplication may seem to be insufficiently characterized by the axioms of PA, since we have for example no axiom stating that addition is commutative—$m + n = n + m$ for all natural number $n, m$—or that the distributive law holds, that is, $n(m + k) = nm + nk$ for all $n, m, k$. In fact these algebraic properties are provable using induction. This was first seen and done by Dedekind, whose set-theoretical characterization of the natural numbers is the basis for the axioms of PA.

Producing long proofs in PA is not something that we have any reason or occasion to do in this book, but we do need a basic grasp of what proofs $P$ of an arithmetical statement $A$ carry over (more or less immediately) to PA, in the sense that a formal deduction of (a formalization of) $A$ from the axioms of PA can be seen to exist on the basis of the more or less informal proof $P$. To this end, we will start this chapter by going through much of the arithmetical reasoning from earlier chapters in order to convince ourselves that it can be carried out in PA. Note that we are not giving any formal derivations in PA, but only ordinary more or less informal mathematical proofs, usually with the accompanying claim—which is sometimes supported by explicit argument, and otherwise assumed unproblematic—that there is a proof in PA formalizing the argument.

In the specialized literature, various subtheories of PA are considered, with restricted induction axioms, and provability in these different subtheories is studied. Such refinements are irrelevant to the concerns of this book, but have connections with corresponding refinements of the notion of computability. (See Hájek and Pudlák [1993] for an extended treatment.)

In convincing ourselves that a particular statement can be proved in PA, we make good use of the completeness theorem for predicate logic, which tells us that any argument for the truth of an arithmetical sentence, even if it involves concepts or rules not formalized in PA, has a counterpart among the derivations in PA, as long as it does not invoke any other properties of the natural numbers than those stated in the axioms of PA. But here we must keep in mind that the axioms of PA only stipulate that the principle of mathematical induction holds for *arithmetical* properties. If we carry out an argument by induction, we don't know that a corresponding proof exists in PA unless the induction is applied to an arithmetical property. For example, we know from the second incompleteness theorem that the consistency of PA is not provable in PA itself, although it is provable by a simple induction: every axiom of PA is true, the rules of logical deduction preserve truth (for all values of the free variables), so every theorem of PA is true, and thus PA is consistent. This proof (which will be seen in Chapter 10 to be formalizable in a suitable theory of second order arithmetic) cannot be carried out in PA, since the property of being a true statement is, as we have seen, not arithmetical. This of course doesn't in itself rule out the possibility of the consistency of PA being provable in PA by other means, although we know from the incompleteness theorem that in fact it is not.

A second caveat concerns the phenomenon of *ω-incompleteness*. That PA is ω-incomplete means that there are formulas $\phi$ with free variable $x$ such that $\phi_x(\overline{n})$ is provable in PA for every $n$, but $\forall x\phi$ is not. For example, PA is a *reflexive* theory, meaning that the consistency of any *finitely axiomatized* subtheory of PA is provable in PA, by a theorem of Mostowski's that will not be proved in this book. Since we can also prove in PA that PA is consistent if and only if every finitely axiomatized subtheory of PA is consistent, why doesn't this imply that the consistency of PA is in fact provable in PA itself? The answer is that "for every theorem $\phi$ of PA, the theory with $\phi$ as its only axiom is consistent" is *not* provable in PA, although for every theorem $\phi$ of PA, "the theory with $\phi$ as its only axiom is consistent" *is* provable in PA.

A final comment on the axioms of PA. It is clear that PA is effectively axiomatized, and we can easily show that the set $Ax_{\mathrm{PA}}$ of axioms of PA is in fact primitive recursive. As noted in §8.6, it follows that the relation $Prf_{\mathrm{PA}}(n,k)$, defined as "$n$ is a derivation from the axioms of PA of the formula $k$", is primitive recursive.

Since arithmetical truth is not arithmetically definable, but provability in PA is, it also follows that the class of true arithmetical sentences cannot coincide with the class of sentences provable in PA. In other words, since all theorems of PA are true, there must be true arithmetical sentences not provable in PA. This is the observation originally made by Gödel, which set him on the path of explicitly formulating a true sentence not provable in PA.

Although PA is not finitely axiomatizable, all but finitely many of the axioms are instances of an easily described axiom schema. This is typically the case for theories formalizing ordinary mathematics, and as a result there is never any doubt about how to arithmetically define the axioms of the theory, in a way which shows the set of axioms to be primitive recursive. The pattern is that "$\phi$ is an axiom of $T$" is defined as "$\phi$ is one of the formulas $\phi_1, \ldots, \phi_n$ or a substitution instance of one of the schemata $\Phi_1, \ldots, \Phi_m$". This will be referred to as the *canonical* definition of the axioms of $T$ (where $T$ is PA or one of the other named theories introduced in the book: ZF, EA, CA, ACA). When we talk about *arbitrary* effectively axiomatizable theories, as in proving a general incompleteness theorem, there is no canonical definition of the axioms of a theory, and various complications ensue, as will emerge in later chapters.

## 9.2. Doing arithmetic in PA

### Basics

Before showing that the proofs in Chapter 2, or suitable variants of them, can be carried out in PA, it is necessary to verify that the basic algebraic properties of addition and multiplication, which were taken for granted in Chapter 2, can be proved to hold in PA. Similarly for the basic properties of the ordering relation between natural numbers.

For the algebraic properties—commutativity and associativity of addition and multiplication, distributivity of addition over multiplication—we basically only need to systematically apply induction and the defining equations for addition and multiplication. Take as a typical example the commutativity of addition. First we prove a lemma, by induction on $n$: for all $n$ and $m$, $m + s(n) = s(m) + n$. For the base case, $m + s(0) = s(m + 0) = s(m) = s(m)+0$, and for the induction step, $m+s(s(k)) = s(m+s(k)) = s(s(m)+k)$ (by the induction hypothesis) $= s(m) + s(k)$, which proves the lemma. Next we prove by induction on $n$ that $m + n = n + m$ for every $m$. For the base case we need to show that $0 + m = 0$ for every $m$, since $m + 0 = 0$ by the first axiom for addition. Here we again use induction. For $m = 0$, $0 + 0 = 0$ follows by the first addition axiom. Assuming $0 + m = m$, we get $0 + s(m) = s(0 + m) = s(m)$. Thus $0 + m = 0 = m + 0$ for every $m$. For the induction step in the main induction, assume $m + n = n + m$ for every $m$. Then $m + s(n) = s(m + n) = s(n + m) = n + s(m) = s(n) + m$ by the lemma. To prove associativity of addition, $m + (n + k) = (m + n) + k$, we similarly use induction on $k$. The base case follows from $m+(n+0) = (m+n)+0 = m+n$. For the induction step, we have $m + (n + s(k)) = m + s(n + k) = s(m + (n + k)) = s((m + n) + k) = (m + n) + s(k)$. Such manipulations establish that the basic algebraic identities follow from the axioms.

As an aside, the kind of proof by induction illustrated by the above arguments can lead quite far. It is characterized by being more algebraic than logical in character, in the sense that the statement proved by induction states that a certain quantifier-free formula is true for all values of its variables, and in proving the statement there is little logical reasoning, but mostly rewriting of formulas using recursion equations. In a variant of formal arithmetic called *primitive recursive arithmetic* or PRA, all induction proofs have this form. The induction axioms are restricted to formulas $\phi$ that contain no quantifiers, and to the axioms for addition and multiplication are added corresponding axioms for *all* primitive recursive functions. The resulting theory is usually regarded as encompassing those arithmetical theorems which can be proved by what is called *finitary* means, which covers most of the classical theorems of arithmetic. Our purpose here is instead to make it plausible that those theorems which have *arithmetical* proofs in the informal sense can also be formalized and proved in PA.

Other properties of multiplication and addition taken for granted in Chapter 2 were the cancellation laws for addition and multiplication: if $x+y = x+z$ then $y = z$, and if $x * y = x * z$ and $x$ is not 0, $y = z$. The easy proofs (by induction) are omitted here.

To prove the basic properties of the ordering relation, we first define $m < n$ as $\exists x(n = m + s(x))$. Defining $m < n$ in PA means to introduce an extension by definitions of PA containing the indicated defining axiom for $<$. In talking about formulas containing $<$ or other defined symbols, we will assume, whenever this simplifies matters, that the defined symbols have been eliminated (by the procedure described in §8.7). Thus we will freely use various defined symbols in talking about what is provable in PA, but also assume all those symbols to have been eliminated in various theoretical observations concerning PA.

With this definition of $<$, we can verify the basic properties of the ordering relation. For example, since $s(x) = x + s(0)$, $x < s(x)$ for all $x$. $x < x$ never holds, since if $x = x + s(z)$ it follows that $x + s(z) = x + 0$, so $s(z) = 0$. Transitivity of $<$ also follows by such algebraic reasoning. To prove that $<$ is a total order we use induction on $m$ to show that for every $n$, either $m = n$ or $m < n$ or $n < m$. The details are left to the reader.

Finally, the proofs in Chapter 2 also use *subtraction* of natural numbers. For this we need to prove in PA that if $n \leq m$, there is a unique $k$ such that $m = n + k$, which we denote by $m - n$. Since function symbols in PA can be applied to any arguments, we must also define $m - n$ for $m < n$, and we define $m - n$ to be 0 in this case. With this definition of subtraction, all the usual algebraic rules involving subtraction and addition can be proved in PA, with suitable assumptions of the form $n \leq m$.

## Classical arithmetic

With these preliminaries out of the way, we can pick up the argument of Chapter 2. The proof in §2.2 of the equivalence of the principle of mathematical induction, the principle of well-founded induction, and the smallest number principle can not be directly formalized in PA, since there are no variables for sets or properties in the language of PA. (It can be directly formalized in elementary analysis, as will be seen in the next chapter.) However, we can use that same proof to show that we get equivalent theories by replacing the induction schema in PA with either an induction schema for well-founded induction or a schema for the smallest number principle. Thus in proving theorems in PA, we can freely use any of these principles applied to arithmetical properties.

The proof in §2.3 of the existence of a unique quotient and remainder in division carries over to PA once some details have been filled in (such as an argument justifying the statement that "since $r'$ and $r$ are both greater than or equal to 0 and strictly smaller than $m$, the same is true of $r' - r$"). Note that this proof is an existence proof and does not explicitly refer to any algorithm for computing the quotient and remainder. An algorithm—that of repeatedly subtracting $m$ from $k$ until a number smaller than $m$ is obtained—is implicit in the proof, and can be extracted by various methods used in proof theory. Since no proof theory is introduced in this book, we will be content with informal observations regarding the relation between proofs and algorithms, but the relation between the *existence* of proofs and the *existence* of algorithms, that is between provability and computability, will be used in formally defining computability.

The existence and uniqueness of the g.c.d. of two numbers $k$ and $n$ is also provable in PA, by the argument of §2.3 cast as an inductive proof of an existence assertion. We use the variant where 0 is admitted, proving by (well-founded) induction on $n$ "for every $k$ there is a largest $m$ such that $m$ divides both $k$ and $n$". In using the induction hypothesis, it is shown that the largest $m$ which divides both $k - n$ and $n$ is also the largest $m$ which divides both $k$ and $n$.

The proof of Euclid's lemma can similarly be adapted to PA. Note that although formulated in terms of the set $I$, references to $I$ can easily be eliminated, as commented on in §5.2. The invocation of Euclid's algorithm and its termination is replaced in the PA version by an inductive argument.

The definitions in §2.4, the proof of the division lemma and the material about congruences carry over directly to PA, except that a formulation in PA cannot refer to congruence classes, but must use the congruence relation. But next (§2.5) we come to the fundamental theorem of arithmetic, for which we need to be able to talk about finite sequences of numbers, and about the product of the numbers in such a sequence. This leads us to definitions in PA of primitive recursive functions and relations.

## Primitive recursion in PA

In Chapter 4, it was established that all primitive recursive functions (and thereby also all primitive recursive relations) can be explicitly defined using the primitive language of arithmetic. This is easily seen in the case of the basic functions and the composition of functions, and the only non-trivial observation concerned replacing primitive recursion equations with an explicit definition:

> $f(b, a_1, \ldots, a_k) = c$ if and only if there is a sequence $c_0, c_1, \ldots, c_b$, where $c_b = c$, $c_0 = h(a_1, \ldots, a_k)$, and for every $i$ smaller than $b$, $c_{i+1} = g(i, a_1, \ldots, a_k, c_i)$.

We want to verify that this definition can be given in PA, and the recursion equations for $f$ derived from it. For this, we need to prove in PA (leaving out the parameters $a_1, \ldots, a_k$ for convenience)

(1) For every $b$, there is a unique sequence number $s$ of length $b+1$ such that $(s)_0 = h$, and for every $i$ smaller than $b$, $(s)_{i+1} = g(i, (s)_i)$.

To prove (1), we must first define sequence numbers and their associated functions in PA. The $\beta$-function, which was defined in §2.6 in terms of addition, multiplication, quotient, and remainder, is easily definable in PA. To define sequence numbers, we also need the $OP$ function of §4.4, defined as

$$OP(n,k) = \frac{(n+k)(n+k+1)}{2} + n$$

We need to verify that it is provable in PA that the function $OP$ is a one-one mapping between pairs of numbers and numbers, or in other words that if $OP(n,k) = OP(n',k')$ then $n = n'$ and $k = k'$, and for every $m$ there are $n$ and $k$ such that $OP(n,k) = m$. In §4.4 a counting argument was used for this: if we define the relation $R$ between pairs of natural numbers $\langle n,k\rangle$ and $\langle n',k'\rangle$ to hold if and only if $n+k < n'+k'$ or $n+k = n'+k'$ and $n < n'$, $OP(n,k)$ counts the number of pairs for which $\langle n',k'\rangle R \langle n,k\rangle$ holds, and therefore gives such a one-one mapping. We can't directly formulate this argument in PA, since we can't even express "the number of pairs for which $\langle n',k'\rangle R \langle n,k\rangle$ holds" in the language of PA. We can however state and prove in PA that if $\langle n,k\rangle R \langle n',k'\rangle$ then $OP(n,k) < OP(n',k')$, by a simple argument here omitted. Since $R$ is also easily shown to be a strict total ordering of pairs, it follows that $OP$ maps different pairs to different numbers. That there are, for every $m$, $n$ and $k$ such that $OP(n,k) = m$ follows by an induction on $m$: 0 is $OP(0,0)$, and if $m$ is $OP(n,k)$ then $m+1$ is either $OP(n+1,k)$ (if $k > 0$) or $OP(0,n+1)$ (if $k = 0$).

Thus we can define the functions $P_1$ and $P_2$ in PA, and we are set to define sequence numbers in PA. We define $length(s)$ and $(s)_i$ using the $\beta$-function, and $seq(x)$ is defined as in §4.6:

$$seq(x) \equiv x > 1 \wedge \neg\exists y < x(length(y) = length(x) \wedge \forall i < length(x)(x)_i = (y)_i)$$

There now remains the final hurdle of proving (1) in PA. That there can be at most one $s$ satisfying the stated condition is easily proved by induction. To prove the existence of $s$ by induction on $b$, we need to be able to prove that for any sequence number $s$ of length $n$ there is a sequence number $s'$ of length $n+1$ such that $(s')_i = (s)_i$ for $i < n$ and $(s')_n = g(n-1, (s)_{n-1})$. For this, what we need to prove is that for every $s$ and every $a$, there are $k$ and $m$ such that $\beta(k,m,i) = (s)_i$ for every $i < n$, where $n$ is $length(s)$, while $\beta(k,m,n) = a$. The proof is carried out by adapting the proof of the $\beta$-lemma. "Adapting" because that proof as formulated in §2.6 rather freely speaks of finite sequences, beginning "let $m$ be the product of all positive integers smaller than or equal to the largest of $n, a_0, \dots, a_n$." Since the possibility of formalizing references to finite sequences is precisely what we want to establish, we need to ensure that the argument can be recast so as not to refer to finite sequences. Inspection of the proof of the $\beta$-lemma shows that this is easily done as far as the $\beta$-lemma is concerned. We need only prove by induction on $l$ that there is an $m$ such that every number smaller than or equal to some $(s)_i$ for $i < l$ divides $m$, and continue the argument as before. We need to invoke the Chinese Remainder Theorem only in a special case involving the numbers $1 + (1 + i)m$, and the reasoning in terms of sequences in the proof of that theorem can then be replaced by an inductive argument. The details are left to the reader.

What is established in this way is that for every primitive recursive function $f$, a corresponding extension by definitions of PA can be constructed in which the defining equations for $f$ are provable, and similarly for predicates. In showing this, we don't need to use the upper bound for $k$ and $m$ in the $\beta$-lemma, which may seem a bit odd—can that upper bound be dispensed with in the formalization of arithmetic? The answer is that we still need to establish the upper bound in PA, in order to be able to reason in PA using sequence numbers as we use sequences in informal arguments. For any given $n$, $\langle a_0, \dots, a_n \rangle$ can be defined in PA, since $\langle a_0, \dots, a_n \rangle$ is a primitive recursive function of $a_0, \dots, a_n$. But in order to prove, for example, that $a = a'$ and $b = b'$ whenever $\langle a, b \rangle = \langle a', b' \rangle$, we need to be able to prove that $\langle a, b \rangle$ is the smallest $m$ such that $(m)_0 = 2$, $(m)_1 = a$, and $(m)_2 = b$, and for this we need the upper bound. Thus, having established that every primitive recursive function is definable in PA, we can go back to the proof of the $\beta$-lemma and verify that the upper bound can be defined in PA (using primitive recursive functions) and proved to be an upper bound.

## Primes and beyond

Once we have established that all primitive recursive functions can be defined in the language of PA, and their defining equations proved in PA, and also that reasoning about finite sequences of numbers can be expressed in PA, it's easy to verify that the fundamental theorem of arithmetic can be expressed and proved in PA. Going on to the material in Chapter 3, Euclid's proof immediately carries over to PA in any of its various forms, since we can express in PA such things as "$s$ is a sequence (number) of primes" and "$q$ is the product of the numbers in $s$". The arithmetical proof of Euler's stronger result in §3.2 also carries over to PA. At this point, one verifies this fact only by checking that the reasoning about infinite series can be reduced to reasoning about finite series, using arithmetical inequalities, and more or less takes it for granted that the resulting argument, which is arithmetical in the informal sense, can be turned into a derivation in PA. In fact PA is seen, on the basis of the general considerations presented here, in combination with experience in making proofs explicit and relating them to formal derivations, to be a good candidate for characterizing arithmetical provability, such as was referred to in §3.3.

## Proofs about PA in PA

Where confusion can arise in talking about provability in PA is in connection with the formalization in PA of arithmetical proofs of statements that are themselves about formulas or derivations in PA. For example, by ordinary informal reasoning we note that if $\phi(x,y)$ is a formula with free variables $x$ and $y$ and $\phi(\overline{n},\overline{n})$ is provable in PA, then $\exists x \exists y(\phi(x,y) \wedge \phi(y,x))$ is provable in PA, since this latter formula is derivable using the rules [L1], [L2], [L12]. Although we are here talking about formulas in PA, the argument is still an informal one. To formalize the argument in PA we must first formalize the conclusion, for example as follows:

$$\begin{aligned}&\forall x \forall y \forall z \forall u (\mathit{free}(x, \langle y, z\rangle) \wedge \\ &\exists w\, \mathit{Prf}_{\mathrm{PA}}(w, \mathit{subseq}(\langle \mathit{num}(u), \mathit{num}(u)\rangle, \langle y, z\rangle, x)) \supset \\ &\exists w\, \mathit{Prf}_{\mathrm{PA}}(w, \mathit{ex}(z, \mathit{ex}(y, \mathit{conj}(x, \mathit{subseq}(\langle z, y\rangle, \langle y, z\rangle, x))))))\end{aligned}$$

Here $\mathit{free}(m,n)$ is the primitive recursive relation meaning that $m$ is a formula and $n$ a sequence number containing those variables that have free occurrences in $m$, while $\mathit{Prf}_{\mathrm{PA}}$, *num*, and *subseq* are the primitive recursive predicate and functions defined earlier. By the general conventions stated in §7.3, we use the same names for predicate or function symbols in extensions by definitions of PA defining primitive recursive relations and functions as we use in the text. *ex* and *conj* are also primitive recursive functions—if $m$ is a variable $x$ and $n$ a formula $\psi$, $\mathit{ex}(m,n)$ is the formula $\exists x\psi$, and similarly for *conj*. By the general identifications of formulas with numbers in §7.1, $\mathit{ex}(m,n)$ is in fact $\langle 6, 5, m, n\rangle$. In this case it would be detrimental to the legibility of the formalization to use the same symbols $\exists$, $\wedge$ as we ordinarily use in talking about formulas when formalizing statements about formulas. Even in the above form, the formalization in PA of the statement proved in the simple argument given is not pleasant to work with, and the introduction of explicit formalizations of this kind will usually be avoided, even when it is essential to observe that a certain argument about PA can be carried out within PA.

## 9.3. Computability and provability

### Defining computability

The concept of a computable function from **N** to **N** has been introduced in earlier chapters: the function $f$ is computable if there is an algorithm which gives the value of $f(n)$ for any given argument $n$. A subset $A$ of $\mathbf{N}$ is said to be *decidable* if its characteristic function (see §4.6) is computable, which is to say that there is an algorithm which given any $n$ decides whether or not $n$ is in $A$. An *undecidable set* is one that is not decidable. These concepts carry over in an obvious way to functions from $\mathbf{N}^n$ to $\mathbf{N}$ and subsets of $\mathbf{N}^n$.

A third concept associated with algorithms has also been introduced in passing, in explaining why the theories considered in this book are assumed to be effectively axiomatizable. This is the concept of an *effectively enumerable* set. A set $A$ is effectively enumerable (sometimes abbreviated e.e.) if it is empty or there is an algorithm which generates the elements of $A$.[1] This can be defined in terms of computable functions: a non-empty set $A$ is effectively enumerable if there is a computable function $f$ such that $A$ is the range of $f$, that is, the members of $A$ are precisely the numbers $f(0), f(1), \ldots$. In these terms, it was observed in Chapter 8 that since the set of derivations from the axioms of a theory is primitive recursive in the axioms, if the theory is effectively axiomatized, its set of theorems is effectively enumerable.

An effectively enumerable set $A$ is not necessarily decidable, since even though we can generate the elements of $A$, we can't necessarily decide whether a given $n$ belongs to $A$. If $n$ does belong to $A$, this can in principle be established, since $n$ will eventually appear in the sequence $f(0), f(1), \ldots$, but if $n$ does not belong to $A$, we can't discover this by computing $f(0), f(1), \ldots$ and it may or may not be possible to discover it by other means. (Effectively enumerable sets are sometimes called *semi-decidable*.) But if the complement $\mathbf{N} \setminus A$ of $A$ is also effectively enumerable, then $A$ *is* decidable. For then there is a computable function $f$ that enumerates $A$ and another computable function $g$ that enumerates $\mathbf{N} \setminus A$, and by computing $f(0), g(0), f(1), g(1), \ldots$ we will eventually come upon $n$ as a value of either $f$ or $g$. A second observation concerning e.e. sets that we shall need is that a set $A$ is effectively enumerable if and only if there is a decidable relation $R$ such that for every $n$, $n$ is in $A$ if and only if there is a $k$ such that $R(k, n)$. In one direction this is clear from the fact that $n$ is in the range of $f$ if and only if there is a $k$ such that $f(k) = n$. For the other direction, if the set $A = \{n \mid \exists k R(k, n)\}$ has at least one member $a$, its members are generated by the function $f$ defined by $f(n) = a$ if $R((n)_0, (n)_1)$ does not hold, and otherwise $f(n) = (n)_1$.

To arrive at mathematical definitions of these concepts, it suffices to give a definition of "computable function from $\mathbf{N}$ to $\mathbf{N}$". The primitive recursive functions are clearly computable, and since all the computable functions commonly used in arithmetic are primitive recursive, it would not be unreasonable to suggest that the computable functions coincide with the primitive recursive functions. A diagonalization argument shows that this is not the case. The set

[1] In the literature, effectively enumerable sets are traditionally called *recursively enumerable* (r.e.). The term *computably enumerable* is also used.

of definitions of primitive recursive functions (considered for example as finite sequences of signs in the language of arithmetic) can be effectively enumerated, and therewith we get an effective enumeration of all primitive recursive functions, in the sense of a computable function $P$ such that $P(m,n)$ is the value of the $m$-th primitive recursive function applied to the argument $n$. The diagonal function $D$ defined by $D(m) = P(m,m) + 1$ is then a computable function which is not primitive recursive. So we must look for a more liberal definition in seeking to characterize the computable functions. Note that if a formally defined class $C$ of functions is in fact to exhaust the class of computable functions, it must not be possible to give an effective enumeration of the functions in $C$, since otherwise the diagonalization argument again applies.

There are many equivalent formal definitions of the computable functions, and we will introduce a definition in terms of provability in PA (which could for this purpose be replaced by a weak subtheory). First an observation regarding the possibility of carrying out computations within PA. The value of $f(k_1,\dots,k_n)$ for a primitive recursive $f$ can be computed by repeatedly making substitutions of numerals for variables in the defining equations for $f$, and in those of any other primitive recursive functions involved in the definition of $f$. Since this is a simple logical operation, and since the defining equations for $f$ are provable in PA, it follows (but will for later purposes need to be proved in detail) that $f(\overline{k}_1,\dots,\overline{k}_n) = \overline{m}$ is provable in PA if $f(k_1,\dots,k_n) = m$.

Some clarifying comments are in order here, to set our implicit conventions straight. Suppose we are talking about a primitive recursive function $f$. We know that there is an extension by definitions of PA in which the equations defining $f$ (and those defining any other primitive recursive functions used in defining $f$) are provable, using newly introduced function symbols. We will use the same letter $f$ in referring to the defined function symbol associated with the function $f$ as we use in speaking of the primitive recursive function $f$ itself. So the observation that $f(\overline{k}_1,\dots,\overline{k}_n) = \overline{m}$ is provable if $f(k_1,\dots,k_n) = m$ means that in a suitable extension by definitions of PA where a function symbol $f$ defining the function $f$ has been introduced, this equality is provable. But we will also, whenever there is any reason to do so, assume that defined symbols have been eliminated, and then references to the formula $f(\overline{k}_1,\dots,\overline{k}_n) = \overline{m}$ are to be understood as referring to a formula in primitive notation. Thus in speaking of $f(\overline{k}_1,\dots,\overline{k}_n) = \overline{m}$ as provable we may equally mean that the translation of the formula into primitive notation is provable in PA.

This applies not only to single letters, but to any names we may introduce for primitive recursive functions and relations. Thus e.g., the formula

$$Prf_{\mathrm{PA}}((y)_0, subseq(\langle num(z), num((y)_1)\rangle, \langle \overline{v}_0, \overline{v}_1\rangle, x))$$

which will be introduced below, is understood to use function symbols *subseq* and *num* and a predicate symbol $Prf_{\mathrm{PA}}$, which are defined in PA in accordance with the general scheme for defining primitive recursive functions and relations in PA, and define just those primitive recursive functions and relations denoted *subseq*, *num*, $Prf_{\mathrm{PA}}$ in the text. The notation $\langle a_0, \ldots, a_n \rangle$ for sequence numbers is also used in formulas.

We will return to the provability in PA of all true equations $f(\overline{k}_1, \ldots, \overline{k}_n) = \overline{m}$ in §9.4 below. Here we will consider how to use the relation between computability and provability in PA to give a general formal definition of "computable function". The reason why PA can prove the equations $f(\overline{k}_1, \ldots, \overline{k}_n) = \overline{m}$ is that these can be established by checking for special cases and making repeated substitutions in equations, which is amply covered by the methods of logical reasoning embodied in PA. But because it can be decided by computation whether something is a formal derivation from the axioms of PA, we can also look at the process of searching for proofs in PA as a special case of computation. We say that a function $f$ from $\mathbf{N}$ to $\mathbf{N}$ is *computably definable* in PA if there is a finite extension by definitions of PA containing a function symbol $f$ such that $f(\overline{k}) = \overline{m}$ is provable whenever $f(k) = m$. The formula defining the function symbol $f$ is said to *computably define* $f$ (both the function and the function symbol) in PA. We define a function to be computable if and only if it is computably definable in PA. Any such function is computable in the informal sense, for since the set of proofs in PA is decidable (and in fact primitive recursive) we can compute $f(k)$ by going through proofs in PA until we come upon a proof of (the translation into primitive notation of) $f(\overline{k}) = \overline{m}$.

Given this definition of computable functions, we find that a set $A$ is decidable if and only if there is a finite extension by definitions of PA containing a predicate symbol $p$ such that $p(\overline{n})$ is provable for $n$ in $A$, and $\neg p(\overline{n})$ is provable for $n$ not in $A$. We say that the formula defining $p$ in PA computably defines the predicate symbol $p$, and the set $A$.

In a way, it is a bit misleading to use PA in this definition, since in fact most of the logical apparatus in PA is not needed in order to prove $f(\overline{k}_1, \ldots, \overline{k}_n) = \overline{m}$ for a computable $f$. (See Shoenfield [1967].)

The fact of incompleteness makes the distinction between a formula that defines a certain function or relation and a formula that *computably* defines that function or relation a significant one. For example, for any sentence $\phi$, the formula $(\phi \wedge v_0 = 0) \vee (\neg\phi \wedge v_0 = 1)$ defines the constant function $f$ for which $f(x) = 0$ if $\phi$ is true and 1 otherwise. The formula does not however computably define $f$ in PA unless $\phi$ is decidable in PA, that is, either $\phi$ or $\neg\phi$ is provable in PA. Since $f$ is a constant function, it is trivially primitive recursive, and therefore computably definable in PA, either by the formula $v_0 = 0$ or by the formula $v_0 = 1$, although we can decide which of these computably defines $f$ in PA only if we can decide whether $\phi$ is true.

Note that what is required for a defined function symbol $f$ in a finite extension by definitions of PA to define a computable function is that for every

$k$ there is an $m$ such that $f(\overline{k}) = \overline{m}$ is provable in PA. Although we can effectively enumerate all definitions of function symbols in finite extensions by definitions of PA, there is no obvious way of effectively generating the $f$ that satisfy this further condition, so there is no obvious way of using diagonalization to define a computable function that is not computably definable in PA.

We extend the definition of computability to the case of $n$-ary functions and relations in the obvious way. Thus a function from $\mathbf{N}^n$ to $\mathbf{N}$ is computably defined by a formula $\phi$ if $\phi$ defines a function symbol $f$ in PA such that $f(\overline{k}_0, \ldots, \overline{k}_{n-1}) = \overline{m}$ is provable whenever $f(k_0, \ldots, k_{n-1}) = m$, and similarly for relations.

Having given a formal definition of the computable functions, we need to verify that the two observations made at the beginning of this section regarding effectively enumerable sets still stand: a set $A$ is decidable if and only if both A and its complement are effectively enumerable, and $A$ is effectively enumerable if and only if $A = \{n \mid \exists k R(k, n)\}$ for some decidable relation $R$. And indeed the arguments given carry over easily to the formally defined concepts, using the fact that the pairing function is primitive recursive, and therefore computably definable in PA.

## The Ackermann function

It is generally accepted that all computable functions are in fact computably definable in PA, although this is not obvious at the outset. There are many other formal characterizations of the computable functions, all known to define the same class of functions. To convince oneself that every function that is computable in the informal sense—that is, for which there is an algorithm that can be used to compute the value of the function for any argument—is in fact computably definable in PA, one will need to look at these equivalence proofs, which show for a wide variety of different types of algorithms how a suitable definition in PA can be found given an algorithm for computing a function from $\mathbf{N}$ to $\mathbf{N}$.

Here we shall look at one special example. A version of what is known as the *Ackermann function* is defined by the equations

$$Ack(0, x, y) = x * y$$
$$Ack(n + 1, x, 0) = 1$$
$$Ack(n + 1, x, y + 1) = Ack(n, Ack(n + 1, x, y), x)$$

Inspection shows that the function taking $x, y$ to $Ack(0, x, y)$ is multiplication, while that taking $x, y$ to $Ack(1, x, y)$ is exponentiation, and the later functions generalize this pattern. To convince ourselves that these equations do define a function, we can argue as follows. We prove by induction on $n$ that for every $x$ and $y$, a computation of $Ack(n, x, y)$ using the above equations in the obvious way will eventually terminate. This is clear for $n = 0$. For the induction step, suppose the assertion holds for $n$. We can then prove by induction on $y$ that the computation of $Ack(n+1, x, y)$ terminates for every $y$ and every $x$. The case $y = 0$ is again immediate, and for the induction step we first get that the computation of $Ack(n+1, x, y)$ terminates with some value $m$ by the induction hypothesis for $y$, and then that the computation of $Ack(n, m, x)$ terminates by the induction hypothesis for $n$.

This argument also shows how we can define $Ack$ in PA and prove the equations $Ack(\overline{n}, \overline{m}, \overline{k}) = \overline{p}$: we can define the function by a formula expressing "if $v_0 = \langle v_2, v_3, v_4 \rangle$, there is a sequence of substitutions in the equations defining $Ack$ which yields $v_1$ as final value for the arguments $v_2, v_3, v_4$; otherwise $v_1$ is 0". The argument above shows that this defines a function in PA, and again the substitutions involved can be carried out in PA, showing that every true equation is provable.

The definition of $Ack$ does not follow the pattern of primitive recursion, but uses what is called a *double recursion*. This of course does not exclude the possibility that $Ack$ is primitive recursive, since it might also have a primitive recursive definition. However, it can be proved that $Ack$ is not primitive recursive, and in fact grows faster than any primitive recursive function.

The argument sketched above can be carried out in a general form for all functions definable using systems of equations, and similar arguments apply to functions computable by Turing machines, and so on for other classes of algorithms and methods of computation. As mentioned in §3.3, the Church-Turing thesis, that these various equivalent approches to computability characterize the algorithmically computable functions, is generally accepted.

### Kleene's normal form and computable partial functions

Although a function computably definable in PA is not necessarily primitive recursive, it is closely associated with certain primitive recursive functions. Given that a unary function $f$ is computably definable in PA, we can define $f$ by the equation (Kleene's normal form)

$$f(n) = U\left(\mu x T^1(\phi, x, n)\right) \qquad (1)$$

where $\phi$ is a formula with free variables $v_0$ and $v_1$, $U$ is a certain primitive recursive function, $T^1$ a certain primitive recursive predicate, and $\mu x$ is an application of the *unbounded $\mu$-operator*, read as "the smallest $x$ such that". Thus $f(n)$ can be computed by checking a primitive recursive condition $T^1$ in the cases $T^1(\phi, 0, n), T^1(\phi, 1, n), \ldots$ until an $m$ such that $T^1(\phi, m, n)$ is found, and then applying the primitive recursive function $U$ to $m$.

So what are $U$, $\phi$, and $T^1$? $\phi$ is a formula computably defining $f$ in an extension by definitions of PA, with $v_1$ as argument and $v_0$ as value, so that in particular $\forall v_1 \exists v_0 \phi$ is provable in PA. The predicate $T^1$ is computably defined in PA by the following formula, using the primitive recursive predicate $Prf_{PA}$ defined above and the primitive recursive functions *subseq* and *num* from Chapter 7:

$$T^1(x,y,z) \equiv Prf_{PA}((y)_0, subseq(\langle num(z), num((y)_1)\rangle, \langle \overline{v}_1, \overline{v}_0\rangle, x))$$

$U(x)$, finally, is simply $(x)_1$. Thus the meaning of $T^1(\phi, x, n)$ is that $(x)_0$ is a proof in PA of $\phi(\overline{k}, \overline{n})$ for $k = U(x)$, which implies that $U(x)$ is $f(n)$ given that $\phi$ computably defines $f$ in PA (with $v_1$ as argument and $v_0$ as value). Conversely, if $k = f(n)$, there are infinitely many pairs $\langle p, k\rangle$ where $p$ is a proof in PA of $\phi(\overline{n}, \overline{k})$, and for the smallest such pair $x$, $f(n)$ is $U(x)$. Thus every computable function can be represented as in (1), using the fixed primitive recursive $U$ and $T^1$ and a parameter $\phi$.

There is no apparent way of deciding in general whether a given formula $\phi$ does define a computable function, and in fact, as pointed out above, there can't be any way of effectively generating the formulas that do, if the identification of computable functions with functions computably definable in PA is not to be shown incorrect through diagonalization. This is an obstacle to a theoretical treatment of computability, and it turns out that the study of computable functions (as pursued in what is known as classical recursion theory) proceeds smoothly only within the larger context of the computable *partial* functions.

A *partial* function from **N** to **N** is a function defined on some subset $A$ of **N**, with values in **N**. A function defined on all of **N**, or in other words a function from **N** to **N**, is a special case of this, and for contrast or emphasis we will speak of functions from **N** to **N** as *total* partial functions, or less confusingly as just total functions. Informally, a partial function $F$ is said to be *computable* if there is an algorithm which, applied to a number $n$, either yields no result at all—goes on forever—or else terminates with $F(n)$ as its result. We can reduce this concept to those already defined as follows: a partial function $F$ is computable if its graph, the set of all pairs $\langle n, F(n)\rangle$ for $n$ in the domain of $F$, is effectively enumerable. For if the graph of $F$ is effectively enumerable we get an algorithm computing $F$ in the sense described by simply generating the elements of the graph until a pair $\langle n, m\rangle$ is found. For the other direction, we need to establish that if $F$ is computable, the informal explanation shows its graph to be effectively enumerable. This follows since $\langle n, m\rangle$ is in the graph of $F$ if and only if there is an $x$ such that the computation of $F(n)$ terminates after $x$ steps with value $m$, and "the computation of $F(n)$ terminates after $x$ steps with value $m$" is a decidable relation between $\langle n, m\rangle$ and $x$. So in terms of the concepts already introduced, we can define a partial function to be computable if its graph is effectively enumerable.

A consequence of this definition is that a set is effectively enumerable if and only if it is the domain of a computable partial function. In one direction this follows from the fact that in generating the graph of $F$ we also generate the elements of its domain. For the other direction, given any e.e. set $A$, we can define a computable partial function $F$ with $A$ as its domain by stipulating that its graph contains exactly the pairs $\langle a, 0\rangle$ where $a$ is in $A$.

Before considering partial functions further, some notational and terminological conventions need to be fixed. $F$, $G$, $H$ (and in a pinch $I$, $J$, $K$, $L$, $P$) will be used for partial functions (of which total functions, as noted above, are a special case). Functional expressions involving partial functions and variables need not be defined or have a value (these expressions will be used interchangeably) for every value of those variables. For example, if 0 is not in the domain of $F$, $F(0)$ does not have a value. An expression $F(t)$ has a value only if $t$ has a value *and* that value belongs to the domain of the function $F$. We also say that a statement $P(t)$ is defined and has a truth value (true or false) only when the term $t$ has a value, and similarly for functions or predicates of arity grater than 1. Equations $S = T$ between expressions that may not have a value are to be understood as meaning that the expression on one side of the equals sign has a value if and only if the expression on the other side also has that same value (for all values of the variables involved).

Note that the need for these conventions arises only because we want to retain the convenience of using functional notation even when working with partial functions. We might in theory use a more cumbersome approach and talk consistently about the graphs of these functions instead, without using functional notation. The conventions in force here, although fairly standard in logic, are not the only possible ones. In computer programming, for example, one may also choose to take an expression like $Z(F(0))$, where $Z$ is the zero function, to have the value 0 whether or not $F(0)$ is defined. In programming, this is known as "lazy evaluation" as opposed to the "eager" or "strict" evaluation presupposed here.

Use of the unbounded $\mu$-operator can also result in expressions without a value. An expression of the form $\mu x P(x)$ has the value $k$ if $P(k)$ is defined and true while $P(m)$ is defined and false for every $m < k$.

Not every formula defines a computable *total* function, but by considering partial functions, we can associate a computable partial function with every formula $\phi$ with free variables $v_0$ and $v_1$. Using the classical notation introduced by Kleene, we define the partial function $\{\phi\}$ by

(2) $$\{\phi\}(n) = U\left(\mu x T^1(\phi, x, n)\right)$$

This equation has the same form as (1), but our conventions about expressions involving partial functions are now in force. The difference is that we are no longer making any assumptions about $\phi$. Instead $\{\phi\}(n)$ is defined for every $n$ such that there is at least one $k$ for which $\phi(\overline{n}, \overline{k})$ is provable in PA, and for any such $n$, $\{\phi\}(n)$ has some such $k$ as value.

We want $\{m\}$ to be a computable partial function for every number $m$, so if $m$ is not a formula with free variables $v_0$ and $v_1$, we define $\{m\}$ to be the constant function $Z$. We call $m$ an *index* of the computable partial function $\{m\}$, and when thinking of natural numbers as indices of partial functions we will often use $e$ as a variable.

To obtain indices for functions with more than one argument, we introduce a predicate $T^n$ and define for every $\phi$ with free variables $v_0, \ldots, v_{n+1}$,

$$\{\phi\}^n(k_1, \ldots, k_n) = U\,(\mu x T^n\,(\phi, x, k_1, \ldots, k_n))$$

where $T^n(\phi, x, k_1, \ldots, k_n)$ means that $(x)_0$ is a proof in PA of $\phi(\overline{m}, \overline{k}_1, \ldots, \overline{k}_n)$ for $m = (x)_1$. If $e$ is not such a formula, $\{e\}^n$ is again the $n$-ary constant zero function. Thus every number is the index of an $n$-ary computable partial function, for every $n$. We will usually leave out the superscript $n$ when it is clear from the number of arguments.

But now we need to verify that every partial function $\{\phi\}$ is in fact computable in the sense defined—that is, has an effectively enumerable graph—and conversely that every computable partial function $F$ has an index $\phi$ for which $F = \{\phi\}$. That $\{\phi\}$ has an effectively enumerable graph is clear since

$$\{\phi\}(n) = m \equiv \exists x\,(T^1(\phi, x, n) \land \forall y < x \neg T^1(\phi, y, n) \land U(x) = m)$$

We see from this equivalence that in fact $\{k\}(n) = m$ is effectively enumerable as a relation between $k, n, m$. Conversely, if $F$ is a computable partial function, there is a decidable relation $R$ such that $F(n) = m$ if and only if $\exists k R(k, \langle n, m\rangle)$, and an index for $F$ is obtained as the formula $\exists x \phi(x, \langle v_0, v_1\rangle)$, where $\phi$ computably defines $R$. The $n$-ary case for $n > 1$ is similar.

It was noted earlier that a set $A$ is effectively enumerable if and only if $A = \{n \mid \exists k R(k, n)\}$ for some decidable relation $R$, and also that $A$ is e.e. if and only if $A$ is the domain of some computable partial function. Since the domain of $\{\phi\}$ is the set $\{n \mid \exists k T^1(\phi, k, n)\}$, we see that the relation $R$ in the first characterization can in fact be taken to be primitive recursive.

### Effectively enumerable undecidable sets and the incompleteness of PA

What happens if we try to produce by diagonalization a computable partial function that does not have an index? The unary computable partial functions are effectively enumerable as $\{0\}, \{1\}, \ldots$ so we can define a computable partial function $D$ by

$$D(n) = \{n\}(n) + 1$$

However, we can not conclude that $D$ is different from every $\{n\}$ but only that if $D = \{n\}$, then $D(n)$ is not defined. We can also conclude that the set $K$ of $n$ such that $\{n\}(n)$ is defined, although effectively enumerable, is not decidable, since if it were, we could define a computable total diagonal function $D'$ by

$$D'(n) = \{n\}(n) + 1 \text{ if } \{n\}(n) \text{ is defined, and } 0 \text{ otherwise.}$$

That $K$ is undecidable also shows that the set of theorems of PA is undecidable, since for every e.e. set $A$ and every $n$, $n$ is in $A$ if and only if "$n$ is in $A$" is provable in PA. That this is so is essentially the same observation as the one made above, that $f(\overline{k}) = \overline{m}$ is provable whenever $f(k) = m$, for primitive recursive $f$. It will be verified in detail below, because we need to know that one half of it is provable in PA.

A theory $T$ is said to be undecidable (or "recursively undecidable") if the set of theorems of $T$ is undecidable, and is otherwise said to be decidable. If $T$ is effectively axiomatized and undecidable, it is also incomplete in the sense of Gödel's incompleteness theorem, that is, there is a sentence $\phi$ in the language of $T$ such that neither $\phi$ nor $\neg\phi$ is provable in $T$. For if $T$ is complete and effectively axiomatized, it is either inconsistent, and thus trivially decidable, or else we can decide whether $\phi$ is a theorem of $T$ or not by generating the theorems of $T$ until either $\phi$ or $\neg\phi$ appears. Thus it follows that PA is incomplete. Another way of seeing that PA is incomplete is to note that the set of theorems of PA of the form "$n$ is not in $K$" (with $K$ defined as above) is effectively enumerable, but the set of true statements of this form is not (since otherwise $K$ would be decidable), so there must be infinitely many true such statements which are not provable in PA. These are but two of the many ways in which the incompleteness of PA (given that all theorems of PA are true) more or less immediately follows from considerations of effective computability.

The term "undecidable" has two common senses in logic, and there is room for confusion. A theory $T$ is said to be undecidable if its set of theorems is undecidable. A formula $\phi$ is said to be undecidable in a theory $T$ if neither $\phi$ nor its negation is a theorem of $T$, and decidable in $T$ otherwise. Thus undecidability of a statement is always relative to some specific theory, whereas undecidability of sets (effective undecidability or recursive undecidability) is not relative to anything, but is an arithmetical property of those sets. The observation made above is that if a theory $T$ is both undecidable and effectively axiomatized (or equivalently, if it is effectively axiomatizable), then there is a sentence $\phi$ undecidable in $T$.

An example of an undecidable e.e. set which is closer to ordinary mathematics than the set $K$ or the set of theorems of PA is given by a remarkable result in recursion theory, the Matiyasevic-Robinson-Davis-Putnam theorem. A set $A$ of natural numbers is said to be Diophantine if there is a polynomial $p$ with integer coefficients such that $A$ is the set of $x$ for which there are natural numbers $y_1, \ldots, y_n$ such that $p(x, y_1, \ldots, y_n) = 0$. An equation of this form, for which solutions in integers or natural numbers are sought, is known as a Diophantine equation, after the Greek mathematician who studied the subject in antiquity. The theorem states that every e.e. set, surprising as this may seem, is Diophantine. A consequence is that the set of equations $p(x, y_1, \ldots, y_n) = 0$ for which there is at least one solution in integers is undecidable.

## Provably total functions

Since $\{\phi\} = \{\psi\}$ if $\phi$ and $\psi$ are equivalent in PA, we see that every computable partial function has infinitely many indices. For a particular example of this, note that $\{\phi\}$ is total if and only if for every $n$ there is at least one $k$ for which $\phi(\overline{n}, \overline{k})$ is provable in PA. Because of $\omega$-incompleteness, this doesn't imply that $\forall v_1 \exists v_0 \phi$ is provable in PA, let alone that $\phi$ defines a function in PA. So we need to verify that every total $\{\phi\}$ is a computable total function in the sense of the definition given earlier, since otherwise we would be in the uncomfortable situation that a total computable partial function is not necessarily a computable total function. Of course what we would do in such a case would be to conclude that the earlier definition was inadequate. However, we can show that any total $\{\phi\}$ is a computable total function, for if $\{\phi\}$ is total, $\{\phi\} = \{\psi\}$, where $\psi$ is the formula

$$\left(\exists x T(\phi, x, v_1) \wedge v_0 = U\left(\mu x T^1(\phi, x, v_1)\right)\right) \vee \left(\neg \exists x T^1(\phi, x, v_1) \wedge v_0 = 0\right)$$

and this formula computably defines $\{\phi\}$ in PA. Note that this does not imply that "$\{\psi\}$ is total", or equivalently $\forall y \exists x T^1(\psi, x, y)$, is provable in PA. $\forall v_1 \exists v_0 \psi$ is indeed provable in PA, but to conclude in PA that $\forall y \exists x T^1(\psi, x, y)$, we need to be able to prove in PA the formalization of "$\psi(\overline{n}, \overline{k})$ is provable in PA whenever $\psi(n, k)$", and this we can't in general do for the $\psi$ defined, even though it is true by assumption, since we can't in general prove that $\neg \exists x T^1(\phi, x, \overline{n})$ is provable whenever true (which in this case is never, by assumption), and this in turn is again because we can't always prove a true statement of the form $\forall y \exists x T^1(\phi, x, y)$ in PA. So once again we come upon the interplay between effective computability and the incompleteness of PA.

A computable partial function $F$ is said to be *provably total* in a theory $T$ if $T$ proves $\forall y \exists x T^1(\phi, x, y)$ for some index $\phi$ of $F$. Since $T$ is assumed to be effectively axiomatized, the theorems of $T$ of this form can be effectively generated, so there is an effective enumeration of the computable partial functions provably total in $T$. This means that a diagonalization can be carried out, yielding a total computable function which is not provably total in $T$.

## The *s-m-n* theorem

Finally, for later applications we will need a lemma which again has a traditional name given it by Kleene:

> ***s-m-n* theorem:** For every pair $m, n$ there is a primitive recursive function $s_n^m$ such that the following is provable in PA:
>
> $$\{s_n^m(e, k_1, \dots, k_m)\}(a_1, \dots, a_n) = \{e\}(k_1, \dots, k_m, a_1, \dots, a_n)$$

In other words, from an index for a computable partial function with $m+n$ arguments, we can get an index for the "specialized" partial function obtained by setting the first $m$ arguments to $k_1, \ldots, k_m$. Just how an index for the specialized partial function is obtained depends on the details of how indices are defined, and in our version it's obtained by substituting numerals for variables. Let's consider the case $m = n = 1$. If $e$ is not a formula $\phi$ with free variables $v_0, v_1, v_2$, $\{e\}^2$ is the zero function, and we can take $s_1^1(e,k)$ to be $e$. So suppose $e$ is the formula $\phi(v_0, v_1, v_2)$. We can then take $s_1^1(\phi, k)$ to be the formula $\phi(v_0, v_1, \overline{k})$, which can clearly be defined as a primitive recursive function of $k$ and the formula $\phi$. (The reason for the convention whereby $v_0$ represents the function value is precisely to smooth the way for this operation.)

## 9.4. Computability and quantifiers

### $\Delta_0$-formulas and $\Sigma$-formulas

A formula $\phi$ in the first order language of arithmetic is a $\Delta_0$-*formula* if all quantifiers in $\phi$ are bounded, that is if universal quantifiers occur only in contexts of the form $\forall x(x < t \supset \phi)$ where $t$ is a term not containing $x$, which we write as $\forall x < t\phi$, and existential quantifiers only in contexts $\exists x(x < t \wedge \phi)$ where again $t$ does contain $x$, which we write $\exists x < t\phi$.

The set of true $\Delta_0$-sentences is clearly decidable in the informal sense, and as we shall see it is in fact primitive recursive. But what we now wish to establish is that every $\Delta_0$-sentence $\phi$ is decidable in PA and then use this fact to conclude that primitive recursive functions and relations are indeed computably definable in PA. In fact we shall prove a stronger assertion. The $\Sigma$-*formulas* are defined by the following inductive definition:

Every atomic formula and negated atomic formula is a $\Sigma$-formula.
If $\phi_1$ and $\phi_2$ are $\Sigma$-formulas, so are $\phi_1 \vee \phi_2$ and $\phi_1 \wedge \phi_2$.
If $\phi$ is a $\Sigma$-formula, so are $\exists x\phi$ and $\forall x < t\phi$, provided $x$ does not occur in $t$.

A $\Delta_0$-formula is not necessarily a $\Sigma$-formula, since negation occurs in $\Sigma$-formulas only applied to atomic formulas, but any $\Delta_0$-formula can be transformed into a logically equivalent $\Sigma$-formula by a primitive recursive operation that consists in rewriting $\neg\forall x < t\phi$ as $\exists x < t\neg\phi$, $\neg\exists x < t\phi$ as $\forall x < t\neg\phi$, $\neg(\phi_1 \vee \phi_2)$ as $\neg\phi_1 \wedge \neg\phi_2$ and $\neg\neg\phi$ as $\phi$. In the other direction, a $\Sigma$-formula is not in general equivalent in PA to a $\Delta_0$-formula, but every $\Sigma$-formula $\phi$ can be similarly transformed by a primitive recursive operation into a $\Sigma_1$-*formula* which is equivalent to $\phi$ in PA. A $\Sigma_1$-formula is a formula $\exists x\phi$ where $\phi$ is a $\Delta_0$-formula, and a procedure for transforming a $\Sigma$-formula into a $\Sigma_1$-formula will be described below. First we will prove the basic completeness theorem in connection with computability, the

> **Σ-completeness theorem:** If $\phi$ is a Σ-formula all of whose free variables are among $v_0, \ldots, v_n$ and if $k_0, \ldots, k_n$ satisfy $\phi$, that is if the formula $\phi(\overline{k}_0, \ldots, \overline{k}_n)$ obtained by substituting $\overline{k}_i$ for $v_i$ (for $i = 0$ to $n$) is true, then $\phi(\overline{k}_0, \ldots, \overline{k}_n)$ is provable in PA.

The proof is by induction on Σ-formulas. For the base case, with $\phi$ an atomic formula or negated atomic formula, we need to establish that equality, the successor function, addition, and multiplication and identity are computably defined in PA by the formulas $v_0 = v_1$, $s(v_1) = v_0$, $v_2 + v_1 = v_0$, $v_2 * v_1 = v_0$ respectively.

First equality. $\overline{n} = \overline{k}$ is provable if $n = k$ by the identity axiom, and to show by induction on $k$ that $\neg\overline{n} = \overline{k}$ is provable if $n \neq k$ we only need to use the axioms for the successor function.

Note that "show by induction" does not here refer to any use in PA of the induction axioms. In fact the Σ-completeness theorem holds for a weak subtheory of PA, which the interested reader can extract from the proof of the theorem. However, in Chapter 12 we will need to convince ourselves that the above argument can itself be formalized in PA, and then of course the induction axioms in PA will be presupposed.

To show that $s(\overline{n}) = \overline{k}$ is provable if $s(n) = k$, we use induction on $k$, and similarly for addition and multiplication.

Having established this, we can prove by induction on $s$ that for any variable-free term $s$, if the value of $s$ is $k$, $s = \overline{k}$ is provable in PA, and for any $n \neq k$, $\neg s = \overline{n}$ is provable in PA. From this follows that if $k_0, \ldots, k_n$ satisfy $s = t$ then $s(\overline{k}_0, \ldots, \overline{k}_n) = t(\overline{k}_0, \ldots, \overline{k}_n)$ is provable, and otherwise $\neg s(\overline{k}_0, \ldots, \overline{k}_n) = t(\overline{k}_0, \ldots, \overline{k}_n)$ is provable, which takes care of the base case in the main induction.

For the induction step, if the numbers $\overline{k}_0, \ldots, \overline{k}_n$ satisfy $\phi_1 \vee \phi_2$, they satisfy $\phi_1$ or $\phi_2$, so $\phi_1(\overline{k}_0, \ldots, \overline{k}_n)$ or $\phi_2(\overline{k}_0, \ldots, \overline{k}_n)$ is provable by the induction hypothesis, so $\phi_1(\overline{k}_0, \ldots, \overline{k}_n) \vee \phi_2(\overline{k}_0, \ldots, \overline{k}_n)$ is provable. Similarly for conjunction.

If $k_0, \ldots, k_n$ satisfy $\exists x\phi$, there is an $m$ such that $k_0, \ldots, k_n, m$ satisfy $\phi_x(v_{n+1})$, so $\phi(\overline{k}_0, \ldots, \overline{k}_n, \overline{m})$ is provable in PA, and thereby $\exists x\phi(\overline{k}_0, \ldots, \overline{k}_n, x)$.

If $k_0, \ldots, k_n$ satisfy $\forall x < t\phi$, finally, let $k$ be the value of $t(\overline{k}_0, \ldots, \overline{k}_n)$. For every $m < k$, $k_0, \ldots, k_n, m$ satisfy $\phi_x(v_{n+1})$, so by the induction hypothesis, $\phi_x(\overline{k}_0, \ldots, \overline{k}_n, \overline{m})$ is provable in PA for every $m < k$. The desired conclusion follows from the fact that $t(\overline{k}_0, \ldots, \overline{k}_n) = \overline{k}$ is provable in PA, given that PA has the following property: for every $m$ and every formula $\psi$, if $\psi_x(\overline{k})$ is provable for $k = 0, \ldots, m-1$, then $\forall x < \overline{m}\psi$ is provable. This property is a consequence of the fact that for every $m > 0$, $\forall x(x < \overline{m} \equiv x = 0 \vee \cdots \vee x = \overline{k})$ is provable in PA for $k = m - 1$. This concludes the proof of the Σ-completeness theorem.

A relation $R$ is called a *Σ-relation* if it is definable by a Σ-formula, or in other words if there is a Σ-formula $\phi(v_0, \ldots, v_n)$ such that $R(k_0, \ldots, k_n)$ holds if and only if $\phi(\overline{k}_0, \ldots, \overline{k}_n)$ is true. A partial function is a *Σ-function* if its graph is a Σ-relation. The close connection between these definitions and computability lies in the following basic

**Σ-characterization of effective enumerability:** A relation is effectively enumerable if and only if it is a Σ-relation.

In one direction this follows from the fact (to be established below) that every Σ-formula is equivalent in PA to a $\Sigma_1$-formula. For the other direction, it suffices to show that every primitive recursive function is a Σ-function, since every e.e. set is $\{n \mid \exists k R(k, n)\}$ for some primitive recursive $R$. The verification of this consists in a proof by induction on primitive recursive functions that the graph of a primitive recursive function is definable by a Σ-formula. This is immediate for the basic functions, and the graph of a function obtained by composition is definable in terms of the graphs of the functions composed, using conjunction and the existential quantifier. So the detailed verification required consists in checking that we can express using an Σ-formula "there is a sequence $c_0, c_1, \ldots, c_b$, where $c_b = c$, $c_0 = h(a_1, \ldots, a_k)$, and for every $i$ smaller than $b$, $c_{i+1} = g(i, a_1, \ldots, a_k, c_i)$", given that "$c_{i+1} = g(i, a_1, \ldots, a_k, c_i)$" and "$c_0 = h(a_1, \ldots, a_k)$" can be thus expressed. This is just a matter of verifying that the functions and predicates associated with sequence numbers are Σ-definable, which is left to the reader.

This proof also shows that primitive recursive functions have the further property of being computably definable in PA by a Σ-formula. This property is not shared by every total computable function. In fact, a computable partial function is computably definable in PA by a Σ-formula if and only if it is provably total in PA. The verification of this, which is left to the reader, uses the provability in PA of the Σ-completeness theorem (to be considered in Chapter 12). Note that any Σ-formula which defines a function in PA computably defines that function, by the completeness theorem.

Σ-formulas and Σ-completeness will play a large role in the later chapters of the book, dealing with Gödel's theorem. The negations of Σ-formulas are equivalent to *Π-formulas*, defined by a dual induction:

Every atomic formula and negated atomic formula is a Π-formula.

If $\phi_1$ and $\phi_2$ are Π-formulas, so are $\phi_1 \vee \phi_2$ and $\phi_1 \wedge \phi_2$.

If $\phi$ is a Π-formula, so are $\forall x \phi$ and $\exists x < t \phi$, if $x$ does not occur in $t$.

If a relation $R$ is decidable, so is its negation, so $R$ is definable both by a Σ-formula and by a Π-formula (but not necessarily by a $\Delta_0$-formula).

A property of provably total functions which makes them very pleasant to work with in many contexts is the following

**Substitution property for provably total functions:** If a unary function symbol $f$ is defined by a $\Sigma$-formula, and $\phi$ is equivalent to a $\Sigma$-formula (or a $\Pi$-formula) in PA, the formula $\phi_x(f(t))$ is also equivalent to a $\Sigma$-formula (or a $\Pi$-formula) in PA. Similarly for function symbols of other arities.

The substitution property is a consequence of the fact that $\phi_x(f(t))$ is equivalent to both $\exists x(x = f(t) \wedge \phi)$ and $\forall x(x = f(t) \supset \phi)$ (given that $t$ is free for $x$ in $\phi$). It follows in particular that primitive recursive functions can be used freely in producing (formulas equivalent in PA to) $\Sigma$-formulas and $\Pi$-formulas.

## The formula hierarchy

A formula $\phi$ is a *prenex* formula, or in prenex form, if it has the form $Q_1x_1 \dots Q_nx_n\psi$, where $\psi$ does not contain any quantifiers, and each $Q_i$ is either $\forall$ or $\exists$. The part $Q_1x_1 \dots Q_nx_n$ is the *quantifier prefix* of $\phi$.

Prenex formulas are usually difficult to understand as soon as there are more than two quantifiers in the prefix, but for theoretical purposes they are often useful. Any formula $\phi$ is logically equivalent to a formula (and in fact infinitely many formulas) in prenex form, and we can define in different ways a primitive recursive function $p$ such that $p(\phi)$ is in prenex form and is logically equivalent to $\phi$. One way of doing this is the following. We first apply to $\phi$ a primitive recursive operation which yields an equivalent formula $\psi$ in which no variable occurs both free and bound, and in which also no variable has more than one occurrence immediately following a quantifier. We then define a primitive recursive $p$ transforming $\psi$ into prenex form as follows, where $Q'$ is $\forall$ if $Q$ is $\exists$ and $\exists$ if $Q$ is $\forall$:

$p(\psi) = \psi$ if $\psi$ is atomic,

If $\psi$ is $\neg q$ and $p(\theta)$ is $Q_1x_1 \dots Q_nx_n\phi$, $p(\psi)$ is $Q'_1x_1 \dots Q'_nx_n\neg\phi$,

If $\psi$ is $Qx\theta$, $p(\psi)$ is $Qxp(\theta)$

If $\psi$ is $\theta_1 \vee \theta_2$, $p(\theta_1)$ is $Q_1x_1 \dots Q_mx_m\phi_1$ and $p(\theta_2)$ is $Q_{m+1}y_1 \dots Q_{m+n}y_n\phi_2$, $p(\psi)$ is $Q_1x_1 \dots Q_mx_mQ_{m+1}y_1 \dots Q_{m+n}y_n(\phi_1 \vee \phi_2)$.

The transformation of a formula into prenex form is purely logical, in the sense that a formula and its prenex version are logically equivalent. (For the last clause in the definition above, this is so only because we have first ensured that all of the quantified variables are different, and no variable that occurs bound in $\theta_1$ has any occurrence in $\theta_2$, or conversely). In arithmetic, we can go further, and transform any first order formula $\phi$ into a formula which is equivalent to $\phi$ in PA and has the form $Q_1x_1 \dots Q_nx_n\psi$ where $\psi$ is a $\Delta_0$-formula, and the quantifier prefix contains strictly alternating quantifiers, that is, each $\forall$ is followed by $\exists$ and conversely. A formula of this form which begins with an existential quantifier is called a $\Sigma_n$-formula, and if it begins with a universal quantifier it is a $\Pi_n$-formula. We also allow $n = 0$, setting both $\Sigma_0$ and $\Pi_0$ equal to $\Delta_0$.

To transform $\phi$ into a $\Sigma_n$-formula or $\Pi_n$-formula, we can start with a prenex form of $\phi$ and apply the following contraction rules to that prenex form:

> $\forall x \forall y \phi$ is rewritten as $\forall z \forall x < z \forall y < z\phi$, where $z$ does not occur in $\forall x \forall y \phi$,
>
> $\exists x \exists y \phi$ is rewritten as $\exists z \exists x < z \exists y < z\phi$, where $z$ does not occur in $\exists x \exists y \phi$,

together with the following rules for moving bounded quantifiers past unbounded ones:

> $\forall x < y \forall z \phi$ is rewritten $\forall z \forall x < y \phi$
>
> $\exists x < y \exists z \phi$ is rewritten $\exists z \exists x < y \phi$
>
> $\forall x < y \exists z \phi$ is rewritten $\exists w \forall x < y \exists z < w \phi$
>
> $\exists x < y \forall z \phi$ is rewritten $\forall w \exists x < y \forall z < w \phi$

The contraction rules and the two latter rules for bounded quantifiers do not result in logically equivalent formulas, but they do yield formulas that can be proved in PA to be equivalent, using basic properties of the ordering relation and mathematical induction. Thus, to show that $\forall x < y \exists z \phi(x, y, z)$ implies $\exists w \forall x < y \exists z < w \phi(x, y, z)$, we prove by induction on $y$ that for every $u$, if $\forall x < y \exists z \phi(x, u, z)$ then $\exists w \forall x < y \exists z < w \phi(x, u, z)$. For $y = 0$ we can take $w = 0$ (since $x < 0$ is never true), while given $\forall x < y \exists z \phi(x, u, z)$ we get $\forall x < s(y) \exists z < w' \phi(x, u, z)$ if $w'$ is larger than $w$ and larger than the smallest $z$ such that $\phi(y, u, z)$. Specializing $u$ to $y$ we then get the desired implication. This implication is similar to set-theoretical postulates (the axiom of collection, the replacement axiom) which give analogous upper bounds in the context of infinite sets, but it also (as pointed out by A.Göransson) resonates as far back as the Diamond Sutra, where an upper bound is envisioned for the total number of grains of sand, assuming there are as many Ganges rivers as there are grains of sand in the Ganges riverbed. The last equivalence above is perhaps most easily proved by applying negation to the formulas in the preceding equivalence.

These same equivalences can be used to transform every $\Sigma$-formula $\phi$ into a $\Sigma_1$-formula equivalent to $\phi$ in PA, and every $\Pi$-formula into an equivalent $\Pi_1$-formula, by a primitive recursive operation.

## 9.5. Partial truth definitions

Tarski's theorem, that arithmetical truth is not definable by an arithmetical formula, was proved in §7.6 using diagonalization. We can recast this argument to obtain the stronger conclusion that it is not possible to have, in any consistent extension $T$ of PA, a predicate symbol *True* such that

$$(1) \qquad True(\overline{\phi}) \equiv \phi$$

is provable in $T$ for every sentence $\phi$ in the language of $T$. For this we need only formalize in PA the argument of §7.6 to obtain the

> **Diagonal lemma:** For every formula $\psi$ with a single free variable $x$ there is a sentence $\phi$ such that $\phi \equiv \psi_x(\overline{\phi})$ is provable in PA. (A formula for which this holds is called a *fixed point* for $\psi$.) Indeed there is a primitive recursive function giving such a $\psi$ as a function of $\phi$.

Recall that we can choose the diagonalization $\phi$ as $\psi_x(sub(num(\overline{\theta}), \overline{x}, \overline{\theta}))$, where $\theta$ is $\psi_x(sub(num(x), \overline{x}, x))$. Since the functions *sub* and *num* are primitive recursive, $sub(num(\overline{\theta}), \overline{x}, \overline{\theta}) = \overline{\phi}$ is provable in PA, so the equivalence follows.

As in §7.6, we can now conclude that the equivalence (1) leads to a contradiction if we take $\phi$ to be a fixed point for the formula $\neg\, True(x)$. Note that the contradiction depends on the assumption that the schema (1) holds for every sentence $\phi$ in a language including *True*. No contradiction results from extending a language $L$ by adding a predicate *True* and postulating or proving (1) for every $\phi$ in the language $L$.

In §7.6, it was also noted that the set $True_{vf}$ of true variable-free sentences of PA is primitive recursive and therefore arithmetically definable. The class *True*-$\Delta_0$ of true $\Delta_0$-sentences extends the class, and can also be shown to be primitive recursive by a similar argument. The variables add a small complication, so we need to give a primitive recursive definition of the more general relation $Sat_0(a, \phi)$, meaning that $a$ is a sequence $\langle k_0, \ldots, k_n \rangle$, all free variables of the $\Delta_0$-formula $\phi$ are among $v_0, \ldots, v_n$, and $\phi(\overline{k}_0, \ldots, \overline{k}_n)$, the formula obtained by substituting $\overline{k}_i$ for $v_i$ in $\phi$, is true. We also need a corresponding function $val(a, t)$ giving the value of a term $t$ for values $k_0, \ldots, k_n$ of the variables $v_0, \ldots, v_n$. *val* and $Sat_0$ can be shown to be primitive recursive by the usual methods on the basis of their natural recursive definitions. For example,

$$Sat_0(a, \phi_1 \vee \phi_2) \equiv Sat_0(a, \phi_1) \vee Sat_0(a, \phi_2)$$
$$Sat_0(a, \exists v_i < t\phi) \equiv \exists x < val(a, t)\, Sat_0(u(a, i, x), \phi)$$

In the second equivalence, $u(a, i, x)$ is the sequence number $a'$ for which $(a')_i = x$ and $(a')_j = (a)_j$ for $j \neq i$. The proof that $u$ is primitive recursive follows the usual lines.

For $n > 0$, the set *True*-$\Sigma_n$ of true $\Sigma_n$-sentences is not primitive recursive, or even decidable (as we see from the fact that the set $K$ defined above is not decidable), but it is arithmetically definable, and indeed definable by a $\Sigma_n$-formula. Similarly the set *True*-$\Pi_n$ of true $\Pi_n$-sentences is definable by a $\Pi_n$-formula, for each $n > 0$. This follows by a simple induction, using the fact that for $n > 1$, a $\Sigma_n$-sentence $\exists x\phi$ is true if and only if there is a $k$ such that $\phi_x(\overline{k})$ is a true $\Pi_{n-1}$-sentence, where "$\phi_x(\overline{k})$ is a true $\Pi_{n-1}$-sentence" is expressible as a $\Pi_{n-1}$-formula by the induction hypothesis, and the fact that $\phi_x(\overline{k})$ is a primitive recursive function of $\phi$ and $k$. For $n = 1$, we use the fact that *True*-$\Delta_0$ is primitive recursive and therefore definable both by a $\Sigma_1$-formula and by a $\Pi_1$-formula.

Thus the set of true $\Sigma_n$-sentences is arithmetically definable for each $n$, although "$\phi$ is a true $\Sigma_n$-sentence" is not arithmetically definable as a relation between $\phi$ and $n$.

## Tarski equivalences

The definition of *True*-$\Sigma_n$ in PA has the important property that for any $\Sigma_n$-sentence $\phi$, the equivalence

(2) $$\textit{True-}\Sigma_n(\overline{\phi}) \equiv \phi$$

is provable in PA. (2) is known as a *Tarski equivalence* for the restricted truth predicate *True*-$\Sigma_n$ (and (1) is a Tarski equivalence for *True*). Similarly for every $\Pi_n$-sentence $\phi$,

(3) $$\textit{True-}\Pi_n(\phi) \equiv \phi$$

is provable in PA.

For the proof, by induction on $n$, consider as a typical case *True*-$\Sigma_2$. Suppose $\phi$ is $\exists x \forall y \psi$, where $\psi$ is $\Delta_0$. By the definitions of the restricted truth predicates,

$$\textit{True-}\Sigma_2(\overline{\phi}) \equiv \exists x \forall y \textit{True-}\Delta_0(\textit{sub}(\textit{num}(y), \overline{y}, \textit{sub}(\textit{num}(x), \overline{x}, \psi)))$$

is provable in PA. So to conclude (2) it suffices to show that for every $\Delta_0$-formula $\psi$ and all variables $x$ and $w$,

(4) $$\textit{True-}\Delta_0(\textit{sub}(\textit{num}(w), \overline{x}, \overline{\psi})) \equiv \psi_x(w)$$

is provable in PA. (4) is proved by induction on $\Delta_0$-formulas. By inspection of this proof we find that in fact it also proves corresponding Tarski equivalences for formulas with free variables. For example, if we have a $\Sigma_n$-formula $\phi$ with the free variable $x$, we can prove in PA

(5) $$\forall x(\textit{True-}\Sigma_n(\textit{sub}(\textit{num}(x), \overline{x}, \overline{\phi})) \equiv \phi)$$

# CHAPTER 10

# ELEMENTARY AND CLASSICAL ANALYSIS

## 10.1. Classical analysis

*Second order arithmetic,* also known as *classical analysis* is a second-order theory which we will denote CA. It includes among its axioms the same axioms for the successor function, addition, and multiplication as in PA. The induction principle is formalized in CA as a single second order sentence:

$$\forall X(0 \in X \wedge \forall x(x \in X \supset s(x) \in X) \supset \forall x(x \in X))$$

Of course for this axiom to have any arithmetical consequences, we must also include axioms for the existence of sets. In CA these axioms are the instances of the *impredicative comprehension principle*:

$$\exists X \forall x(x \in X \equiv \phi)$$

where $\phi$ is any formula in the language of CA which does not contain the variable X. ("Impredicative" refers to the fact that the formula $\phi$ may itself contain second order quantifiers.) In other words, sets can be defined by abstraction in CA using any formula of the language. Thus proofs in CA can use induction on any property definable in the language.

The induction principle in CA is prima facie stronger than that of PA, since PA only formalizes the induction principle for arithmetical properties, while CA does so for all properties definable in the second order language of arithmetic. Of course there are also properties of the natural numbers that cannot be defined in this language (such as the property of being a true sentence in the language of CA), so it is not to be expected that the induction principle of CA covers every use of mathematical induction, but we still expect CA to be a stronger theory than PA.

But what are these properties expressible in the second order language? This is the same as asking how the set quantifiers in CA are to be interpreted. The comprehension axiom of CA is problematic if we don't accept the notion of an arbitrary subset of N and allow the quantifiers to vary over such arbitrary subsets. We have no reason to suppose that the impredicative comprehension principle is valid if the set variables vary over, say, arithmetical sets, since there is no reason to assume that every set definable by a condition involving quantifiers over arithmetical sets can also be defined by a condition quantifying only over the natural numbers. And indeed as we shall see, the impredicative comprehension principle is false if we restrict the set quantifiers to arithmetical sets. Is there any language $L$ such that the impredicative comprehension principle can be seen to be valid for the sets definable in $L$? The only obvious candidate is the language $L$ of second order arithmetic itself, since the impredicative comprehension principle asserts the existence of sets definable in this language. This is of no help, however, in specifying how the set variables in CA are to be understood, since it is precisely the interpretation of that language that is at issue. A different suggestion might be that the variables vary over the "definable sets" in the absolute sense of "set definable in any language", but then we are faced with the problems surrounding this notion, touched on in §5.2 To most people, the impredicative comprehension principle is evident only if we accept the notion of sets as "extensional totalities", independent of all definitions. Since this notion is itself considered doubtful or problematic by many, the impredicative comprehension principle is also commonly seen as problematic.

So just what can be proved in CA? From a logical point of view, CA is a very strong theory, in which large parts of mathematics can be formalized. In fact large parts of mathematics can be formalized in subtheories of CA, and there is an extensive logical literature on such subtheories (see Simpson [1999] for an extended treatment). In this book we are only concerned with a few basic properties of CA and of the subtheories EA and ACA (presented later in this chapter). The basic relations between these three theories are as follows. CA is a logically strong theory, which we can justify (in the strong sense of presenting its axioms as true) only on the basis of a mathematical concept—that of an arbitrary subset of N—which is usually seen as more abstract, more problematic than the notion of a natural number, and which cannot reasonably be said to be implicit in anything we say about the natural numbers themselves. EA is a formally weak second order theory which gives much of the convenience of talking about sets, without going beyond PA as far as arithmetic is concerned. ACA is a theory of arithmetic embodying the notion of arithmetical set, formally equivalent, as far as arithmetical statements are concerned, to a theory extending PA with the notion of truth for arithmetical sentences. ACA points the way to an indefinite sequence of extensions of PA which can still be held to be justified on the basis of the truth of the axioms of PA, while CA represents those extensions of arithmetic which can by no stretch of the (mathematical) imagination be thought of as based only on the truth of the axioms of arithmetic.

## 10.2. Elementary analysis

Elementary analysis, EA, differs from CA in having only a weak comprehension principle, requiring the formula $\phi$ to contain no bound set variables.

Thus the only sets which can be proved to exist are arithmetical sets. But there is a further essential restriction: since EA has the same induction axiom as CA, induction can be used only for *arithmetical* formulas, formulas that contain no bound set variables. For it is only for such formulas that we can prove the existence of a corresponding set and then use the induction axiom.

A consequence of this is that EA is a *conservative* extension of PA. This can be easily proved using the completeness theorem. Suppose $\phi$ is an arithmetical sentence not provable in PA. There is then an interpretation $I$ in which the axioms of PA are true but $\phi$ is false. From this interpretation $I$ we can get an interpretation $I'$ of the second order language by stipulating that the set quantifiers refer to the subsets of the domain of $I$ which are definable using the language of arithmetic, with $+$ and the other symbols given the interpretation they have in $I$. This interpretation $I'$ will be a model of the axioms of EA, and thus $\phi$ is not a theorem of EA either.

So EA is a theory in which every use of sets can be regarded as just a figure of speech, as far as proving arithmetical theorems are concerned. To be sure, the proof that EA conservatively extends PA given above does not give any method for converting a proof of an arithmetical formula $\phi$ in EA into a proof of $\phi$ in PA. In a proof-theoretical treatment, such a method can be given.

Although not stronger than PA as far as proving arithmetical theorems is concerned, EA is much more convenient in formalizing arithmetical proofs. Thus for example the proofs of the equivalence of the various versions of the induction principle in Chapter 2 carry over directly to EA. Also, the formalizations of many arithmetical proofs in EA will be much shorter than they are in PA, because we can use, applied to different arithmetical conditions, results proved for arbitrary sets instead of having to repeat the proof for each condition.

That EA is a conservative extension of PA has been proved here using the completeness theorem, and the proof of that theorem made free use of inductive reasoning involving sets. Thus, although we have proved that every arithmetical theorem provable in EA is provable in PA, we have not shown that references to sets can be regarded as a figure of speech in proving the general statement "every arithmetical statement provable in EA is provable in PA". When one speaks of proof-theoretical reductions of a theory $T$ to a theory $T'$, it is also required that the relevant relation between $T$ and $T'$ (for example that $T$ is a conservative extension of $T'$) can be proved in a weak theory. In the present case, it can in fact be proved in PA, using a different proof than that given, that EA conservatively extends PA, so EA is said to be reducible to PA as far as arithmetical theorems are concerned. In proof theory, various such reductions are established and studied for different theories.

But now let us note that if we do accept the notion of arithmetical sets of natural numbers, and therewith take the theory EA to have a model, interpreting the set variables as varying over arithmetical sets, it seems an arbitrary restriction not to allow induction for formulas that quantify over sets. This leads to a stronger theory.

## 10.3. Arithmetical analysis and arithmetical truth

The theory which is here called "Arithmetical analysis" has no agreed name in the vernacular, but is often designated ACA, standing for "arithmetical comprehension axiom". ACA has the same comprehension axioms as EA, but instead of the induction axiom of EA it has an induction schema, allowing induction on any condition in the language. Thus ACA differs from EA—which is also referred to as $ACA_0$ - in allowing induction over arbitrary properties (expressible in the language). And indeed if we accept the notion of an arithmetically definable set of natural numbers, there is no obvious reason why we should not accept proofs by induction involving quantification over such sets.

Quite a large part of analysis can be formalized in ACA, but in this book the chief role of ACA is as a theory for formalizing the semantics of arithmetic. Let us first consider how we can define arithmetical truth using only arithmetical sets.

What we need to do is to define $True(\phi)$ so that we can prove in ACA

(T1) If $\phi$ is an atomic sentence, $True(\phi)$ if and only if $True_{vf}(\phi)$.
(T2) If $\phi$ is $\psi_1 \vee \psi_2$, $True(\phi)$ if and only if $True(\psi_1)$ or $True(\psi_2)$.
(T3) If $\phi$ is $\neg\psi$, $True(\phi)$ if and only if not $True(\psi)$.
(T4) If $\phi$ is $\exists x\psi$, $True(\phi)$ if and only if $True(\psi_x(\overline{n}))$ for some $n$.

Once we have defined *True* in such a way that (T1)–(T4) are provable in ACA, an induction proof shows that every instance of the Tarski schema

$$True(\overline{\phi}) \equiv \phi$$

is provable in ACA. In fact we can prove corresponding equivalences for formulas containing free variables. Thus if $\phi$ contains the free variable $x$, we can prove the parametrized Tarski equivalence

$$\forall x(True)(sub(num(x), \overline{x}, \overline{\phi}) \equiv \phi)$$

Note that proving the Tarski equivalences presupposes that we can use the predicate *True* in induction proofs, and since *True* is defined using quantification over sets, this can only be done in ACA, not in EA. Again, since quantification over sets is used in the definition of *True*, we can't prove the existence of the *set* of true arithmetical statements in ACA. This is as it should be, since we know that this set is not arithmetical, whereas only the existence of arithmetical sets can be proved in ACA.

To carry out the definition of *True* in ACA, we first define a binary predicate $True(X, m)$ meaning that $X$ is the set of true sentences in the set $M_m$ of sentences with at most $m$ occurrences of $\neg$, $\vee$, $\exists$, by the following explicit definition, where $X(n)$ stands for the set $\{y \mid \langle n, y\rangle \in X\}$:

> *True*$(X, m)$ if and only if there is a set $Y$ such that $X$ is $Y(m)$, $Y(0)$ is the set of true atomic sentences, and for every $i < m$, and every arithmetical sentence $\phi$, $\phi \in Y(i+1)$ if and only if
>
> $\phi \in Y(i)$, or $\phi$ is in $M_{i+1}$ and
>
> $\phi$ is $\neg\psi$ where $\psi \notin Y(i)$, or
>
> $\phi$ is $\psi_1 \vee \psi_2$ where $\psi_1 \in Y(i)$ or $\psi_2 \in Y(i)$, or
>
> $\phi$ is $\exists x\psi$ where $\psi_x(\overline{n}) \in Y(i)$ for some $n$.

In ACA, we can now prove by induction that for every $m$ there is a unique $X$ such that *True*$(X, m)$, and we can then define *True*$(\phi)$ as $\exists X(\mathit{True}(X, f(\phi)) \wedge \phi \in X)$, where $f(\phi)$ is the number of occurrences of $\wedge$, $\vee$, $\exists$ in $\phi$. (T1)–(T4) then follow by induction on $f(\phi)$.

It is instructive to consider what happens if we try to define *True* in ACA using the set-theoretic version of the inductive definition of arithmetical truth given in §7.6. There should be a problem with this, because as noted in §6.2, the inductive definition of $M_R$ (here the set of true arithmetical sentences) is clearly correct only if $M_R$ itself is included among the sets $A$ quantified over in "for every set $A$ closed under $R$, $x$ is in $A$", and in ACA we cannot prove the existence of any non-arithmetical set. So suppose we define in ACA *True*$(x)$ to mean that $\langle x, 1\rangle$ belongs to every set $A$ such that

> $\langle \phi, 1\rangle$ belongs to $A$ for every atomic $\phi$ in $\mathit{True}_{Vf}$,
>
> $\langle \phi, 0\rangle$ belongs to $A$ for every atomic sentence $\phi$ not in $\mathit{True}_{Vf}$,
>
> $\langle \phi_1 \vee \phi_2, 1\rangle$ belongs to $A$ whenever $\langle \phi_1, 1\rangle$ or $\langle \phi_2, 1\rangle$ belongs to $A$,
>
> $\langle \phi_1 \vee \phi_2, 0\rangle$ belongs to $A$ whenever $\langle \phi_1, 0\rangle$ and $\langle \phi_2, 0\rangle$ belong to $A$,
>
> $\langle \neg\phi, 1\rangle$ belongs to $A$ whenever $\langle \phi, 0\rangle$ belongs to $A$,
>
> $\langle \neg\phi, 0\rangle$ belongs to $A$ whenever $\langle \phi, 1\rangle$ belongs to $A$,
>
> $\langle \exists x\phi, 1\rangle$ belongs to $A$ whenever $\langle \phi_x(\overline{n}), 1\rangle$ belongs to $A$ for some $n$,
>
> $\langle \exists x\phi, 0\rangle$ belongs to $A$ whenever $\langle \phi_x(\overline{n}), 0\rangle$ belongs to $A$ for every $n$.

This definition can be given in ACA, but in proving e.g., (T3) we need to carry out a proof by induction that no sentence is both true and false. To carry out this proof, we need to be able to prove the existence of the set of true sentences, which can't be done in ACA.

### Proving PA sound in ACA

We can formalize in ACA the proof by induction on theorems that every sentence provable in PA is true: all axioms are true, and the inference rules preserve truth, so every theorem is true.

That every axiom is true only requires using (T1)–(T4) to unwind what it means for a formula to be true. For example, an induction axiom

$$\phi_x(0) \wedge \forall x(\phi \supset \phi_x(s(x))) \supset \forall x \phi$$

is true if and only if

if $\phi_x(0)$ is true and for every $k$, if $\phi_x(\overline{k})$ is true then $\phi_x(s(\overline{k}))$ is true, then for every $k$, $\phi_x(\overline{k})$ is true

and this is itself easily proved by induction.

The proof that the inference rules preserve truth must take into account that the formulas involved in an inference are often not sentences. Thus what is proved by induction on derivations is a formalization of the statement

For every $\Gamma$ and $\phi$ such that all free variables occurring in $\phi$ or in any formula in $\Gamma$ are among $v_0, \ldots, v_n$, and for every sequence $\langle k_0, \ldots, k_n \rangle$, if $\Gamma \Rightarrow \phi$ and $\psi(\overline{k}_0, \ldots, \overline{k}_n)$ is true for every $\psi$ in $\Gamma$, then $\phi(\overline{k}_0, \ldots, \overline{k}_n)$ is true.

The inductive proof of this in ACA is straightforward, but requires that we formalize and prove by induction formalizations of such observations as

If the value of $t$ is $n$ and $\phi_x(t)$ is true then $\phi_x(\overline{n})$ is true.

While requiring some care to set out properly, these proofs are more or less automatic.

## Extending PA by a truth predicate

In a sense, the definability in ACA of arithmetical truth is its only essential property as an extension of PA. To see just what sense this is, we consider a new theory PA$^{True}$, obtained from PA by adding to the language of PA a unary predicate *True*, with *True*$(x)$ interpreted as "$x$ is a true arithmetical sentence", and to the axioms of PA the principles (T1)–(T4) for arithmetical sentences, together with all induction axioms for formulas in the extended language. We also add an axiom stating that *True*$(x)$ does not hold if $x$ is not an arithmetical sentence.

Since we can define *True* in ACA so as to make the new axioms of PA$^{True}$ provable in ACA, we see that every sentence in the language of arithmetic provable in PA$^{True}$ is also provable in ACA. But the converse also holds: every sentence in the language of arithmetic that is provable in ACA is provable in PA$^{True}$. This is a consequence of the fact that ACA can be interpreted using the language and axioms of PA$^{True}$, just as PA$^{True}$ can be interpreted using the language and axioms of ACA. For any formula $\phi$ of ACA, replace every occurrence in $\phi$ of a formula of the form $t \in X$ by a formalization of "$x$ is a formula in the language of PA with one free variable, and $x(t)$ is true". Also replace "$X = Y$" by a formalization of "$x$ and $y$ are formulas in the language of PA with one free variable, and for every $k$, $x(\overline{k})$ is true if and only if $y(\overline{k})$ is true". In other words, we interpret the set variables of ACA as ranging over formulas in the language of PA with one free variable, and we interpret "$k$ is a member of $X$" as "$k$ satisfies the formula $X$". On this interpretation, all the axioms of ACA become provable in $\mathrm{PA}^{True}$, and as a consequence every first order arithmetical theorem of ACA is provable in $\mathrm{PA}^{True}$.

The decisive step in verifying that this is the case concerns the comprehension axioms:

$$\text{(C)} \qquad \exists X \forall x (x \in X \equiv \phi)$$

Recall that in general $\phi$ may contain free first and second order variables, and the corresponding comprehension axiom is then the universal closure of the formula (C). To show that (C) is provable in $\mathrm{PA}^{True}$ when we interpret set variables as ranging over arithmetical formulas with one free variable, we use a straightforward induction on formulas, proving (C) first for atomic formulas $s = t$ or $t \in X$ and then for compound formulas $\neg\psi$, $\psi_1 \vee \psi_2$, $\exists x\psi$. Note that the inductive proof would break down if we admitted second-order quantification in $\phi$, since that would entail applying the truth predicate to sentences that themselves contain the truth predicate, whereas the truth predicate of $\mathrm{PA}^{True}$ applies only to arithmetical sentences.

## Defining truth for other languages

The proof of Tarski's theorem in §9.5 applies not only to theories with the language of PA, but equally to ACA and $\mathrm{PA}^{True}$. So if we need a formal treatment of truth for sentences in the language of ACA or $\mathrm{PA}^{True}$, we need to extend that language. This can again be done by starting with an inductive definition of truth for sentences in the language. If we consider the language of ACA, there is a difference compared to the language of PA or $\mathrm{PA}^{True}$, since in the latter case, sentences quantify only over natural numbers, and every natural number has a name in the language. Sentences in the second order language also quantify over sets, and there are no names for sets in the language. Indeed, if we take the set quantifiers to refer to arbitrary subsets of **N**, we can't introduce a name in the language for every set, since there are uncountably many such subsets. So in defining truth for the second order language, we need to go via a definition of the satisfaction relation (rather than the other way around, as has been done in the arithmetical case).

We define the relation "$\langle X_0, \ldots, X_n \rangle$ satisfies $\phi$", where $X_0, \ldots, X_n$ are sets and $\phi$ is a formula which has no free number variables and all of whose set variables free or bound are among $V_0, \ldots, V_n$, by the following clauses, where for simplicity (but without logical loss of generality) we restrict ourselves to formulas in which negation is applied only to atomic formulas:

If $s = t$ is a true atomic sentence, $\langle X_0, \ldots, X_n \rangle$ satisfies $s = t$.

If $s = t$ is a false atomic sentence, $\langle X_0, \ldots, X_n \rangle$ satisfies $\neg s = t$.
If $X_i = X_j$, $\langle X_0, \ldots, X_n \rangle$ satisfies $V_i = V_j$.
If $X_i \neq X_j$, $\langle X_0, \ldots, X_n \rangle$ satisfies $\neg V_i = V_j$.
If the value of $t$ is $n$ and $n$ is a member of $X_i$, $\langle X_0, \ldots, X_n \rangle$ satisfies $t \in V_i$.
If the value of $t$ is $n$ and $n$ is not a member of $X_i$, $\langle X_0, \ldots, X_n \rangle$ satisfies $\neg t \in V_i$.
If $\langle X_0, \ldots, X_n \rangle$ satisfies $\psi_1$ or $\psi_2$, $\langle X_0, \ldots, X_n \rangle$ satisfies $\psi_1 \vee \psi_2$.
If $\langle X_0, \ldots, X_n \rangle$ satisfies $\psi_x(\overline{n})$ for some $n$, $\langle X_0, \ldots, X_n \rangle$ satisfies $\exists x \psi$.
If $\langle X_0, \ldots, X_n \rangle$ satisfies $\psi_x(\overline{n})$ for all $n$, $\langle X_0, \ldots, X_n \rangle$ satisfies $\forall x \psi$.
If $\langle X_0, \ldots, X_n \rangle$ satisfies $\psi$ for some $X_i$, $\langle X_0, \ldots, X_n \rangle$ satisfies $\exists V_i \psi$.
If $\langle X_0, \ldots, X_n \rangle$ satisfies $\psi$ for every $X_i$, $\langle X_0, \ldots, X_n \rangle$ satisfies $\forall V_i \psi$.

Note that this definition applies whatever we take to be the range of the set quantifiers in the second order language, as long as we use the same range in the quantifiers "for some $X_i$", "for every $X_i$" in the truth definition.

When this inductive definition is formalized in the set-theoretical style, we need to quantify not only over subsets of **N**, but over sets of such subsets. Thus we need a *third order language*, in which we can talk about sets of subsets of **N**. This is not surprising, since we know that truth for sentences in the second order language cannot be defined in that language.

However, different definitions of truth are possible in the second order case, depending on how we interpret the set variables. In particular, if we restrict our attention to the interpretation of the set variables as ranging over *arithmetical* sets, we don't need to introduce any third-order variables in our language in order to define truth, but can distinguish instead between arithmetical sets, which we call sets of *order one*, and sets definable using the arithmetical language and quantification over arithmetical sets, which we call sets of *order two*. If we introduce corresponding set variables of order one and order two into our language for referring to sets of order one and sets of order two, we can define truth for sentences that contain only set variables of order one by quantifying over sets of order two, by just the same method that we used to define truth of arithmetical sentences by quantifying over sets of order one. The formalism becomes simpler if we consider instead the corresponding iterated truth theory $\mathrm{PA}^{True_1}$ obtained by extending the language of $\mathrm{PA}^{True}$ with a truth predicate $True_1$ for sentences in the language of $\mathrm{PA}^{True}$, and adding the induction axioms for the extended language, together with new truth axioms stating that for all sentences $\phi$ in the language of $\mathrm{PA}^{True}$,

If $\phi$ is an atomic arithmetical sentence, $True_1(\phi)$ if and only if $True(\phi)$.

If $\phi$ is an atomic sentence of the form *True*($t$), $True_1(\phi)$ if and only if the value of $t$ is a sentence $\psi$ in the language of PA, and *True*($\psi$).

If $\phi$ is $\psi_1 \vee \psi_2$, $True_1(\phi)$ if and only if $True_1(\psi_1)$ or $True_1(\psi_2)$.

If $\phi$ is $\neg\psi$, $True_1(\phi)$ if and only if not $True_1(\psi)$.

If $\phi$ is $\exists x\psi$, $True_1(\phi)$ if and only if $True_1(\psi_x(\overline{n}))$ for some $n$.

As in the case of $PA^{True}$ and ACA, the theory $PA^{True_1}$ is equivalent, as far as arithmetical theorems are concerned, to the second order theory obtained by having set variables of order one and order two, with corresponding comprehension axioms for sets of order one and order two. The process can be continued, and such iterated truth theories will be described in the final chapters of the book, where some complications are introduced into the formal description because the iteration is continued into the transfinite.

# CHAPTER 11

# THE RECURSION THEOREM AND ORDINAL NOTATIONS

## 11.1. The recursion theorem

In a definition of a function $f$ by course-of-values recursion (which we know can be reduced to primitive recursion), $f(n)$ is defined in terms of $f(k)$ for $k < n$. $f$ may have further arguments, but these are only carried along as parameters.

As noted in Chapter 9, not all definitions by recursion follow this pattern. The Ackermann function was used as an example of a definition by "double recursion". Another example, which can't be read as either primitive recursion or double recursion, is J. McCarthy's definition of "the 91 function":

(1) $F(x) = x - 10$ if $x > 100$, and otherwise $F(x) = F(F(x + 11))$

This looks pleasantly loopy, and it's not obvious what partial function it defines. In fact, although the definition of the 91 function is not a definition by either primitive or double recursion, it can be proved by an inductive argument (or by a lengthy computation) that the function it defines is primitive recursive: $F(x) = x - 10$ for $x > 100$, and $F(x) = 91$ for $x \leq 100$.

But now we need to consider just what we mean by saying that a set of equations defines a function. In §4.6, a pair of primitive recursion equations were said to define a unique function $f$, given $g$ and $h$:

$$f(0, a_1, \ldots, a_k) = h(a_1, \ldots, a_k)$$
$$f(b + 1, a_1, \ldots, a_k) = g(b, a_1, \ldots, a_k, f(b, a_1, \ldots, a_k))$$

That these equations define a unique function was explained as meaning that there *exists* a unique function satisfying the equations (for all values of $b$ and the parameters $a_1, \ldots, a_k$). But at the same time it was emphasized that the equations can be used to *compute* the value of $f(b, a_1, \ldots, a_k)$ for any given arguments. Similarly, the equations defining the Ackermann function in §9.3 were presented as defining in the sense of characterizing a particular function from $\mathbf{N}^3$ to $\mathbf{N}$, but also as a set of equations that can be used to compute the value of the function for given arguments. In considering more general definitions such as (1), which in general define *partial* functions, we need to separate these two aspects.

Considered as a description of an algorithm, a definition such as (1) is interpreted as an instruction to pick, given an argument $n$, the appropriate branch in the definition, and after a test either compute $F(n)$ as $n - 10$ (if $n > 100$) or else first compute, using the same algorithm, $F(n + 11) = m$ and then compute $F(n)$ as $F(m)$. This is a *recursive* algorithm, meaning that it contains an instruction to apply the very algorithm that is being defined. Most programming languages accept recursively formulated algorithms like (1). Euclid's algorithm was presented in §2.3 both in an explicitly recursive formulation—with equations such as $(k, m) = (k, m - k)$—and in a formulation using the phrase "continue in this way until ... ", which also corresponds to a standard construction in programming languages.

Considered as a specification of an algorithm, then, (1) defines a certain partial function, namely the function $F$ for which $F(n)$, if defined, is the value computed by the algorithm described in (1). As it happens, this function is in the case of (1) also the unique function that satisfies (1) when it is read as an *assertion* about $F$. On this second reading, the meaning of (1) is the assertion "for every $x$, if $x > 100$, $x$ is in the domain of $F$ and $F(x) = x - 10$, while if $x > 100$, $x$ is in the domain of $F$ if and only if $x + 11$ is in the domain of $F$ and $F(x + 11)$ is in the domain of $F$, and in that case $F(x) = F(F(x + 11)))$". This assertion is true of the function computed by the algorithm described in (1), and it is the only function for which the assertion is true.

In general, however, it is not the case that the function computed by a set of equations (with conditions, such as "if $n > 100$", giving different possible branches in the computation) read as an algorithm is the unique function satisfying those equations read as an assertion. Consider the following:

(2) $G(x) = 2$ if $x = 0$, and otherwise $G(x) = G(x + 1) * (G(x + 1) + 1)$

There is a unique total function satisfying (2), namely the function $G$ for which $G(n) = 0$ for all $n > 0$, and $G(0) = 2$. However, the function computed by (2) read as an algorithm is undefined for all $x > 0$, since the attempt to compute $G(x)$ by first computing $G(x + 1)$ never terminates.

Another example is

(3) $$G(x) = G(x + 1)$$

Clearly infinitely many total functions (those that have a constant value) satisfy (3), but the function computed by (3) is not defined anywhere.

The function computed by the algorithms associated with (1), (2), (3) is in fact in each case the *smallest* partial function satisfying the equations for all values of the variable $x$. "Smallest" here means smallest in the sense of set inclusion, so that $F \leq G$ if the domain of $F$ is a subset of the domain of $G$, and $F(n) = G(n)$ for every $n$ in the domain of $F$. In other words, the graph of $F$

is a subset of the graph of $G$. Thus this is the function $G$ given by an inductive definition of the graph of $G$. This inductive definition can be extracted from the corresponding equations and conditions in a systematic way.

The specifications (1), (2), and (3) have only been used by way of example. To generalize these examples, and put the whole discussion on a firm footing, we would need to give a proper definition of the allowable specifications and explain how to use such a specification to define a computable partial function by induction. This is a natural procedure to follow in many contexts, for example in discussing the semantics of programming languages. Here we will take a different route, and prove instead a basic existence theorem, Kleene's (second) recursion theorem, which implies the *existence* of at least one computable partial function satisfying assertions of the kind exemplified by (1), (2), (3), and many more besides.

> **Recursion theorem:** For any computable partial $F$ there is an $e$ such that
>
> (4) $$\{e\}(k) = F(e,k) \text{ for every } k,$$
>
> and in fact there is a primitive recursive function $K$ such that if $F = \{f\}$, the equation (4) is satisfied for $e = K(f)$.

Note that the meaning of (4) is that for any $k$ such that $F(e,k)$ is defined, $\{e\}(k)$ is also defined, and has the value $F(e,k)$, and for other $k$, $\{e\}(k)$ is undefined. We can apply the recursion theorem to the formulations used in (1), (2), and (3) to conclude that there is at least one $G$ satisfying (1), (2), (3) read as assertions, but we don't learn anything from the recursion theorem about the number of such $G$, and no particular algorithm for computing such a $G$ is given as part of the formulation of the recursion theorem. Thus we must give special arguments using the definition of $F$ and the equation (4) if we want to draw any further conclusions about $G$.

To see how the recursion theorem yields partial functions satisfying such descriptions as (1), (2), (3), take (1) as a typical example. To say that $F = \{f\}$ satisfies (1) read as a condition on $F$ is to say that $\{f\}(k) = H(f,k)$ for every $k$, where $H$ is defined by

$$H(e,n) = n - 10 \text{ if } n > 100, \text{ and otherwise } \{e\}(\{e\}(x+11))$$

But the recursion theorem can be used with any arbitrary computable $F$. For example, we can conclude from the recursion theorem that there is a computable total function $\{e\}$ satisfying the equation

(5) $$\{e\}(k) = e + k$$

The equation (5) can't be read as an algorithm in any immediately apparent way, and it isn't obvious why there should be any computable function $F$ which for a given argument $k$ outputs the sum of $k$ and an index for $F$. But the recursion theorem tells us that there is such an $F$.

The proof of the theorem is short but baffling. Choose $h$ so that $\{h\}(m,n) = F(s_1^1(m,m),n)$ and let $e = s_1^1(h,h)$. Then $\{e\}(k) = \{s_1^1(h,h)\}(k) = \{h\}(h,k) = F(s_1^1(h,h),k) = F(e,k)$. $h$ can itself be chosen as $s_1^1(b,f)$ where $\{b\}(k,m,n) = \{k\}(s_1^1(m,m),n)$, so we get $e$ as a primitive recursive function of $f$.

We will have occasion to use the fact that this proof is formalizable in PA. More precisely we will use the following version of this observation: There is a primitive recursive function $K$ such that

$$\forall x \forall y(\{K(x)\}(y) = \{x\}(K(x),y))$$

is provable in PA.

The recursion theorem has been formulated and proved above for the case of unary partial functions, but the formulation and proof carry over with obvious changes to the case of $n$-ary partial functions. There are many other versions of the recursion theorem (see Smullyan [1993]), but the above will suffice for the arguments in this book.

The recursion theorem is often used in combination with the $s_n^m$-functions. As a typical example, we may observe in the context of an argument that by the recursion theorem, there is, for any $a, b$, a function $\{e\}$ where $e = E(a,b)$ for some primitive recursive $E$, such that

$$\{e\}(n) = F(a,b,e,n) \text{ for every } n.$$

This typically occurs in a context in which we treat $a, b$ as inert parameters—they are carried along in the argument, but nothing is really done with them—and are only interested in $e$, which we say is obtained "primitive recursively" from $a$ and $b$, and in the argument $n$. The justification for this treatment derives from the recursion theorem. The function $E$ is defined by $E(a,b) = K(s_1^2(f,a,b))$, where $K$ is as above, and $f$ is an index for the function $F$. Thus it is provable in PA that

$$\begin{aligned}\{E(a,b)\}(n) &= \{K(s_1^2(f,a,b))\}(n)\\ &= \{s_1^2(f,a,b)\}(E(a,b),n) = F(a,b,E(a,b),n).\end{aligned}$$

In this book, it is only in some proofs in Chapter 14 that the recursion theorem will be used intensively. In Chapter 13 it will be used to establish the existence of recursive progressions of theories. Several of these applications will use the recursion theorem for ordinal notations to be formulated below, by which we can define primitive recursive functions.

## 11.2. Ordinal notations

Chapter 6 gave an axiomatic characterization of the countable ordinals as a generalization of the natural numbers. In order to formalize in arithmetic assertions about specific infinite sequences of theories, counted using ordinals, we also need to generalize the *notation* used to refer to natural numbers. For each natural number $k$, we have a numeral $\overline{k}$ which is used in formulas to refer to $k$. For infinite ordinals, the whole subject of notations becomes more complicated.

In the case of the formal numerals denoting natural numbers, we began by speaking of the terms "0", "$s(0)$", "$s(s(0))$" as expressions built up from symbols, and only in Chapter 7 were these expressions identified, for theoretical purposes, with certain natural numbers. In introducing notations for countable ordinal numbers, we go directly to the theoretical treatment, and define a system of notation for ordinals—Kleene's system **O**—in which the notations are certain natural numbers rather than expressions in any ordinary sense. It is not immediately clear how more familiar systems of notation, using symbols like "$\omega$" and "+", can be related to the system **O**. This should become clear in §11.3 when we consider in detail how to represent ordinals smaller than $\varepsilon_0$ and (using more complicated notations) ordinals smaller than $\Gamma_0$ and show how the resulting notations can be regarded as a part of the system **O**.

The definition of **O** involves the particular formal representation of computable functions used in this book. Thus if we were to use some other such representation, a different set of natural numbers would result as notations for ordinals. However, as in other similar cases (such as the precise mathematical definition of the terms and formulas of a language), the relevant *structure* of **O** is independent of these choices. It's irrelevant whether 23940293844293498049348888880001 is a member of **O** or not; what matters are the mathematical and logical properties of **O**, and these are the same whatever standard representation of the computable functions we use.

Deviating somewhat from tradition,[1] we will use $suc(a)$, which we define simply as $\langle a, 0\rangle$ and $lim(e)$, which is defined as $\langle e, 1\rangle$, in defining by simultaneous induction the class **O** of ordinal notations, a partial ordering relation $<_{\mathbf{O}}$ between such notations, and the ordinal $|a|$ denoted by the notation $a$:

> 0 is in **O**, and $|0| = 0$.
>
> If $a$ is in **O** then $suc(a)$ is in **O**, $a <_{\mathbf{O}} suc(a)$ and $|suc(a)| = |a| + 1$.
>
> If $e$ is the index of a total function and $\{e\}(n) <_{\mathbf{O}} \{e\}(n+1)$ for every $n$, then $lim(e)$ is in **O**, $\{e\}(n) <_{\mathbf{O}} lim(e)$ for every $n$, and $|lim(e)|$ is the supremum of the ordinals $|\{e\}(n)|$ for $n = 0, 1, \ldots$.
>
> If $a <_{\mathbf{O}} b$ and $b <_{\mathbf{O}} c$ then $a <_{\mathbf{O}} c$.

To see what all this means, consider first the finite ordinals. These have unique notations in **O**, namely the numbers $0, suc(0), suc(suc(0)), \ldots$. We will denote the notation in **O** for a natural number $n$ by $n_{\mathbf{O}}$. For these notations, $m_{\mathbf{O}} <_{\mathbf{O}} n_{\mathbf{O}}$ just means that $m < n$. So for finite ordinals, their notations and the relation $<_{\mathbf{O}}$ are isomorphic to the ordinals themselves and $<$. In general,

[1] Kleene uses 1 instead of 0, $2^a$ instead of $suc(a)$, and $3 * 5^e$ instead of $lim(e)$.

however, we can only prove, by induction, that $<_{\mathbf{O}}$ is a strict *partial* ordering of the elements in **O**. If $a <_{\mathbf{O}} b$ then $|a| < |b|$, but the converse does not hold.

The reason for this complication emerges when we consider limit ordinals. As was emphasized in §6.3, every limit ordinal is the supremum of infinitely many fundamental sequences of ordinals. In giving a notation to a limit ordinal $\alpha$ in the system **O**, we choose an index $e$ for an effective enumeration $\{e\}(0), \{e\}(1), \ldots$ of notations for the ordinals in some such fundamental sequence, and give $\alpha$ the notation $lim(e)$. There will always be infinitely many choices of $e$, both because there are infinitely many fundamental sequences for $\alpha$ and because every computable function has infinitely many indices. Thus $\omega$, the first limit ordinal, is named in the system **O** by $lim(e)$ for infinitely many $e$. For each such $e$, we get a corresponding series of notations $lim(e), suc(lim(e)), suc(suc(lim(e))), \ldots$ for the ordinals $\omega, \omega + 1, \omega + 2, \ldots$, and after that we again have infinitely many choices of notation for $\omega + \omega$. If $lim(e)$ and $lim(e')$ are two different notations for $\omega$, the notations $lim(e)$ and $suc(lim(e'))$ are *incomparable* with respect to $<_{\mathbf{O}}$, even though for the corresponding ordinals, $|lim(e)| < |suc(lim(e'))|$. Thus the elements of **O**, partially ordered by $<_{\mathbf{O}}$, form what is called a *tree*, branching off into infinitely many directions at each notation for a limit ordinal.

Given a natural number $n$, there is no algorithm for deciding whether or not it belongs to **O**. We can check whether $n$ is 0, or has the form $suc(m)$ or $lim(e)$, but when it comes to checking whether a number $e$ is the index of a total function that yields a sequence of notations, there is no algorithm for doing this. Indeed, although we will have no occasion to prove or use this fact in the book, the set **O** is in a sense more undecidable than any set that we've encountered so far, and there is for example a primitive recursive function $f$ such that an arithmetical sentence $\phi$ is true if and only if $f(\phi)$ is in **O**. Thus a way of determining whether a number belongs to **O** would give a way of determining whether a given arithmetical statement is true. It follows that the relation $<_{\mathbf{O}}$ is not decidable either, since $a$ is in **O** if and only if $a <_{\mathbf{O}} suc(a)$. This undecidability is a necessary consequence of **O** being a *universal* system of notation, in a sense that will be illustrated below. More restricted systems of notation can differ from **O** in these respects: they can assign a unique notation to ordinals, and they can have a decidable set of notations and a decidable relation between notations meaning "the ordinal named by $a$ is smaller than the ordinal named by $b$". These systems of notation can, for theoretical purposes, be regarded as parts of **O**, as we shall see.

An inductive argument shows that every ordinal $\alpha$ that is smaller than $|b|$ for some $b$ in **O** is itself $|a|$ for some $a$ in **O** such that $a <_{\mathbf{O}} b$. On the other hand, since **O** is countable, there are only countably many ordinals that are given a notation by **O**, that is, that are $|a|$ for some $a$ in **O**. Thus the ordinals given at least one notation (and therefore, except for the finite ordinals, infinitely many notations) by **O** are precisely those that are smaller than a certain countable ordinal $\omega_1^{CK}$ (for "Church-Kleene"). They are known as the *constructive* ordinals.

The above simultaneous inductive definition included **O**, $<_{\mathbf{O}}$ and the operation taking $a$ to $|a|$ in a single package. It is instructive to also consider these components separately. An important property of $<_{\mathbf{O}}$ is that there is an effectively enumerable relation $<_{\mathbf{K}}$ such that $<_{\mathbf{K}}$ and $<_{\mathbf{O}}$ coincide for members of **O**. $<_{\mathbf{K}}$ is inductively defined by

$a <_{\mathbf{K}} suc(a)$ for every $a$,

$\{e\}(n) <_{\mathbf{K}} lim(e)$ for every $n$ such that $\{e\}(n)$ is defined,

if $a <_{\mathbf{K}} b$ and $b <_{\mathbf{K}} c$ then $a <_{\mathbf{K}} c$.

To see that $<_{\mathbf{K}}$ is effectively enumerable, note that $a <_{\mathbf{K}} b$ if and only if there is a sequence $a_0, \ldots, a_n$ beginning with $a$ and ending with $b$, where for every $i < n$ either $a_{i+1}$ is $suc(a_i)$ or $a_{i+1}$ is $lim(e)$ and $a_i$ is $\{e\}(n)$ for some $n$. Since this definition uses only effectively enumerable relations, conjunction, disjunction, existential quantification, and bounded universal quantification, it can be expressed by a $\Sigma$-formula, and so itself defines an e.e. relation. By an easy inductive argument (in two directions), we find that for $a, b$ in **O**, $a <_{\mathbf{O}} b$ if and only if $a <_{\mathbf{K}} b$.

Using the relation $<_{\mathbf{K}}$, we can define **O** by induction. Through the general set-theoretical interpretation of inductive definitions this shows that "$n$ is in **O**" can be defined by a condition of the form $\forall X \psi(X, x)$, where $\psi$ is an arithmetical formula, that is, has no bound set variables. Specifically, $n$ is in **O** if and only if $n$ belongs to every set $X$ such that

0 is in $X$,

if $a$ is in $X$ then $suc(a)$ is in $X$,

if $\{e\}$ is total and for every $n$, $\{e\}(n) <_{\mathbf{K}} \{e\}(n+1)$ and $\{e\}(n)$ is in $X$, then $lim(e)$ is in $X$.

Thus $<_{\mathbf{K}}$ can be defined in arithmetic, and **O** can be defined in the language of second order arithmetic. The relation $a <_{\mathbf{O}} b$ can now be defined as "$a$ and $b$ are in **O** and $a <_{\mathbf{K}} b$", or equivalently as "$b$ is in **O** and $a <_{\mathbf{K}} b$".

To define the ordinal $|a|$ named by $a$ notation $a$, we need an inductive definition involving a bit more set theory, namely the theory of ordinals:

$|0|$ is 0,

if $|a|$ is $\alpha$ then $|suc(a)|$ is $\alpha + 1$,

if $lim(e)$ is in **O** and for every $n$, $|\{e\}(n)|$ is $\alpha_n$, $|lim(e)|$ is the supremum of the $\alpha_n$.

Now let's take a closer look at the set $\mathbf{O}_b$ of $a$ such that $a <_{\mathbf{O}} b$, for $b$ in **O**. This subset of **O** gives a notation to every ordinal smaller than $|b|$, and in fact a unique notation, for we can prove by induction that for $a_1, a_2$ in $\mathbf{O}_b$, $a_1 <_{\mathbf{O}} a_2$

if and only if $|a_1| < |a_2|$. So $<_{\mathbf{O}}$ (and thus $<_{\mathbf{K}}$) restricted to $\mathbf{O}_b$ is an effectively enumerable total ordering, and in fact a well-ordering, and $\mathbf{O}_b$ is an effectively enumerable set (since it is identical with the set of $a$ such that $a <_{\mathbf{K}} b$). The relation $<_{\mathbf{K}}$ is not well-founded as a relation on $\mathbf{N}$, and indeed we know from the recursion theorem that there are $e$ such that $\{e\}(n) = lim(e)$, so that $<_{\mathbf{K}}$ is not even irreflexive.

Can we in some way obtain unique notations for all constructive ordinals by choosing suitable members of $\mathbf{O}$? Yes, but only by a more abstract set-theoretical approach. A *branch* of $\mathbf{O}$ is a subset $B$ of $\mathbf{O}$ that is totally ordered by $<_{\mathbf{O}}$, that is, for which $a <_{\mathbf{O}} b$ or $b <_{\mathbf{O}} a$ for any different elements $a, b$ of $B$, and which is closed under $<_{\mathbf{O}}$, so that $a \in B$ whenever $a <_{\mathbf{O}} b$ and $b \in B$. A branch gives a unique notation to every ordinal smaller than the length of the branch, which is the supremum of the ordinals $|b|$ for $b$ in $B$. A *maximal* branch is a branch $B$ which is not a proper subset of any branch $B'$. It can be proved that there are maximal branches of length $\omega_1^{CK}$ (although they can not be effectively enumerable), and such a branch will thus give a unique notation in $\mathbf{O}$ to every constructive ordinal. Generally, a maximal branch has length at least $\omega^\omega$, and by simple cardinality considerations we can verify that there must be maximal branches of no greater length than $\omega^\omega$. However, in this book we will not study any of the intricate structure of $\mathbf{O}$ and its ordering, but only make use of some basic facts set out in this chapter.

## Defining functions on O by recursion

We know that if an ordinal $\alpha$ has a notation $a$, $\alpha + 1$ has the notation $suc(a)$. But can we assert more generally that whenever ordinals $\alpha$ and $\beta$ have a notation, so does their sum $\alpha + \beta$, and can we define an addition operation on notations corresponding to addition of ordinals?

For this and similar applications, we need to be able to define $F(a)$ using the values of $F(b)$ for $b <_{\mathbf{O}} a$. This can be done using the following

> **Recursion theorem for ordinal notations.** For any primitive recursive $G, H, J$ there is a primitive recursive $F$ such that it is provable in PA that for all $a, e, k_1, \ldots, k_n$
>
> $F(0, k_1, \ldots, k_n) = G(k_1, \ldots, k_n)$,
>
> $F(suc(a), k_1, \ldots, k_n) = H(a, k_1, \ldots, k_n, F(a, k_1, \ldots, k_n))$,
>
> $F(lim(e), k_1, \ldots, k_n) = J(e, b, k_1, \ldots, k_n)$,
>
> where $\{b\}(m) = F(\{e\}(m), k_1, \ldots, k_n)$ for every $m$ and $b$ is given as a primitive recursive function of $e, k_1, \ldots, k_n$.

Of course, since $F$ is primitive recursive, $F(n)$ is not only defined for ordinal notations $n$, and the above equations apply to $lim(e)$ for any number $e$. For a number $n$ which is not 0 and is not $suc(a)$ or $lim(e)$ for any $a$ or $e$, any convenient value can be given to $F(n)$. Note that $\{b\}(m)$ is not necessarily defined for every $m$, since $\{e\}(m)$ is not necessarily defined, but $F$ is always defined, since the index $b$ is given by a primitive recursive function.

To prove the theorem using the general recursion theorem, we first note that there is a partial recursive function $I$ such that for all $e, f, m, k_1, \dots k_n$

$$I(e, f, k_1, \dots, k_n, m) = \{f\}(\{e\}(m), k_1, \dots, k_n)$$

By setting $L(e, f, k_1, \dots, k_n) = s_1^{n+1}(i, e, f, k_1, \dots, k_n)$, where $i$ is an index for $I$, we then get a primitive recursive $L$ such that

$$\{L(e, f, k_1, \dots, k_n)\}(m) = \{f\}(\{e\}(m), k_1, \dots, k_n)$$

To define $F$, we conclude from the recursion theorem that there is an $f$ such that for all $m, k_1, \dots k_n$

$$\{f\}(m, k_1, \dots, k_n) = P(f, m, k_1, \dots, k_n) \tag{1}$$

where the partial function $P$ is defined by

$P(u, m, k_1, \dots, k_n) = H(a, k_1, \dots, k_n, \{u\}(a, k_1, \dots, k_n))$
if $m = suc(a)$,

$P(u, m, k_1, \dots, k_n) = J(e, L(e, u, k_1, \dots, k_n), k_1, \dots, k_n)$
if $m = lim(e)$,

$P(u, m, k_1, \dots, k_n) = G(k_1, \dots, k_n)$ otherwise.

Given that $f$ satisfies (1), we can prove by induction on $m$ that $F = \{f\}$ is total (since $a < suc(a)$). But how do we conclude that $F$ is primitive recursive? For this, we need to observe that $F$ satisfies

$F(m, k_1, \dots, k_n) = H(a, k_1, \dots, k_n, F(a, k_1, \dots, k_n))$
if $m = suc(a)$

$F(m, k_1, \dots, k_n) = J(e, L(e, f, k_1, \dots, k_n), k_1, \dots, k_n)$
if $m = lim(e)$

$F(m, k_1, \dots, k_n) = G(k_1, \dots, k_n)$ otherwise

and this is a definition by course-of-values recursion, using primitive recursive functions, and thus defines a primitive recursive function.

As an application, we can define an operation $a \oplus b$, by

$a \oplus 0 = a$,

$a \oplus suc(b) = suc(a \oplus b)$,

$a \oplus lim(e) = lim(b)$, where $\{b\}(n) = a \oplus \{e\}(n)$.

We can then prove that for $a$ and $b$ in $\mathbf{O}$, $a \oplus b$ is in $\mathbf{O}$, $|a \oplus b| = |a| + |b|$, and $a <_{\mathbf{O}} a \oplus b$ if $|b|$ is not 0. Other applications will be made in later chapters.

In the logical literature, there are many highly technical treatments of systems of ordinal notations, and the subject can get very hairy indeed. Fortunately, because the results needed for the philosophical conclusions of this book are simple and qualitative rather than difficult and quantitative, we don't need to know more about ordinal notations than has been stated in this section. But to get some feeling for the intricacies involved in notation systems it is a good idea to see how a couple of more restricted systems of notation can be regarded as subsystems of **O**. The systems presented have both been used and studied in classical work in proof theory, and working with them will also give the reader some appreciation of how the abstract set-theoretical notion of ordinal number, through systems of notation that can be defined in elementary terms, relates to the logical complexity of reasoning and the logical strength of theories.

## 11.3. Two restricted ordinal notation systems

### Notations for ordinals below $\varepsilon_0$

Cantor proved that any ordinal can be written in a unique way as a sum

$$\omega^{\gamma_1} + \omega^{\gamma_2} + \cdots + \omega^{\gamma_n}$$

where $\gamma_1 \geq \gamma_2 \geq \cdots \geq \gamma_n$. This is known as the *Cantor normal form* for ordinals, and is the foundation for the first restricted ordinal notation system to be defined in this section, which gives notations to the ordinals smaller than $\varepsilon_0$, the first $\alpha$ for which $\omega^\alpha = \alpha$. The argument below shows how to obtain the Cantor normal form for $\alpha < \varepsilon_0$ and use it as a notation for $\alpha$. Note that it is only for these $\alpha$ that we can use the Cantor normal form itself as a notation for $\alpha$. If we consider larger ordinals $\alpha$, we may find that the Cantor normal form for $\alpha$ is just $\omega^\alpha$, which is of no help in giving a notation to $\alpha$.

First we note that if $\beta \leq \alpha < \beta + \gamma$, there is a unique $\delta < \gamma$ such that $\alpha = \beta + \delta$. This is proved by a simple induction on $\gamma$.

So suppose $\alpha$ is smaller than $\varepsilon_0$. We will define a notation for $\alpha$ by recursion. Let $\beta$ be the smallest $\gamma$ such that $\alpha < \omega^\gamma$. Since $\alpha < \varepsilon_0$, $\beta \leq \alpha$. $\beta$ can't be a limit ordinal, so it's 0 or a successor. If $\beta$ is 0, $\alpha$ must be 0, and we give $\alpha$ the notation 0. Otherwise, let $\beta = \gamma + 1$. We then have that $\omega^\gamma \leq \alpha < \omega^{\gamma+1} = \omega^\gamma * \omega$. Now let $m$ be the smallest $k$ such that $\alpha < \omega^\gamma * k$. $m$ is greater than 1, so $m = n + 1$ for some positive $n$, and we have that $\omega^\gamma * n \leq \alpha < \omega^\gamma * n + \omega^\gamma$. It follows that there is a unique $\beta < \omega^\gamma \leq \alpha$ such that $\alpha = \omega^\gamma * n + \beta$. $\gamma$ and $\beta$ are both smaller than $\alpha$ and are uniquely determined, so assuming recursively that $\gamma$ and $\beta$ have been given notations $a$ and $b$, we can denote $\alpha$ by the expression $\omega^a \cdot n + b$ (now simply regarded as syntactic object), or by $\omega^a \cdot n$ if $\beta$ is 0. In this way every ordinal smaller than $\varepsilon_0$ is given as notation a unique expression which is either 0 or is of the form

$$\omega^{a_1} \cdot n_1 + \omega^{a_2} \cdot n_2 + \cdots + \omega^{a_m} \cdot n_m$$

where $n_1, \ldots, n_m$ are positive natural numbers and each $a_1, \ldots a_m$ is in turn an expression of this form, or 0. Thus these notations, from a typographical point of view, may contain exponential towers of any height.

Now let's consider how to treat these notations in arithmetic. For greater theoretical simplicity, we do away with the factors $n_1, \ldots, n_m$ and just consider the above expression as a sequence number $\langle a_1, a_1, \ldots, a_m \rangle$ where each $a_i$ occurs $n_i$ times in succession. A first attempt to arithmetically define the set $E$ of notations in this system might be to define it by primitive recursion as follows:

> $a$ is in $E$ if and only if $a$ is 0 or is a sequence $\langle a_1, \ldots, a_k \rangle$, where each $a_i$ is in $E$.

But here we must take into account that in the representation of an ordinal $\alpha$ as $\omega^{a_1} \cdot n_1 + \omega^{a_2} \cdot n_2 + \cdots + \omega^{a_m} \cdot n_m$ in this system, the ordinals represented by $a_1, \ldots, a_m$ form a strictly decreasing sequence. So not every element of $E$ as defined above does in fact represent a notation for an ordinal.

To get around this, we define $E$ as above but call it the set of *prenotations*. We then define an ordering relation $<_E$ on prenotations by primitive recursion:

> For $a$ and $b$ in $E$, $a <_E b$ if and only if $a$ is 0 and $b$ is not, or $a$ is $\langle a_1, \ldots, a_m \rangle$ and $b$ is $\langle b_1, \ldots, b_n \rangle$ and either $m < n$ and $a_i = b_i$ for $1 \leq i \leq m$, ($a$ is a *proper initial subsequence* of $b$) or else $a_i <_E b_i$, where $i$ is the smallest $j$ such that $a_j \neq b_j$.

We then define the set $E_0$ of notations as containing 0 together with those prenotations $\langle a_1, \ldots, a_k \rangle$ for which $a_i \geq_E a_{i+1}$ holds for every $i$ from 1 to $k-1$. The motivation for the definition of $<_E$ lies of course in the fact that for notations $a$ and $b$, the ordinal denoted by $a$ is smaller than the ordinal denoted by $b$ if and only if $a <_E b$.

The procedure may be easier to understand as a definition by simultaneous induction of the set $E_0$ and the relation $<_E$:

> 0 is in $E_0$.
>
> If $a_1, \ldots, a_k$ are in $E_0$ and $a_i \geq_E a_{i+1}$ for every $i$ from 1 to $k-1$, $\langle a_1, \ldots, a_k \rangle$ is in $E_0$ and $0 <_E \langle a_1, \ldots, a_k \rangle$.
>
> If $\langle a_1, \ldots, a_k \rangle$ and $\langle a_1, \ldots, a_k, a_{k+1}, \ldots, a_m \rangle$ are in $E_0$, $\langle a_1, \ldots, a_k \rangle <_E \langle a_1, \ldots, a_k, a_{k+1}, \ldots, a_m \rangle$.
>
> If $\langle a_1, \ldots, a_k, a_{k+1}, \ldots, a_m \rangle$ and $\langle a_1, \ldots, a_k, b_{k+1}, \ldots, b_n \rangle$ are in $E_0$ and $a_{k+1} <_E b_{k+1}$,
> $$\langle a_1, \ldots, a_k, a_{k+1}, \ldots, a_m \rangle <_E \langle a_1, \ldots, a_k, b_{k+1}, \ldots, b_n \rangle.$$

The first version, which first defines prenotations, then the ordering of prenotations, and then the notations proper, has the advantage of making it easy to show that $E_0$ and $<_E$ are primitive recursive.

We have, then, a system of notations for the ordinals smaller than $\varepsilon_0$, which assigns an ordinal $|a|$ smaller than $\varepsilon_0$ to every notation $a$, and a unique notation to every ordinal smaller than $\varepsilon_0$. Further, both the set $E_0$ of notations and the relation $<_E$ are primitive recursive, and thus in particular definable in the language of arithmetic. $<_E$ is a well-ordering of $E_0$, since $a <_E b$ if and only if $|a| < |b|$. (However, $<_E$ is not a well-ordering of the set $E$ of prenotations.)

If we forget about the origin of $<_E$ and just look at its arithmetical definition, how can we prove that $<_E$ is a well-ordering of $E_0$? That $<_E$ is a *total* ordering of $E$ and therefore of $E_0$ can be proved in PA by ordinary well-founded induction: given two different prenotations $a = \langle a_1, \ldots, a_k \rangle$ and $b = \langle b_1, \ldots, b_n \rangle$ such that neither is a proper initial segment of the other, find the first $i$ for which $a_i \neq b_i$. By the induction hypothesis, $a_i <_E b_i$ or $b_i <_E a_i$, implying that $a <_E b$ or $b <_E a$, by the definition of $<_E$. But how do we prove the further condition that $<_E$ is well-founded when restricted to the set $E_0$?

Since the statement "$<_E$ is well-founded on $E_0$" cannot be formulated in the language of first order arithmetic, we cannot ask whether it is provable in PA. But we can ask whether every application of $<_E$-induction can be uniformly reduced to ordinary mathematical induction, the way well-founded induction on $<$ can be uniformly reduced to ordinary mathematical induction. Equivalently, we can ask whether "$<_E$ is well-founded on $E_0$" is provable in EA, since EA is a conservative extension of PA.

In ACA, we can prove that $<_E$ is well-founded on $E_0$ as follows. Reverting to the notation $\omega^{a_1} \cdot n_1 + \omega^{a_2} \cdot n_2 + \cdots + \omega^{a_m} \cdot n_m$, we define the *height* of a notation $a$ in $E_0$ by $height(0) = 0$ and $height(\omega^a \cdot n + b) = height(a) + 1$. Thus the height of $a$ is the number of levels of exponentiation in the first term of the formal sum $a$. If $height(a) < height(b)$, $a <_E b$, although the converse does not hold. We now prove by induction on the smallest height of an element in $X$ that every non-empty $X$ contains a $<_E$-smallest element. If the smallest height of an element in $X$ is 0, $X$ certainly contains a $<_E$-smallest element, namely 0. Otherwise, by the induction hypothesis, there is a $<_E$-smallest notation $a_0$ among the $a$ such that

$$\exists m(\omega^a \cdot m \in X \vee \exists b(\omega^a \cdot m + b \in X)),$$

and by the smallest number principle there is a smallest number $m_0$ among the $m$ such that

$$\omega^{a_0} \cdot m \in X \vee \exists b(\omega^{a_0} \cdot m + b \in X).$$

If $\omega^{a_0} \cdot m_0 \in X$, this is the $<_E$-smallest element of $X$. Otherwise we choose, again by the induction hypothesis, $a_1$ and $m_1$ similarly minimal so that

$$\omega^{a_0} \cdot m_0 + \omega^{a_1} \cdot m_1 \in X \vee \exists b(\omega^{a_0} \cdot m_0 + \omega^{a_1} \cdot m_1 + b \in X).$$

If $\omega^{a_0} \cdot m_0 + \omega^{a_1} \cdot m_1 \in X$, we again have the $<_E$-smallest element of $X$. We define inductively in this way a strictly $<_E$-decreasing sequence $a_0, a_1, \ldots$. Since $a_0$ has height smaller than the smallest height of an element in $X$, there is by the induction hypothesis a $<_E$-smallest $a_i$. For this $a_i$, we must have

$$\omega^{a_0} \cdot m_0 + \omega^{a_1} \cdot m_1 + \cdots + \omega^{a_i} \cdot m_i \in X,$$

and this will be the $<_E$-smallest element of $X$.

This proof can't be carried out in EA, since it uses induction to prove a statement involving quantification over sets. However, we can use the proof to see that for every $n$, induction with respect to the notations of height smaller than $n$ can be carried out in EA (and thus in a uniform way in PA). That is, for every $n$, "for every $X$, if $X$ contains a notation of height $n$, then $X$ contains a $<_E$-smallest notation" can be proved in EA. To prove this by induction on $n$, we define $a_0, a_1, \ldots$ as above, but instead of invoking an induction hypothesis in EA to conclude that there is a $<_E$-smallest $a_i$, we apply in EA the proof of $<_E$-induction for smaller heights which exists by the induction hypothesis in our reasoning about provability in EA.

Although the proof given cannot be formalized in EA, it is of course conceivable that some other proof would establish general $<_E$-induction in EA, and even if there is no such proof in EA, it is conceivable that everything provable in PA extended with a principle of $<_E$-induction is provable in PA itself, although not uniformly. In fact this is not the case. There are arithmetical statements provable using $<_E$-induction that cannot be proved in PA. This fact will not be used or proved in this book, but two examples will be given.

One example is the statement "PA is consistent", which can be proved using $<_E$-induction in combination with some elementary arithmetical reasoning. More precisely, through methods pioneered by Gerhard Gentzen, the consistency of PA can be proved in primitive recursive arithmetic (described in §9.2) extended with the principle of $<_E$-induction. Adding the principle of $<_E$-induction to PA is equivalent, by a theorem of Kreisel and Lévy, to adding the uniform reflection principle which will be introduced in Chapter 14.

Another example of a theorem provable using $<_E$-induction which is not provable in PA is *Goodstein's theorem*, the demonstration of which can be relied on as a successful party trick in a suitable environment. Here is the recipe.

To write a natural number $q$ in *hereditary base k* is to write it as the value of an expression

$$n_1 \cdot k^{a_1} + n_2 \cdot k^{a_2} + \cdots + n_m \cdot k^{a_m} = q$$

where $a_1, a_2, \ldots, a_m$ are also such expressions, and $a_1 > a_2 > \cdots > a_m$ (when we compute the ordinary arithmetical value of the expressions), while $n_1, n_2, \ldots, n_m$ are positive integers smaller than $k$. The implicit base case in this inductive definition occurs when $q$ is 0, and 0 expressed in hereditary base $k$ is just the expression 0. (Thus other numbers $q$ smaller than $k$ are written $q \cdot k^0$)

The *Goodstein sequence* with initial base $k$ and starting value $n$ is a sequence of numbers determined by the following rules. The first number in the sequence is $n$, and the first base is $k$. To compute further bases and numbers in the sequence, two alternating operations are applied. The first consists in subtracting 1 from the latest number in the sequence (as long as this number is greater than 0), and the result of this operation becomes the next number (with the base unchanged). The second operation consists in writing the latest number in the sequence in hereditary base $k$, where $k$ is the current base, and then replacing $k$ everywhere in the resulting expression by $k+1$. The value of this expression is the next number in the sequence, and $k+1$ is the new base.

For example, the Goodstein sequence with initial base 2 and starting value 10 begins with the numbers

$$10, 9, 82, 81, 1024, 1023, 9843, 9842, 140744, 140743, \ldots$$

and the complete Goodstein sequence with initial base 3 and starting value 5 is

$$5, 4, 5, 4, 5, 4, 4, 3, 3, 2, 2, 1, 1, 0.$$

This latter sequence ends, since it reaches 0. The reason why 0 is reached is that the base after a few steps is larger than the latest value in the sequence, so that the only operation affecting the value of the last number in the remaining steps is subtraction by 1. The result of replacing $k$ by $k+1$ in the number written in hereditary base $k$ does not give a new value, since the base no longer appears in the expression (except with exponent 0).

Since the value of an expression written in hereditary base $k$ increases considerably when $k$ is replaced by $k+1$ in every other case, we see that the only case in which a Goodstein sequence terminates is when the base becomes larger than the latest number in the sequence. After this, a sequence of subtractions will eventually lead to 0.

It's easy to see that this "eventually" must mean "after a great many steps" in many cases. For example, the first six numbers in the Goodstein sequence with starting value 20 and initial base 2 are

20,

19,

7625597484991,

7625597484990,

13407807929942597099574024998205846127479365820592393377723561443721764030073546976801874298166903427690031858186486050853753882811946569946433649006084100,

13407807929942597099574024998205846127479365820592393377723561443721764030073546976801874298166903427690031858186486050853753882811946569946433649006084099, and the next number in the sequence has more than two thousand digits. That nevertheless every Goodstein sequence does eventually terminate can be seen by assigning notations in $E_0$ to numbers written in hereditary base $k$. Suppose the numerical expression is

$$n_1 \cdot k^{a_1} + n_2 \cdot k^{a_2} + \cdots + n_m \cdot k^{a_m}$$

To this we assign the corresponding ordinal notation

$$\omega^{b_1} \cdot n_1 + \omega^{b_2} \cdot n_2 + \cdots + \omega^{b_m} \cdot n_m$$

where the $b_i$ are similarly obtained from the $a_i$. 0 is assigned the ordinal notation 0. Note that the ordinal notation is the same whatever the base $k$, so when $k$ is replaced by $k+1$ in the numerical expression, the corresponding ordinal notation remains the same. What happens when we subtract 1 from the value of the numerical expression? If $a_m = 0$, the numerical expression in the next step will be

$$n_1 \cdot k^{a_1} + n_2 \cdot k^{a_2} + \cdots + (n_m - 1) \cdot k^{a_m}$$

and the corresponding ordinal notation is

$$\omega^{b_1} \cdot n_1 + \omega^{b_2} \cdot n_2 + \cdots + \omega^{b_m} \cdot (n_m - 1)$$

which is smaller in the $<_E$-ordering than the first notation. The decisive case is when $a_m$ is different from 0. We will then subtract 1 from a power of $k$, and need to write the result in hereditary base $k$ (and in the next step replace $k$ by $k+1$ in the resulting expression). The relevant algebraic identity is

$$\begin{aligned} n \cdot k^a - 1 &= (n-1) \cdot k^a + k^a - 1 \\ &= (n-1) \cdot k^a + (k-1)k^{a-1} + (k-1)k^{a-2} + \cdots + (k-1) \cdot a^0 \end{aligned}$$

where the term $(n-1) \cdot k^a$ falls away if $n = 1$. When we write the result of the subtraction in hereditary base $k$ we apply the above rule, first to $n \cdot k^a - 1$, then recursively to the exponents $a-1, a-2 = (a-1)-1, \ldots$. The corresponding ordinal notation is

$$\omega^b \cdot (n-1) + \omega^{b_1} \cdot (k-1) + \omega^{b_2} \cdot (k-1) + \cdots + \omega^0 \cdot (k-1)$$

which again is smaller in the $<_E$-ordering than the notation for the original. Thus Goodstein's theorem follows by $<_E$-induction.

## $E_0$ as a part of O

There is a notation $a$ in **O** for $\varepsilon_0$ such that $E_0$ and $<_E$ can be identified with $\mathbf{O}_a$ and $<_\mathbf{O}$ restricted to $\mathbf{O}_a$, as we shall now see.

First note that "$a$ is a notation in $E_0$ for 0", "$a$ is a notation in $E_0$ for a successor ordinal", and "$a$ is a notation in $E_0$ for a limit ordinal" are all primitive recursive conditions

Next, there is a primitive recursive function *pred* such that for any $a$, if $a$ is the notation for a successor ordinal $\beta + 1$ in $E_0$, $pred(a)$ is a notation in $E_0$ for $\beta$. We can define *pred* by

$$pred(a) = \langle a_1, \ldots, a_k \rangle \text{ if } a = \langle a_1, \ldots, a_k, 0 \rangle,\ pred(a) = 0 \text{ otherwise.}$$

Similarly there is a primitive recursive function *fundseq* such that if $a$ is the notation for a limit ordinal $\alpha$ in $E_0$, $fundseq(a)$ is an index for a computable total function $F$ such that $F(0), F(1), \ldots$ is a sequence of notations in $E_0$ corresponding to a fundamental sequence for $a$, that is, $|F(0)|, |F(1)|, \ldots$ is a strictly increasing sequence of ordinals with $\alpha$ as limit. *fundseq* is defined by course-of-values recursion: if $a = \langle a_1, \ldots, a_k \rangle$, where $a_k$ is the notation for a successor ordinal, $fundseq(a)$ is an index for the $F$ defined by

$$F(0) = \langle a_1, \ldots, pred(a_k) \rangle,$$
$$F(n+1) = F(n) * \langle pred(a_k) \rangle$$

(where $*$ is the concatenation operator defined in §4.6). If $a_k$ is the notation for a limit ordinal, $fundseq(a)$ is an index for the $F$ defined by

$$F(n) = \langle a_1, \ldots, \{fundseq(a_k)\}(n) \rangle.$$

An index for $F$ is obtained as a primitive recursive function of $a$ as in the proof of the recursion theorem.

Finally, we note that $E_0$ is what is called a *univalent* system of notation, meaning that no ordinal is assigned more than one notation in $E_0$.

Using these functions we can associate each notation $a$ in $E_0$ with a notation $G(a)$ in $\mathbf{O}$ for the same ordinal as follows. By the recursion theorem, there is a $G$ such that

$G(0) = 0$,

if $a$ is the notation for a successor ordinal, $G(a) = suc(G(pred(a)))$,

if $a$ is the notation for a limit ordinal, $G(a)$ is $lim(e)$, where $e$ is an index for the total $F$ such that $F(n) = G(\{fundseq(a)\}(n))$,

otherwise, $G(a) = 0$.

A proof by $<_E$-induction shows that $G$ is total, $|a| = |G(a)|$, and for $a, b$ in $E_0$, $a <_E b$ if and only if $G(a) <_{\mathbf{O}} G(b)$. Thus we can regard $E_0$ as a part of $\mathbf{O}$.

The general definition of a system of notation for ordinal numbers (as given by Kleene) only requires that we can effectively enumerate the notations for 0, for successor ordinals, and for limit ordinals, and that there are functions corresponding to *pred* and *fundseq* above. The embedding of $E_0$ in $\mathbf{O}$ given also makes use of the fact that $E_0$ is univalent. For suppose we have a system of notation which gives infinitely many notations $a_0, a_1, a_2, \ldots$ to $\omega$. We can then have a notation for $\omega + \omega$ with a corresponding fundamental sequence

$$a_0, suc(a_1), suc(suc(a_2)), \ldots$$

and this will not yield a corresponding notation for $\omega + \omega$ in **O**. However, the construction can be modified in this case to yield an embedding of an arbitrary system of notation in **O**, which is therefore called a *universal* system of notation for the constructive ordinals.

## Notations for ordinals below $\Gamma_0$

Recall from §6.3 that if we define $F$ by

$F_0(\alpha) = \omega^\alpha$,

for $\beta > 0$, $F_\beta(\alpha) =$ the $\alpha$-th common fixed point of all the functions $F_{\beta'}$ for $\beta' < \beta$,

$\Gamma_0$ is defined as the smallest ordinal $\beta$ such that $F_\beta(0) = \beta$. Note that $\varepsilon_0$ is $F_1(0)$, while $F_2(0)$ is the first $\alpha$ such that $\alpha$ is the $\alpha$-th epsilon number. After taking a few more steps in this direction, we are bound to feel that $\Gamma_0$ is "much larger" than $\varepsilon_0$. From a logical point of view, this is reflected in the greater complexity of the ordering of the ordinals smaller than $\Gamma_0$, and the greater amount of mathematics needed to prove that it is a well-ordering. To put this in more concrete terms, we need to introduce a system of notations for the ordinals smaller than $\Gamma_0$.

This is actually not much more difficult than for $\varepsilon_0$. We can no longer rely only on the Cantor normal form, but as will be seen we can express every ordinal smaller than $\Gamma_0$ in a suitable way as a sum of the form

$$F_{\beta_1}(\gamma_1) + \cdots + F_{\beta_n}(\gamma_n)$$

where

$$F_{\beta_1}(\gamma_1) \geq \cdots \geq F_{\beta_n}(\gamma_n)$$

So suppose $\alpha < \Gamma_0$. If $\alpha$ is 0 we again give it the denotation 0. Otherwise, since $\alpha < F_\alpha(0)$ and is thus not a fixed point of $F_\beta$ for every $\beta < \alpha$, let $\beta$ be the smallest $\beta$ such that $\alpha$ is not a fixed point of $F_\beta$. This means that $\alpha < F_\beta(\alpha)$. We now have two cases. If $\beta = 0$, $\alpha < \omega^\alpha$, and we can argue as in the $\varepsilon_0$-case to find uniquely determined $k > 0$, $\gamma < \alpha$ and $\delta < \omega^\gamma \leq \alpha$ such that $\alpha = \omega^\gamma * k + \delta = F_0(\gamma) + F_0(\gamma) + \cdots + \delta$. By recursion applied to $\delta$ and $\gamma$, we get a notation for $\alpha$ as a formal sum. In the case $\beta > 0$, it follows that $\alpha$ is a fixed point of $F_\gamma$ for every $\gamma < \beta$, which means that $\alpha = F_\beta(\delta)$ for some $\delta < \alpha$. In either case, we get a unique well-defined notation for $\alpha$ as a formal sum of notations for a decreasing sequence of smaller ordinals. For example, $\varepsilon_0 + 1$, the first ordinal not given a notation in $E_0$, is given the notation $F_0(F_1(0)) + F_0(0)$.

Again reducing these notations to numbers, as in the $E_0$-case, we get the following set $G$ of prenotations:

0 is in $G$,

if all of $a_1, \ldots, a_n, b_1, \ldots, b_n$ are in $G$, $\langle\langle a_1, b_1\rangle, \ldots, \langle a_n, b_n\rangle\rangle$ is in $G$.

To define the set $G_0$ of notations proper, we now need to define the ordering relation $<_G$ between notations corresponding to the ordering of the ordinals they denote. For this, we need to consider when $F_\beta(\gamma) < F_{\beta'}(\gamma')$ holds. If $\beta = \beta'$, since each $F_\beta$ is a normal function, this reduces to $\gamma < \gamma'$. If $\beta < \beta'$, $F_\beta(F_{\beta'}(\gamma')) = F_{\beta'}(\gamma')$, since by definition $F_{\beta'}(\gamma')$ is a fixed point of $F_\beta$ for every $\beta < \beta'$, so $F_\beta(\gamma) < F_{\beta'}(\gamma')$ if and only if $F_\beta(\gamma) < F_\beta(F_{\beta'}(\gamma))$, and thus if and only if $\gamma < F_{\beta'}(\gamma)$. Similarly if $\beta > \beta'$, $F_\beta(\gamma) < F_{\beta'}(\gamma')$ if and only if $F_\beta(\gamma) < \gamma'$. This gives us the following definition by course-of-values recursion of $<_G$ for prenotations $a$ and $b$:

> $a <_G b$ if and only if $a = 0$ and $b \neq 0$, or $a$ is $\langle\langle a_1, b_1\rangle, \ldots, \langle a_n, b_n\rangle\rangle$ and $b$ is $\langle\langle c_1, d_1\rangle, \ldots, \langle c_m, d_m\rangle\rangle$ and $a$ is a proper initial subsequence of $b$, or for the first elements $\langle a_i, b_i\rangle, \langle c_i, d_i\rangle$ where they differ, either $a_i = c_i$ and $b_i <_G d_i$ or $a_i <_G c_i$ and $b_i <_G \langle c_i, d_i\rangle$ or $c_i <_G a_i$ and $\langle a_i, b_i\rangle <_G d_i$.

$G_0$ can now be defined as the set of prenotations $\langle\langle a_1, b_1\rangle, \ldots, \langle a_n, b_n\rangle\rangle$ for which $\langle a_{i+1}, b_{i+1}\rangle \leq_G \langle a_i, b_i\rangle$ for $i = 1, \ldots, n-1$.

As in the case of $E_0$ and $<_E$, $<_G$ can be proved in PA to be a total ordering of the prenotations, though by a more involved inductive argument. That $<_G$ is a well-ordering of the notations in $G_0$ is provable in CA, but not in ACA. In a sense to be clarified in later chapters, the proof that $<_G$ is a well-ordering of $G_0$ requires us to go beyond anything that is implicit in our accepting the axioms of PA as true.

Like the notations in $E_0$, those in $G_0$ can be represented as the notations a in **O** that satisfy $a <_\mathbf{O} b$ for a particular $b$ in **O**. The method of doing this is just the same as in the case of $E_0$. We need only observe that it is decidable whether a notation $a$ in $G_0$ is the notation for 0, a successor ordinal, or a limit ordinal, and that primitive recursive functions can be defined that yield a notation in $G_0$ for the predecessor of $|a|$, or an index for a computable function that enumerates notations in $G_0$ for a fundamental sequence with $|a|$ as limit.

CHAPTER 12

# THE INCOMPLETENESS THEOREMS

## 12.1. Gödel's first incompleteness theorem without the Gödel sentence

Gödel's proof of the first incompleteness theorem for a theory $T$ uses what is known as a *Gödel sentence* for $T$. This is a sentence which formalizes "This sentence is not provable in $T$". The proof that a Gödel sentence for $T$ is not decidable in $T$ (assuming some soundness condition on $T$) is memorable and pleasing to the imagination. However, this proof has contributed to a fairly widespread impression that incompleteness is essentially tied to such rather strange self-referential sentences, and that the incompleteness theorem applies only to theories in which self-referential sentences can be formalized. This impression is incorrect. In this book, for reasons explained in §8.6, we are only concerned with effectively axiomatizable theories. For such theories, it is easy to see on very general grounds that they cannot correctly decide every arithmetical statement. For we know (see §9.3) that there are effectively enumerable sets $A$ of natural numbers which are not decidable. (By the Matiyasevic-Robinson-Davis-Putnam theorem, we know in fact that $A$ can be chosen as the set of numbers $k$ such that a particular Diophantine equation $p(k, x_1, \ldots, x_m) = 0$ has a solution.) This means that the set of true statements of the form "the natural number $k$ does not belong to $A$" is not effectively enumerable. Since the set of statements of this form which can be proved in $T$ *is* effectively enumerable for effectively axiomatizable $T$, it follows that infinitely many statements of this form are not correctly decided in $T$, but are either false and provable or true and unprovable. So if $T$ is $\Sigma$-*sound*, meaning that it does not prove any false $\Sigma$-statements, there are infinitely many such statements that are undecidable in $T$. Note that this conclusion does not depend on any assumption that self-referential statements can be formalized in $T$.

In this chapter we will nevertheless begin by giving Gödel's original proof of the first incompleteness theorem. The reason is that this proof leads directly to the second incompleteness theorem, which is the foundation for all that follows in the rest of the book.

Throughout the chapter, it will be assumed unless otherwise stated that $T$ is an effectively axiomatizable first or second order extension of PA. The incompleteness theorems can be proved in much greater generality, and we could in particular define a simple finitely axiomatized subtheory Q of PA such that the incompleteness theorem can be proved for every effectively axiomatizable extension of Q. In this book, however, our concern is only with theories at least as strong as PA.

## 12.2. Provability predicates

### Formalizing consistency statements

A derivation has been defined (in §8.6) as a sequence number $\langle s_0, \ldots, s_n \rangle$ where each $s_i$ is a pair $\langle t_i, \phi_i \rangle$ with $t_i$ the sequence number of a sequence of formulas, such that every $s_i$ in the sequence is related by one of the logical rules presented in Chapter 8 to zero, one, or two earlier pairs in the sequence. A proof of a formula $\phi$ in a theory $T$ is a derivation in which $s_n$ is $\langle t_n, \phi \rangle$, where every formula in the sequence $t_n$ is an axiom of $T$. So given an arithmetical formula $\psi$ defining the axioms of $T$, we can express "$x$ is a theorem of $T$" by a formula $Thm_T(x)$, formalizing "there is an $s$ such that $s$ is a derivation and the last element in $s$ is $\langle t, x \rangle$, where $\psi((t)_i)$ holds for every $i < length(t)$". We can then formalize "$T$ is consistent" as a formula $Con_T$, defined as $\neg\exists x(Thm_T(x) \wedge Thm_T(neg(x)))$, where $neg(x)$ is the negation of the formula $x$. (In our identification of formulas and numbers, as set out in §7.1, $neg(x)$ is $\langle 7, 7, x \rangle$.)

That $\psi$ defines the axioms of $T$ means only that $\psi(\overline{k})$ is true if and only if $k$ is an axiom of $T$—for emphasis, we sometimes express this by saying that $\psi$ *extensionally* defines the axioms of $T$—so all we can say about $Thm_T(x)$ given that $\psi$ defines the axioms of $T$ is that $Thm_T(x)$ in the same extensional sense defines the theorems of $T$. It is more difficult to say in what sense $Con_T$ is a statement expressing that $T$ is consistent. Because of the phenomenon of incompleteness, there are always different definitions of the axioms of $T$ which are not provably equivalent in $T$. This means that we get different definitions of the theorems of $T$ which are not provably equivalent in $T$, and thereby different formulations of $Con_T$ which are not in general provably equivalent in $T$. So which of these different $Con_T$ should we take to be the formalization of the assertion that $T$ is consistent?

In some cases the answer is determined by what description of the axioms of $T$ we ourselves have in mind in asserting that $T$ is consistent. As was noted in §9.1, for the theories that we actually use in formalizing mathematics there is a *canonical* description of their axioms, and there is then also a corresponding canonical or standard formula $\psi$ defining the axioms of $T$, yielding a standard definition $Thm_T(x)$ of the theorems of $T$ and a standard formalization $Con_T$ of the statement "$T$ is consistent". Any definition of the theorems or axioms of $T$ (and the resulting formulation of $Con_T$) which, although it defines the theorems or the axioms of $T$ in the extensional sense, is not equivalent in $T$ to the standard definition is called non-standard.

When we are talking about theories in general, there are no such canonical definitions of their axioms, and we are at once faced with a basic and recurring difficulty of the subject. The statement "for every effectively axiomatizable consistent extension $T$ of PA, there is a true $\Pi$-sentence which is not provable in $T$" does not have this problematic aspect. We can state and prove this assertion without having to consider any choice of definition of the axioms of an arbitrary consistent extension $T$ of PA. The statement "for any consistent extension $T$ of PA, it is not provable in $T$ that $T$ is consistent" is another matter. Here we must formulate "$T$ is consistent" as an arithmetical sentence, and this requires us to choose some definition of the axioms of $T$. If we allow any arbitrary definition, the statement is false, since it is then possible for $T$ to prove "$T$ is consistent". A trivial example is the following definition of the axioms of a theory $T$, modifying $A(x)$, which is any condition defining the axioms of $T$: "$A(x)$, and there is no derivation of a contradiction from the formulas $x$ such that $A(x)$". If $T$ is in fact consistent, this formula defines the axioms of $T$, and it is then provable in $T$ that $T$ is consistent (using the modified definition of the axioms of $T$). Since the mere existence of an axiom of $T$ logically implies that $T$ is consistent, we can argue in $T$ as follows: "Given a derivation of a contradiction from the axioms of $T$, this derivation can in fact use no axiom of $T$, but then pure predicate logic (the theory with the language of PA and no axioms) is inconsistent." But the consistency of pure predicate logic is easily proved in PA.

The definition of the axioms of $T$ used in this example is distinctly odd, since it is not even provable in $T$ for any axiom $\phi$ of $T$ that $\phi$ satisfies this definition. Since we are only concerned with effectively axiomatized theories, it is reasonable to introduce the requirement that the formula $\psi$ must *computably* define the axioms of $T$. However, an example given in Feferman [1961] shows that this is not sufficient to exclude pathological cases. Consider the following non-standard definition $A(x)$ of the axioms of PA: "$x$ is an axiom of PA (according to the standard definition) and there is no proof of any contradiction from those axioms of PA that are $\leq x$". $A(x)$ computably defines the axioms of PA in PA, because (as was commented on in §9.1) PA is *reflexive*, meaning that the consistency of any finitely axiomatized subtheory of PA is provable in PA, although this fact will not be proved in this book. Furthermore, the corresponding statement "PA is consistent" is again provable in PA, since we can now argue in PA as follows. Any derivation of an inconsistency in PA will use an axiom $x$ such that all other axioms used are $\leq x$. But then there is a derivation of an inconsistency from axioms of PA that are $\leq x$, so $x$ isn't an axiom (in the sense of the non-standard definition) after all.

Fortunately the situation is not as bad as it might seem on the basis of the above example, for it turns out that a sufficient condition for Gödel's theorem to be provable for a particular choice of $Thm_T(x)$ (and therewith of $Con_T$) is that the definition $\psi$ of the axioms of $T$ which is used in $Thm_T(x)$ is a *Σ-formula*. (This was established in Feferman [1961], using a less perspicuous characterization of these formulas.) The reason why this is so is, as we shall see, that the Σ-completeness theorem of §9.4 is not only true, but provable in PA. This suffices for the proof of Gödel's theorem (and extensions like Löb's theorem), but a Σ-definition of the axioms of a theory may still be "intensionally incorrect" in the sense that it includes irrelevant conditions. We will have occasion to consider this possibility closer in connection with Turing's completeness result for consistency extensions and Feferman's extension of that result, presented in Chapter 14.

Henceforth, when we are talking about theories $T$ in general, we assume that $Thm_T(x)$ and $Con_T$ are formalized as above, using some Σ-formula which defines the axioms of $T$. When we are talking about theories $T$ for which there is a canonical definition of the axioms of $T$, this definition will be assumed to be used if nothing is said to the contrary. Sometimes we will use the notation $Con(\psi)$, meaning $Con_T$ where the formula $\psi$ has been used to define the axioms of $T$.

## The deducibility conditions

We will use a special notation (borrowed from modal logic) for formalizations in PA of provability statements. If $\phi$ is a sentence, we write $\Box\phi$ for the sentence $Thm_T(\overline{\phi})$, where the theory $T$ is implicit. As always, it is assumed whenever necessary or convenient that $Thm_T(\overline{\phi})$ is the primitive formula obtained by eliminating defined symbols. $\Box\Box\phi$ accordingly is the formula $Thm_T(\overline{\psi})$, where $\psi$ is $Thm_T(\overline{\phi})$. In such iterated uses of the square, it is always assumed that the same theory is referred to throughout, and the same formula used to define the axioms of the theory.

Instead of $Con_T$, we will sometimes use the formula $\neg\Box\bot$, where $\bot$ is a special notation for the formula $\exists x(s(x) = 0)$. $Con_T$ and $\neg\Box\bot$ are equivalent in $T$, and in fact in PA, by a formalization of the proof by inspection of the rules of deduction: if $\phi$ and $\neg\phi$ are both provable in $T$, $\bot$ is provable by the rule [L9] of §8.2, and conversely, if $\bot$ is provable in $T$, $\neg\, Con_T$ follows since $\neg\bot$ is also provable, by the first successor axiom. (The proof of this equivalence does not depend on the axioms of $T$ being defined by a Σ-formula.)

We will also need to consider formalizations of statements about the provability of substitution instances of formulas. For this, we extend the interpretation of the square. If the formula $\phi$ is not a sentence, but contains free variables $x_1, \ldots, x_n$, $\Box\phi$ is the formula

$$Thm_T(subseq(\langle num(x_1), \ldots, num(x_n)\rangle, \langle \overline{x}_1, \ldots, \overline{x}_n\rangle, \overline{\phi}))$$

Thus $\Box\phi$ is a formalization in PA of the property of $n$ numbers $k_1, \ldots, k_n$ that we would informally express as "$\phi(\overline{k}_1, \ldots, \overline{k}_n)$ is provable in $T$".

Using this notation, we can express the basic properties of the provability predicate needed for Gödel's theorem (including the unprovability of consistency). Essentially these properties, known as the *deducibility conditions*, were formulated by Hilbert and Bernays [1939] in their careful working out of the details of the proof of the second theorem.

(D1) If $\phi$ is a theorem of $T$, so is $\Box\phi$, for every sentence $\phi$.

(D2) $\Box\phi \supset \Box\Box\phi$ is a theorem of $T$ for every sentence $\phi$.

(D3) If $\Box(\phi \supset \psi)$ is a theorem of $T$, so is $(\Box\phi \supset \Box\psi)$, for all sentences $\phi$ and $\psi$.

## Proving the deducibility conditions

Given that the axioms of $T$ are defined using a $\Sigma$-formula $\psi$, (D1)–(D3) all hold. To verify this, we first need to note that if $\psi$ is a $\Sigma$-formula, $\Box\phi$ is equivalent in PA (and therefore in $T$) to a $\Sigma$-formula, and we can therefore assume that $\Box\phi$ is a $\Sigma$-formula and $Con_T$ a $\Pi$-formula. (D1) is then a consequence of the $\Sigma$-completeness theorem. (D3) is proved by a formalization in $T$ of the obvious argument: if $s$ is a proof in $T$ of $\phi \supset \psi$ and $s'$ of $\phi$, we get a proof of $\psi$ by suitably combining $s$ and $s'$ and applying modus ponens.

The proof of (D2) follows the proof of the $\Sigma$-completeness theorem in §9.4. Although the proof of that theorem is formalizable in PA, we actually only need, for (D2), to show that for each $\Sigma$-formula $\phi$, $\phi \supset \Box\phi$ is provable in PA. (D2) then follows by specializing this to $\Sigma$-sentences of the form $\Box\phi$.

Note that $\phi$ is not necessarily a sentence, so $\phi \supset \Box\phi$ is the formula

(1) $\quad \phi \supset \mathit{Thm}_T(\mathit{subseq}(\langle \mathit{num}(x_1), \ldots, \mathit{num}(x_n)\rangle, \langle \overline{x}_1, \ldots, \overline{x}_n\rangle, \overline{\phi}))$

To show that (1) is provable in PA (and thereby in $T$) for every $\Sigma$-formula $\phi$, we proceed by induction on $\Sigma$-formulas, following the proof in §9.4 and using the fact that primitive recursive functions like *subseq* are computably defined in PA. Thus for the base case we need to show that (1) is provable in PA for atomic and negated atomic $\phi$. This is proved by induction in PA, as in the proof given in §9.4. Similarly for the induction step in the main induction. Consider for example the case of disjunction. Here we need to establish that the following is provable in PA:

(2) $\quad \phi_1 \vee \phi_2 \supset \mathit{Thm}_T(\mathit{subseq}(\langle \mathit{num}(x_1), \ldots, \mathit{num}(x_n)\rangle, \langle \overline{x}_1, \ldots, \overline{x}_n\rangle, \overline{\psi}))$

where $\psi$ is $\phi_1 \vee \phi_2$.

By the induction hypothesis, the following are provable:

(3) $\phi_1 \supset Thm_T\,(subseq\,(\langle num(\boldsymbol{x}_1),\ldots,num(\boldsymbol{x}_n)\rangle, \langle\overline{\boldsymbol{x}}_1,\ldots,\overline{\boldsymbol{x}}_n\rangle, \overline{\phi}_1))$

(4) $\phi_2 \supset Thm_T\,(subseq\,(\langle num(\boldsymbol{x}_1),\ldots,num(\boldsymbol{x}_n)\rangle, \langle\overline{\boldsymbol{x}}_1,\ldots,\overline{\boldsymbol{x}}_n\rangle, \overline{\phi}_2))$

To get (2) from (3) and (4) in PA we also need to prove in PA, where $\psi$ is $\phi_1 \vee \phi_2$:

(5)
$$\begin{aligned} &subseq\,(\langle num(\boldsymbol{x}_1),\ldots,num(\boldsymbol{x}_n)\rangle, \langle\overline{\boldsymbol{x}}_1,\ldots,\overline{\boldsymbol{x}}_n\rangle, \overline{\psi}) \\ &\quad = dis\Big(subseq\,(\langle num(\boldsymbol{x}_1),\ldots,num(\boldsymbol{x}_n)\rangle, \langle\overline{\boldsymbol{x}}_1,\ldots,\overline{\boldsymbol{x}}_n\rangle, \overline{\phi}_1), \\ &\qquad\qquad subseq\,(\langle num(x_1),\ldots,num(x_n)\rangle, \langle x_1,\ldots,x_n\rangle, \overline{\phi}_2)\Big) \end{aligned}$$

and

(6)
$$Thm_T(x) \vee Thm_T(y) \supset Thm_T(dis(x,y))$$

where $dis(x,y) = \langle 7,8,x,y\rangle$ is the disjunction of formulas $x$ and $y$. The proofs of (5) and (6) in PA are left to the reader, together with the remaining details. This proof by vigorous handwaving of (D2) is traditional, as was commented on in §1.2, and it should be clear at this point how the argument can be fleshed out ad libitum.

## 12.3. The Gödel sentence

A Gödel *sentence* $G_T$ for $T$ is the diagonalization of $\neg\, Thm_T(v_0)$. Thus, by §9.4,

(7)
$$G_T \equiv \neg\Box G_T \text{ is provable in PA.}$$

Since the definition of $Thm_T$ uses an arbitrary $\Sigma$-formula $\psi$ to define the axioms of $T$, there is no uniquely defined Gödel sentence for a theory $T$. Nor, as we shall see, are Gödel sentences associated with different such $\psi$ equivalent in $T$ in general. It is thus strictly speaking inappropriate to refer to "the" Gödel sentence , but we will do so anyway when differences between Gödel sentences are immaterial, and also when there is a canonical definition of the axioms of a theory.

We can now prove

**Gödel's first incompleteness theorem for $T$:** If $T$ is consistent, $G_T$ is not provable in $T$, and if $T$ is $\Sigma$-sound, $G_T$ is undecidable in $T$.

The proof is short. If $G_T$ is provable in $T$, so is $\Box G_T$ by (D1), and by (7) it follows that $\neg G_T$ is provable in $T$, so $T$ is inconsistent. If $\neg G_T$ is provable in $T$, either $T$ is inconsistent and thus not $\Sigma$-sound, or if $T$ is consistent, $\neg G_T$ is false, and again $T$ is not $\Sigma$-sound, since $\neg G_T$ is (equivalent in $T$ to) a $\Sigma$-formula.

The proof only uses the fact that $G_T$ is a fixed point of $\neg\, Thm_T(v_0)$, that is, that (7) holds. That a sentence $\phi$ is a fixed point of a formula $\psi(x)$ does not necessarily mean that $\phi$ is a formula which "says of itself that it satisfies $\psi(x)$". For example, if $\phi$ is the fundamental theorem of arithmetic and $\psi(x)$ the formula $x = \overline{\phi}$, $\phi$ is a fixed point of $\psi(x)$, but one wouldn't want to say that the fundamental theorem of arithmetic "says of itself that it is the formula $\phi$". The diagonal lemma produces a formula $\phi$ that does "say of itself that it satisfies $\psi(x)$" in the strong sense that $\phi$ is (the translation into primitive notation of) a sentence $\psi(t)$, where $t$ is a closed term such that the value of $t$ is $\phi$. Although the proof of the first incompleteness theorem only uses the fact that $G_T$ is a fixed point of $\neg\, Thm_T(v_0)$, it is only through the diagonal construction that we know of the existence of such a fixed point.

Gödel, instead of $\Sigma$-soundness, used a property called $\omega$-consistency in his formulation of the theorem. This original formulation will not play any role in the following. Rosser proved a significant strengthening of the first incompleteness theorem, by defining a variant $R_T$ of the Gödel sentence which is undecidable in $T$ on the assumption that $T$ is consistent. (See Smullyan [1992].) This again does not have any role to play in the main argument of this book, concerning inexhaustibility, and so will be set aside.

The deducibility conditions imply that the reasoning in the first incompleteness theorem can be carried out in $T$ itself:

$\Box(G_T \supset \neg\Box G_T)$ is provable in $T$ by (7) and (D1), so
$\Box G_T \supset \Box\neg\Box G_T$ is provable in $T$ by (D3), and
$\Box G_T \supset \Box\Box G_T$ is provable in $T$ by (D2), so
$\Box G_T \supset \Box\Box G_T \wedge \Box\neg\Box G_T$ is provable in $T$, so
$\Box G_T \supset \neg\, Con_T$ is provable in $T$.

Thus $Con_T$ implies $\neg\Box G_T$ in $T$, and thereby also $G_T$. So from the first incompleteness theorem we can conclude

**Gödel's second incompleteness theorem for $T$:** If $T$ is consistent, $Con_T$ is not provable in $T$.

The implication $G_T \supset Con_T$ is also provable in $T$, since "if $T$ is inconsistent, every formula is provable in $T$" is provable in $T$ by a simple argument. So in fact $G_T$ and $Con_T$ are equivalent in $T$.

Like the first incompleteness theorem, the second theorem has been strengthened, and in this case the strengthening will be very relevant in the remainder of the book:

**Löb's theorem:** If $\phi$ is a sentence for which $\Box\phi \supset \phi$ is provable in $T$, then $\phi$ is provable in $T$.

The original proof of Löb's theorem will be given below, but Kreisel found a simple argument using the second incompleteness theorem. If $\Box\phi \supset \phi$ is provable in $T$, the theory $T + \neg\phi$ obtained by adding $\neg\phi$ as a new axiom to $T$ proves $\neg\Box\phi$, which is equivalent in $T$ to $Con_{T+\neg\phi}$. Thus $T + \neg\phi$ proves its own consistency, and so is inconsistent, implying that $\phi$ is a theorem of $T$.

Implicit in this argument is that if the axioms of $T$ are defined by a formula $\psi$, those of $T + \neg\phi$ are defined by the formula $\psi \vee v_0 = \overline{\neg\phi}$. This kind of argument, showing that a theory is inconsistent by showing that it proves its own consistency, will be used in several later proofs.

Löb's theorem generalizes the second incompleteness theorem, since if we choose $\phi$ as $\perp$, the second incompleteness theorem results.

The proof of Löb's theorem given above makes essential use of the fact that theorems in $T$ can be proved using the classical rule of reductio ad absurdum. In this it differs from the proof of the second incompleteness theorem, which can be carried out for theories $T$ restricted to intuitionistic or minimal logic. It would be odd if this were not the case for Löb's theorem, and in fact Löb's original proof makes no use of negation. Suppose $\Box\phi \supset \phi$ is provable in $T$, and choose $\psi$ by the diagonal lemma so that

(8) $\psi \equiv (\Box\psi \supset \phi)$ is provable in $T$.

We'll show that $\psi$ is provable in $T$. From (8) we get that

(9) $\psi \supset (\Box\psi \supset \phi)$ is provable in $T$

and by (9), (D1), and two applications of (D3) it follows that

(10) $\Box\psi \supset (\Box\Box\psi \supset \Box\phi)$ is provable in $T$.

(D2) then gives that

(11) $\Box\psi \supset \Box\phi$ is provable in $T$

so by the assumption on $\phi$,

(12) $\Box\psi \supset \phi$ is provable in $T$,

and by (8) it follows that $\psi$ is provable in $T$. Therefore, $\Box\psi$ is provable in $T$, and (12) gives that $\phi$ is provable in $T$.

## Consistency proofs

The second incompleteness theorem tells us that $Con_T$ cannot be proved in $T$ itself. This is sometimes taken to mean that $Con_T$ can only be proved in a logically *stronger* theory than $T$, but a moment's reflection tells us that this is not the case. The consistency of CA, for example, is provable in the theory obtained by adding $Con_{CA}$ as a new axiom to PA, and the resulting theory is neither weaker, nor stronger than CA.

Of course a consistency proof for $T$ that consists in producing the axiom $Con_T$ is pretty pointless. In the history of logic in the twentieth century, consistency proofs have in general not been thought of as pointless, but as important for one reason or another. The point of view adopted here is the following. The consistency of the theories considered in this book is not problematic, and the problem regarding these theories posed by the second incompleteness theorem does not lie in finding convincing consistency proofs, but in coming to grips with the peculiar impossibility of formulating a theory of arithmetic that consistently asserts its own consistency. But in other cases, consistency proofs have an obvious interest. A consistency proof for $T$ shows that $T$ is *Π-sound*, that is, that every Π-sentence provable in $T$ is true. This is a consequence of the Σ-completeness theorem: if a Π-sentence $\phi$ is false, $\neg\phi$ is provable in $T$, so if $\phi$ is provable in $T$, $T$ is inconsistent. (However, as the second incompleteness theorem shows, consistency does not imply Σ-soundness, since PA $+\neg$ $Con_{\mathrm{PA}}$ is consistent although it proves a false Σ-sentence.) Thus one point of consistency proofs, in those cases where $T$ is not known to be sound, is to establish a limited soundness of $T$. When a proof that $T$ is consistent can be carried out within a theory $T'$, it also follows that every Π-sentence provable in $T$ is provable in $T'$, since the above argument showing that consistent theories are Π-sound is formalizable in $T'$ (for the $T'$ considered here). Thus, whether or not the soundness or consistency of a theory $T$ is in doubt, a consistency proof for $T$ can give us interesting information about what it takes to prove the Π-theorems of $T$.

The latter aspect of consistency proofs has been much generalized in proof theory, through the investigation of various possible ways of interpreting a theory $T$ in a theory $T'$ and thereby establishing that for some particular class of formulas $\phi$, $\phi$ is provable in $T'$ whenever $\phi$ is provable in $T$. (For a semi-popular survey, see Feferman [1993].) Here $T$ and $T'$ may be theories involving different kinds of abstractions—for example, in what is known as Gödel's Dialectica interpretation, first order arithmetic is interpreted in a theory of higher-order computable functions with only a limited form of logical reasoning.

In this book, except in the proof of Goryachev's theorem in Chapter 14, the only consistency proofs to be considered will be trivial consistency proofs whereby the consistency of $T$ follows from the observation that the axioms of $T$ are all true.

## 12.4. "Going outside the system"

It is sometimes said that the Gödel sentence $G_T$ for a theory $T$ has the peculiar property that it can only be seen to be true by "going outside the system". While a natural metaphor, this is in some ways misleading.

$G_T$ is a formalization of an odd self-referential sentence, and another such is the corresponding *Henkin sentence* $H_T$, which formalizes "This sentence is provable in $T$". That is, we get $H_T$ by applying the diagonalization lemma to $Thm_T(v_0)$ instead of $\neg\, Thm_T(v_0)$. By Löb's theorem, $H_T$ is in fact provable in $T$. It is not, then, necessary to "go outside the system" to prove $H_T$. To understand why $G_T$ is not provable in $T$ we are not helped by the reflection that $G_T$ formalizes an odd self-referential statement about provability in $T$

and for this reason can only be proved "from the outside", since we might say as much about $H_T$.

We might take the view that it is necessary to "go outside the system" even to *understand* $H_T$ or $G_T$, since they are only arithmetical sentences in virtue of a definition of syntactic objects as natural numbers. But in this they are no different from any of innumerable other mathematical statements which we assume formalized in PA or in CA or in axiomatic set theory ZFC, such as the fundamental theorem of arithmetic. We never actually write out these formalizations—which often simply can't be done in practice—or make any use of the identification of rational numbers with certain natural numbers, or of ordered pairs with certain sets of sets, in our mathematical reasoning. If "going outside the system" means to refer to sentences like $G_T$ only within a wider context, presupposing an understanding of concepts not explicitly axiomatized in PA, we are simply never "inside the system" in any realistic mathematical context.

$G_T$ is equivalent in $T$ to $Con_T$, and "$T$ is consistent" is not self-referential at all. If we forget about $G_T$ and consider instead $Con_T$, how are we to understand the unprovability of $Con_T$ in $T$? Here it is often emphasized that the resources of a theory $T$ do not themselves suffice to enable a proof of the consistency of $T$. Again it is only by "going outside the system" that one can prove that $T$ is consistent.

A weakness of this emphasis is that it doesn't take into account that the relevant concept of "proof" is a very liberal one. The consistency of $T$ is provable in the theory $T + Con_T$. This is not because any new fundamental principle has been introduced or because the theory $T + Con_T$ incorporates any new insight that goes beyond those expressed in $T$, but simply because the consistency of $T$ has been *postulated*. We don't require any more of a proof, as the term is used in logic. Accordingly, the second incompleteness theorem makes a stronger statement than one might naturally suppose. The consistency of $T$ not only cannot be derived from the basic principles embodied in $T$, it cannot even be consistently asserted in $T$. A theory cannot consistently *postulate* its own consistency. By the diagonal lemma, we can produce a formula $\phi$ formalizing "This sentence is consistent with $T$", but since $T + \phi$ then proves its own consistency, we know that in fact it is inconsistent.

Why is it impossible for $T$ to consistently postulate $Con_T$? Because a paradox results from such a postulate, or so Gödel's proof of the second theorem suggests. If $T$ asserts its own consistency, it must both assert and deny the provability of the sentence formalizing "This sentence is not provable in $T$". It's not just a matter of $T$ lacking the resources to establish a particular truth (that $T$ is consistent) but of it being impossible to consistently sneak in this truth as an assertion or postulate in the theory itself. Saying that one must "go outside the system" to prove the consistency of $T$ conveys the suggestion that $T$ metaphorically speaking has a kind of "blind spot", that it cannot reflect on or understand or inspect itself sufficiently to establish its own consistency—and indeed in extrapolations from the incompleteness theorem to other fields (religion, physics, psychology) this suggestion is frequently made explicit. The fact that $T$ cannot even consistently assert its own consistency, without attempting any inspection or justification whatever, would seem to indicate that this suggestion is a bit of a red herring. A better analogy to the inability of $T$ to truly affirm its own consistency might be our own inability to truly affirm that we never speak about ourselves. This inability is not due to any difficulty in analyzing, penetrating or justifying our own capacities or predilections, but stems from the logical fact that such an affirmation invalidates itself.

Another weakness of "going outside the system" is that it is rather too positive-sounding, suggesting some general method or approach. It is true that $Con_T$ can only be proved by "going outside the system" in the straightforward sense that it cannot be proved *within* the system, that is in $T$ itself. But in general we have no idea at all of any "going outside the system" by which $Con_T$ can be proved in any non-trivial sense. Goldbach's conjecture (that every even number greater than 2 is the sum of two primes) is true if and only if the theory obtained by adding the conjecture to PA as a new axiom is consistent, but this does not mean that we have any idea of any "going outside the system" whereby the consistency of $PA +$ Goldbach's conjecture, or equivalently the truth of the conjecture, might be proved.

It appears that the phrase "going outside the system" derives its impact from a particular case, namely that in which we accept the axioms of $T$ as true, and therefore conclude that $T$ is consistent and its Gödel sentence true. It is doubtful whether "going outside the system" is a very good metaphor in such a case. It may be argued that we are not bringing to bear any new insight not contained in the axioms of $T$ when concluding that $T$ is consistent, but rather only making explicit what is implicit in accepting the axioms, and cannot be explicitly asserted in the system itself on pain of paradox. This point will be considered further in connection with reflection principles in Chapter 14.

The impossibility for a theory $T$ to correctly postulate its own consistency seems to contrast with the ability of humans to affirm their own consistency with impunity. The contrast is only apparent. The mathematical concepts used in the formulation and proof of the incompleteness theorem have no application to ordinary informal discourse. Formal theories $T$ have their formally defined theorems, and if among those theorems is a formalization of "$T$ is consistent", every formula is a theorem. Humans have no formally defined theorems at all, but argue, utter, question, assert, deny statements in response to a multitude of circumstances, and even the solemn affirmation of a logical contradiction one minute need not have any influence whatever on what they say the next minute. There is no reason to believe that computers can't be programmed to exhibit the same kind of behavior, but of course neither computers, nor humans, can formally specify a consistent $T$ that proves its own consistency.

The attempt is sometimes made to confront people with a logical conundrum in the form of a statement corresponding to the Gödel sentence which they cannot consistently either affirm or deny. For example, a "Gödel sentence" for the mathematician $P$ might be put forward:

$(G)$ $\quad$ $P$ will never, referring to $(G)$, describe it as a true sentence.

Suppose $P$ accepts the challenge, and begins by stating "I hereby describe $(G)$ as a true sentence, and thus refute it." In response to the claim that he is now being inconsistent, he can say that there is no inconsistency whatever in his statement. He has not said that $(G)$ is and is not true; he has "described it as a true sentence" in the sense of uttering the words "I hereby describe $(G)$ as a true sentence", but he has asserted no contradictory statements. To deal with this pettifoggery, his interlocutors might modify $(G)$:

$(G')$ $\quad$ $P$ will never accept $(G')$ as true.

$P$ may now ask just what counts as accepting a sentence as true. If this is explained in terms that do not refer to any overt action of $P$'s, but rather to his beliefs or internal state, indefinite wrangling will ensue, because of the fuzziness and doubtful aspects of the notions of belief and inner state, and because $P$ cannot even start to believe or disbelieve $(G')$ until the matter of the meaning of "accept" has been resolved. The only way of getting around this is by tying "accept" to some overt action independent of the confusion engendered by going into the meaning of $(G')$. Thus his interlocutor might switch to

$(G'')$ $\quad$ $P$ will never assert $(G'')$

where to assert a sentence is to utter it in the form of a declarative statement, perhaps prefacing it by "I assert the following". $P$ responds by uttering the words: "I assert the following: $P$ will never assert $(G'')$", also affirming that by "$(G'')$" he does indeed mean the sentence "$P$ will never assert $(G'')$". When now formally charged with inconsistency, $P$ blithely responds that his uttering those words was merely a means of refuting $(G'')$, and had nothing to do with any actual claim of his. His interlocutor might at this point wax legalistic, stipulating that to "assert $(G'')$" means to utter $(G'')$ in a context where it is understood that one will assume full responsibility for the content of the utterance as conventionally understood, on pain of fine or imprisonment, as in a court of law. $P$ may now in all sincerity affirm $(G'')$, since the circumstances of the present discussion do not conform to this description.

Details apart, the point of the above is to underline that it is not necessarily problematic, in ordinary discourse, to make apparently contradictory or absurd assertions that are themselves frivolous or metaphorical or devious or legalistic or otherwise guided by some modality other than the straightforwardly alethic. The Gödel sentence for a formal theory is not associated with any problem of just what is meant by "belief" or "acceptance" or "assertion", since the relevant concepts are formally defined in purely arithmetical terms, and there is only one modality involved, that of provability in $T$. In the case of $(G)$ and its variants, there simply are no such formal definitions. Whatever confusion a sentence like $(G)$ may engender in the minds of people, it does not serve to indicate any lacuna in our mathematical knowledge or understanding, and indeed has no apparent theoretical significance at all.

The inexhaustibility conundrum, as formulated in the introduction to this book, is not tied to any idea that the incompleteness theorem or its proof can sensibly be applied to human beings and their mathematical discourse (or to computers simulating such discourse). In particular, it is not assumed that there is any well-determined range of human arithmetical knowledge, or any well-defined totality of those arithmetical truths that are in principle accessible to human knowledge, to which Gödel's theorem can be applied. Rather, the significance in this context of the impossibility for a formal theory to consistently assert its own consistency lies in the resulting impossibility for us to come up with any formally defined set of arithmetical sentences which we can assert exhausts our arithmetical knowledge. This difficulty remains whether or not there is any such thing as the totality of humanly provable arithmetical truths.

At the same time, the inexhaustibility phenomenon does not automatically entail any way in which human beings "transcend Gödel's theorem" in the sense of having ways of arriving at the truth of arithmetical statements which cannot be simulated by algorithmic procedures. This aspect will be commented on in greater detail in Chapter 15.

## CHAPTER 13

# ITERATED CONSISTENCY

### 13.1. Consistency extensions

The second incompleteness theorem gives us a way to extend PA to a stronger theory $PA_1$, obtained by adding to PA a formula $Con_{PA}$, formalizing "PA is consistent". More precisely, what is added is $Con(\phi)$, with $\phi$ chosen as the canonical $\Sigma$-formula defining the axioms of PA. In general there is no canonical definition of the axioms of a theory $T$, and we say that $T + Con_T$ is a *consistency extension* of $T$ for any $Con_T$ formulated as in §12.2 using any $\Sigma$-formula that defines the axioms of $T$. Unless otherwise noted, $T$ will be assumed to be a consistent first order extension of PA with an effectively enumerable set of axioms, which we also refer to as a *numerical* theory, so such a $\Sigma$-formula exists, and a consistency extension of $T$ is always logically stronger than $T$, for consistent $T$. By an *arithmetical* theory will be meant a numerical theory which is an extension by definitions of a theory with the language of PA (and so, for theoretical purposes, may be assumed to be a theory formulated in the primitive language of arithmetic). Thus both arithmetical and numerical theories quantify only over the natural numbers, but in a numerical theory we may introduce functions and relations that are not arithmetically definable.

When we look at the axioms of PA from the point of view of ordinary mathematics, it is not at all obvious that there is any way of extending them, without extending the language, so as to get a logically stronger but still mathematically correct theory. And indeed the extension $PA_1$ goes off in a very odd direction from a mathematical point of view. It doesn't add any mathematical operations or properties of such operations that we've overlooked, and it doesn't give any sharper characterization of the natural numbers than that expressed in the induction schema of PA. Instead it adds a statement about the formal system PA itself, that it is consistent, or equivalently (as noted in §12.3) that it is $\Pi$-sound. It's far from clear whether this statement, even though unprovable in PA, has any new implications about the distribution of primes, the solutions of Diophantine equations, or anything else of traditional mathematical interest. In presenting his theorem, Gödel emphasized that the consistency statement did in fact imply new theorems of a simple and traditional mathematical *form*, and this observation has since been greatly strengthened through the Matiyasevic-Robinson-Davis-Putnam theorem (see §9.3), which shows that $Con_{PA}$ is equivalent in PA to the statement that a certain Diophantine equation has no solution. Such observations, however, only concern the form of the new theorems obtained, and leave it open whether there exist any examples of statements of actual independent mathematical interest that are implied by $Con_{PA}$. This question has been given much attention in recent years, but it will not be our topic here. Instead we will be concerned with the very features of the extension $PA_1$ that make it mathematically odd.

$PA_1$ is far from being a bold or striking extension of PA by new principles or interesting conjectures. It seems more correct to say (as in the comments by Gödel quoted in Chapter 1) that in accepting PA as a formalization of part of our mathematical knowledge we have already implicitly accepted $PA_1$. This view will be considered further in Chapter 14, in the discussion of reflection principles. There are two contrary and fairly widespread lines of thought which will be commented on here.

The first of these is the view that we don't strictly speaking *know* that PA is consistent, that the consistency of PA is problematic, and the step from PA to $PA_1$ is therefore not to be taken lightly. Such skeptical views will just be set aside in the following. A reader who is skeptically inclined should still be able to make good sense of the argument presented in the following chapters, as dealing with the apparent inexhaustibility of our mathematical knowledge that results if we *do* accept the axioms of arithmetic as true

The second contrary line of thought will be very relevant in the following. It consists in its basic form in taking the step from PA to $PA_1$ too easily, by remarking that since PA is consistent, $Con_{PA}$ is true, and so can be added as a new axiom. This may sound somewhat convincing until we reflect that the incompleteness theorem shows that there are consistent theories that prove their own inconsistency. Thus if we accept $PA_1$ on the basis of our accepting PA, our accepting PA cannot consist only in our accepting that it is consistent, since we cannot from "PA is consistent" conclude "$PA_1$ is consistent". (This is also emphasized in Pudlák [1999], the conclusions of which are at least in part similar to those presented in this book.)

So if not mere consistency, what property of PA justifies our accepting $PA_1$? There is a simple answer: PA is *sound*, that is, every theorem of PA is true. This being so, $PA_1$ is also sound, since the axiom added is true. Similarly for any theory $T$, if $T$ is sound, so is every consistency extension of $T$. In other words, soundness is *inherited* by consistency extensions—this in contrast to consistency, which by the second incompleteness theorem is not in general inherited by consistency extensions. Further, if we *know* that $T$ is sound and that the formula $\phi$ used in $Con_T$ defines the axioms of $T$, we also know that $T + Con_T$ is sound.

Our concern in this book is primarily with theories that we accept as sound, and with the consequences of accepting them as sound. Although consistency is a much weaker condition than soundness, we will begin by considering extensions of a sound theory $T$ obtainable by repeated addition of consistency statements.

## 13.2. Iterated consistency extensions

By a *consistency sequence* will be meant a sequence of theories $T_\alpha$ indexed by the ordinals $\alpha$ smaller than some limit ordinal $\lambda$, such that

> For every $\alpha < \lambda$, $T_{\alpha+1}$ is $T_\alpha + Con(\phi)$, where $\phi$ is some $\Sigma$-formula defining the axioms of $T_\alpha$. That is, the axioms of $T_{\alpha+1}$ are the axioms of $T_\alpha$ together with the formula $Con(\phi)$.
>
> For every limit ordinal $\beta < \lambda$, the set of axioms of $T_\beta$ is the union of the sets of axioms of $T_\alpha$ for $\alpha < \beta$.

By an *iterated consistency extension* of $T$ will be meant a theory which is identical with some $T_\alpha$ in a consistency sequence beginning with $T = T_0$. Thus an iterated consistency extension of $T$ is a theory which is obtainable from $T$ by iterating ad infinitum the addition of consistency statements, using ordinals to count the iterations and collecting all earlier theories into a single new theory at every limit ordinal. The question is, how can we define iterated consistency extensions, and how can we recognize a theory as an iterated consistency extension of a given $T$?

Note that the definition of "consistency sequence" doesn't show us how to define consistency sequences, or tell us for which ordinals $\lambda$ at least one consistency sequence exists. This is because we need some $\Sigma$-formula defining the axioms of $T_\alpha$. If we set out to define a consistency sequence on the basis of the general definition, we are faced with the difficulty of introducing some particular formulas that define the axioms of earlier theories in the sequence.

The problem appears only with limit ordinals. We can easily define a short consistency sequence of length $\omega$, given a theory $T$ and some $\Sigma$-definition $\phi$ of its axioms. $T_0$ is defined as $T$, and given a theory $T_i$ in the sequence with axioms described by $\phi_i$ we define $T_{i+1}$ as the theory $T_i + Con(\phi_i)$, and take $\phi_{i+1}$ to be the $\Sigma$-formula $\phi_i \vee v_0 = \overline{\psi}$, where $\psi$ is $Con(\phi_i)$. We can go on to define $T_\omega$ as the union of the theories $T_i$ for $i < \omega$, but to get a formula defining the axioms of $T_\omega$ and thus be able to define $T_{\omega+1}$ we need to use a notation for $\omega$ or some equivalent construction.

Generally, in defining consistency sequences we need to be able to express in the language of arithmetic that a theory in the sequence is consistent, that is, we need an arithmetical definition of its axioms, and so we have to deal with *notations* for ordinals. Further, since all theories in the sequence are to be effectively axiomatizable, to specify a theory $T$ is the same as specifying a $\Sigma$-formula $\phi$ such that the axioms of $T$ are the sentences satisfying $\phi$.

We accordingly define an *explicit consistency sequence* as a mapping of notations $a$ to $\Sigma$-formulas $\phi_a$, for $a$ in some branch $B$ of $\mathbf{O}$, such that the mapping of $|a|$ to $T_a$ is a consistency sequence, where $T_a$ is the theory whose axioms are the sentences satisfying $\phi_a$. We also impose the condition that $\phi_{suc(a)}$ is the formula $\phi_a \vee v_0 = \overline{\psi}$ where $\psi$ is $Con(\phi_a)$, so that the axioms of $T_{suc(a)}$ are described in a canonical manner, given a description of the axioms of $T_a$. The point of this condition is to make explicit consistency sequences more epistemologically relevant than consistency sequences in general, where any $\Sigma$-formula may be used in defining the axioms of a theory, even if there is no apparent way of knowing that it does define those axioms. However, this restriction does not exclude non-standard descriptions of the axioms of $T_0$ or $T_{lim(e)}$, as we shall see. So the question now is, how can we define explicit consistency sequences?

## Consistency progressions

One way of defining explicit consistency sequences is by way of *recursive consistency progressions*, as first introduced by Turing and further developed by Feferman.

A consistency progression is not a consistency sequence, but is instead a primitive recursive mapping taking any natural number $a$ to a $\Sigma_1$-formula $\phi_a$ with free variable $v_0$, such that the following statements, which we will refer to as the characteristic conditions on progressions, are provable in PA:

> For every $a$ and $n$, $n$ satisfies $\phi_{suc(a)}$ if and only if $n$ satisfies $\phi_a$ or is the formula $Con(\phi_a)$.
>
> For every $e$ and $n$, $n$ satisfies $\phi_{lim(e)}$ if and only if $n$ satisfies $\phi_{\{e\}(k)}$ for some $k$.

To show the existence of consistency progressions, we use the recursion theorem for ordinal notations (§11.2). Let $\phi$ be a formula with free variable $v_0$ which defines the axioms of $T$. By the theorem cited, we define the formula $\phi_a$ as a primitive recursive function of $a$ so that (the formalization of) the following is provable in PA for some primitive recursive function $H$:

$\phi_0 = \phi$

$\phi_{suc(a)}$ = the formula $\phi_a \vee v_0 = \overline{\psi}$ where $\psi$ is $Con(\phi_a)$.

$\phi_{lim(e)}$ = the formula $\exists v_1 Sat\text{-}\Sigma_1(v_0, \{\overline{m}\}(v_1))$, where $\{m\}(n) = \phi_{\{e\}(n)}$ for every $n$, and $m = H(e)$.

Here $Sat\text{-}\Sigma_1$ is the effectively enumerable relation for which $Sat\text{-}\Sigma_1(n,k)$ ($n$ satisfies the $\Sigma_1$-formula $k$) holds if and only if $k$ is a $\Sigma_1$-formula $\phi$ with one free variable $x$, and $\phi_x(\overline{n})$ is a true $\Sigma_1$-sentence. For arguments $a$ not covered by any of the above equations, $\phi_a$ is taken to be $\phi$. The mapping of $a$ to $\phi_a$ is called a *recursive consistency progression based on* $\phi$. If the axioms of a theory $T$ are given through the formula $\phi$, we also speak of this as a consistency progression based on $T$.

The proof of the characteristic conditions on progressions in PA is straightforward, using the defining equations for the progression. Thus in the case of $lim(e)$, $n$ satisfies $\phi_{lim(e)}$ if and only if $n$ satisfies $\exists v_1 Sat\text{-}\Sigma_1(v_0, \{\overline{m}\}(v_1))$, or in other words if and only if $\exists v_1 Sat\text{-}\Sigma_1(\overline{n}, \{\overline{m}\}(v_1))$ is true. This holds if and only if there is a $k$ such that $Sat\text{-}\Sigma_1(\overline{n}, \{\overline{m}\}(\overline{k}))$ is true, and by the condition on $m$ this means that there is a $k$ such that $Sat\text{-}\Sigma_1(n, \phi_{\{e\}(k)})$, that is, such that $n$ satisfies $\phi_{\{e\}(k)}$.

In a progression, $\phi_a$ is defined for every $a$, but in general the corresponding theory $T_a$, which has as axioms the formulas satisfying $\phi_a$, has no particular connection with what can or cannot be proved in an iterated consistency extension of $T$. For example, we know by the recursion theorem that there exist $e$ such that $\{e\}(0) = suc(lim(e))$. For such an $e$, $T_{lim(e)}$ proves everything provable in $T_{suc(lim(e))}$ and thus in particular proves the consistency of $T_{lim(e)}$, and so is inconsistent. (This construction is used below, in the proof of Turing's completeness theorem.) So in general, $T_a$ may well be an inconsistent theory. However, if we restrict $a$ to some branch $B$ of $\mathbf{O}$, the mapping of $a$ to $\phi_a$ for $a$ in $B$ is an explicit consistency sequence. In particular, the mapping of $a$ to $\phi_a$ for $a$ in $\mathbf{O}_b = \{a \mid a <_\mathbf{O} b\}$ is an explicit consistency sequence of length $|b|$ given that $b$ is in $\mathbf{O}$. This follows easily by induction on $|a|$.

Thus consistency progressions are a way of obtaining explicit consistency sequences. The maximal length of a sequence that can be obtained in this way is the maximal length of a branch in $\mathbf{O}$, and thus $\omega_1^{CK}$. The structure and properties of progressions is an interesting subject in its own right, but in this book consistency progressions are of interest only as a means of defining explicit consistency sequences. Indeed the focus is even narrower, for we are interested primarily in explicit sequences that can be regarded as justified by our acceptance of the original theory. In general, sequences obtained from progressions need not have any such character. This is illustrated in a striking way by a completeness theorem proved by Turing.

### Turing's completeness result

Turing proved the following

> **Completeness theorem for consistency progressions.** For any consistency progression and any true $\Pi_1$-sentence $\phi$, there is an $a \in \mathbf{O}$ with $|a| = \omega + 1$ such that $\phi$ is provable in $T_a$.

Thus in terms of sequences, any progression yields a consistency sequence such that $\phi$ is provable in $T_{\omega+1}$.

This result seems rather puzzling at first sight. $\phi$ is provable in the theory asserting the consistency of $T_\omega$, where $T_\omega$ itself has infinitely many consistency statements as axioms. Just what reasoning leads from the consistency of $T_\omega$ to $\phi$, and how are the consistency axioms in $T_\omega$ used?

In fact the consistency axioms in $T_\omega$ are not used at all. As the proof of the theorem shows, the only reason why $\phi$ can only be proved at stage $\omega + 1$ in a consistency sequence based on an arbitrary consistency progression is that it is only at stage $\omega$ that a non-standard definition of the axioms of $T_a$ can be introduced in such a sequence.

For the proof, given any progression and assuming $\phi$ to be $\forall x\psi$ where $\psi$ is a $\Delta_0$-formula, we define $e$ by the recursion theorem so that provably in PA, for every $n$,

$\{e\}(n) = no$ if $\psi(\overline{k})$ is true for every $k \leq n$,
and otherwise $\{e\}(n) = suc(lim(e))$.

Since $\phi$ is true, $lim(e)$ is in fact a notation in **O** for $\omega$, but an oddly defined one. The point of the definition is that it is provable in PA that if $T_{lim(e)}$ is consistent, $\phi$ is true. For $T_a$, where $a = suc(lim(e))$, proves the consistency of $T_{lim(e)}$, and if $\phi$ is false, $T_{\{e\}(n)}$ is $T_a$ for all sufficiently large $n$, and so $T_{lim(e)}$ also proves the consistency of $T_{lim(e)}$, implying that $T_{lim(e)}$ is inconsistent by the second incompleteness theorem. This argument can be carried out within PA, and so shows that it is provable in $T_a$ that $\phi$ is true, since it is provable in $T_a$ that $T_{lim(e)}$ is consistent.

If we take a closer look at the argument, we see that the essential point is that the axioms of the theory $T_\omega$ defined on page 187 are given a non-standard description in $T_a$, corresponding to the definition of $e$. If our concern is not with consistency progressions, but only with consistency sequences, the consistency sequence obtained from the above $e$ is quite unnecessarily long. If we introduce a non-standard definition of the axioms of $T_\omega$ we may as well use a non-standard definition of the axioms of $T_0$. Simplifying the construction in Turing's argument accordingly, we come up with

> **The $\Pi$-completeness observation** (Turing): For any true $\Pi_1$-sentence $\phi$, there is a $\Sigma$-formula $\theta$ defining the axioms of $T$ such that $Con(\theta)$ implies $\phi$ in PA.

For the proof of this, assume as above that $\phi$ is $\forall x\psi$ and take $\phi_1$ to be any $\Sigma$-formula with free variable $v_0$ defining the axioms of $T$. We then take $\theta$ to be the formula $\phi_1 \vee (\exists x \neg\psi \wedge v_0 = \overline{0 = s(0)})$.

Thus there is, for any true $\Pi$-sentence $\phi$, a consistency extension of $T$ in which $\phi$ is provable. When our concern is with the epistemology of iterated consistency extensions rather than with the structure of progressions, the $\Pi$-completeness observation is more relevant than Turing's completeness theorem, since it exhibits directly the role played by non-standard definitions and does not give the misleading impression that the consistency axioms of $T_\omega$ play a role in the proof of $\phi$.

### Autonomous progressions

The completeness observation does not of course tell us anything about what can be *known* on the basis of iterated consistency assertions, for we know that the definition $\theta$ of the axioms of $T$ defines the axioms of $T$ only if we know that $\phi$ is true. Thus although $\phi$ is provable in a consistency extension of $T$, it is not necessarily provable in any theory that we can convince ourselves *is* a consistency extension of $T$. In terms of progressions, we don't know in general that the notation $lim(e)$ for $\omega$ defined in Turing's argument is a notation for $\omega$ unless we already know that $\phi$ is true.

So how can we obtain from a progression beginning with $T$ iterated consistency extensions of $T$ that we know to be such? The easy answer is that if we know, on whatever grounds, that $a$ is in $\mathbf{O}$, we also know that $T_a$ is an iterated consistency extension of $T$. However, as noted in §11.2, knowing that $a$ is in $\mathbf{O}$ can involve any amount of arithmetical knowledge, since there is a primitive recursive function $f$ for which we can prove that an arithmetical sentence $\phi$ is true if and only if $f(\phi)$ is in $\mathbf{O}$. In order to be able to say something informative about how we can come to accept that $T_a$ is an iterated consistency extension of $T$, we need to specify some source of knowledge for the information that $a$ is in $\mathbf{O}$. The most immediately interesting possibility is if we can prove in $T$ itself that $a$ is in $\mathbf{O}$, for then we know on the basis of $T$ alone that $T_a$ is an iterated consistency extension of $T$, and can therefore in a reasonable sense take accepting $T_a$ to be justified by our accepting $T$.

But then we are faced with the problem how to even express in $T$ that $a$ is in $\mathbf{O}$ when $T$, as in the current discussion, is a numerical theory. We know that for $a$ and $b$ in $\mathbf{O}$, $a <_{\mathbf{O}} b$ if and only if $a <_{\mathrm{K}} b$ where $<_{\mathrm{K}}$ is the e.e. relation defined in §11.2, so we might try to define $\mathbf{O}$ in PA by an arithmetical condition $O(a)$ expressing

> $<_{\mathrm{K}}$ restricted to $\{b \mid b \leq_{\mathrm{K}} a\}$ is a strict total ordering, and for every $b \leq_{\mathrm{K}} a$, either $b$ is 0 or $b$ is $suc(c)$ for some $c$, or $b$ is $lim(e)$ for some $e$ such that $\{e\}$ is total and $\{e\}(n) <_{\mathrm{K}} \{e\}(n+1)$ for every $n$.

Although $O(a)$ seems rather restrictive, it is insufficient to constrain $a$ to be a member of $\mathbf{O}$, for as implied (but not proved) in §11.2, the set $\mathbf{O}$ is not arithmetically definable. Any arithmetical condition satisfied by all $a$ in $\mathbf{O}$ will also be satisfied by infinitely many numbers not in $\mathbf{O}$. We can formulate stronger arithmetical conditions on $a$ that do constrain $a$ to be in $\mathbf{O}$, for example "$a <_{\mathrm{K}} b$" where $b$ is in $\mathbf{O}$. However, that $\overline{k} <_{\mathrm{K}} \overline{n}$ is provable in PA, where $n$ is in $\mathbf{O}$, does not mean that we can know, on the basis only of PA, that $k$ is in **O**. We need a way of proving in PA that $a$ is in **O** even though we cannot express "$a$ is in **O**" in PA.

This can be achieved in a sense, as we shall see. The class of $a$ satisfying the condition $O(a)$ defined above is larger than the class of ordinal notations, but if we add the condition that the relation $<_K$ is *well-founded* on $\{b \mid b \leq_K a\}$, we get a condition equivalent to being in **O**. That this is so follows in one direction from the fact that $<_K$ coincides with $<_O$ for $a$ and $b$ in **O**, and for the other direction we can easily prove by well-founded induction on $<_K$ that $b$ is in **O** for every $b \leq_K a$. Now, we can't express "$<_K$ is well-founded on $\{b \mid b \leq_K a\}$" by an arithmetical formula, but we can introduce a sufficient condition, formulated in terms of provability in PA, for this to be true. Expand the language of PA by adding a new unary predicate symbol $P$, and extend the axioms of PA to include all induction axioms in which the new predicate symbol appears. If in the resulting theory $\text{PA}^P$ the formula

$$(1)\quad \exists x(x \leq_K \bar{n} \wedge P(x)) \supset \exists x(x \leq_K \bar{n} \wedge P(x) \wedge \neg\exists y(y \leq_K \bar{n} \wedge y <_K x \wedge P(y)))$$

is provable, $<_K$ is well-founded on $\{b \mid b \leq_K n\}$. This is so since the axioms of $\text{PA}^P$ do not constrain the interpretation of $P$ in any way, so anything proved about $P$ in $\text{PA}^P$ is true of every subset of **N**. Thus if $\Omega(\bar{n})$ is provable in $\text{PA}^P$, where $\Omega(\bar{n})$ is the conjunction of (1) and a formula expressing $O(n)$, this can be interpreted as meaning that $n$ can be seen to be an ordinal notation on the basis of the mathematical knowledge contained in PA. (For convenience, we will drop the superscript $P$ and just say that $\Omega(\bar{n})$ is provable in PA.)

To say that the provability of $\Omega(\bar{n})$ in PA "can be interpreted" in this way is only to underline that there is no obviously correct way of characterizing the ordinal notations that can be recognized as such on the basis of the mathematical knowledge contained in PA. The idea of iterating the adding of consistency statements ad infinitum, up to any given ordinal, involves concepts that go beyond arithmetic. The only obvious argument for the assertion that $n$ is in **O** whenever $\Omega(\bar{n})$ is provable in PA uses the soundness of PA, plus a bit of set-theoretical reasoning. The justification for nevertheless regarding the $n$ for which $\Omega(\bar{n})$ is provable in PA as ordinal notations recognizable as such on the basis of PA is that these $n$ can be proved to belong to **O** on the basis of a uniform bit of set-theoretical reasoning together with reasoning that is purely arithmetical (even if applied to an unspecified predicate $P$) and can be carried out in PA. The same justification applies for any theory $T$, and if $\Omega(\bar{n})$ is provable in $T$—that is, in a theory $T^P$ corresponding to $\text{PA}^P$—we will interpret this as meaning that $n$ can be seen to be an ordinal notation on the basis of the mathematical knowledge contained in $T$.

If $\Omega(\bar{n})$ is provable in a theory $T$, proofs by induction on $<_K$ restricted to $\{b \mid b \leq_K n\}$ can be carried out in $T$ with respect to any arithmetical formula $\phi(x)$. This is so because we can take a proof of (1) and replace $P(x)$ everywhere by $\phi(x)$ to obtain a proof of the relevant induction principle.

Now suppose $T_a$ can be seen to be an iterated consistency extension of PA on the basis of PA, in the above sense, and that $T_b$ can in the same sense be seen to be an iterated consistency extension of PA on the basis of $T_a$, that is, suppose $\Omega(\overline{b})$ is provable in $T_a$. Since PA is sound, $T_a$ is also sound, so we can conclude that $T_b$ is also an iterated consistency extension of PA. This leads us to the concept of the autonomous part of a progression. Given any progression, we define its *autonomous part* to be the restriction of the progression to the set $A$ of *autonomous notations* inductively defined by

0 is in $A$,
if $a$ is in $A$ and $\Omega(\overline{b})$ is provable in $T_a$, $b$ is in $A$.

(This is essentially equivalent to Feferman's definition in Feferman [1962].)

An *autonomous iterated consistency extension* of $T$ is a theory identical with $T_a$ for some $a$ in the autonomous part of some consistency progression based on $T$. If $T$ is a sound theory, every $a$ in A is in **O**, and every autonomous iterated consistency extension of $T$ is in fact an iterated consistency extension of $T$.

## Autonomous extensions and arithmetical knowledge

To recapitulate: if a theory $T$ is sound, so is every iterated consistency extension of $T$. Thus if we accept $T$ as sound, we will also accept as sound every iterated consistency extension of $T$ that we recognize as such. We arrive at iterated consistency extensions by defining explicit consistency sequences, and if we know $a$ to be a limit notation in **O**, we know that the theories $T_b$ for $b <_K a$ constitute an explicit consistency sequence. Regarding the autonomous notations as those that can be recognized as notations in **O** on the basis of our accepting $T$ (and thereby its iterated consistency extensions) as sound, we arrive at a characterization of the arithmetical statements that can be recognized as true on the basis of $T$ and the repeated adding of consistency statements: they are the theorems of the union of all autonomous iterated consistency extensions.

At this point we note that the inductive definition of $A$ shows it to be effectively enumerable, by the same kind of argument used in §11.2 to show that $<_K$ is effectively enumerable. Thus the union of all $T_a$ for $a$ in $A$ is itself an effectively axiomatizable theory $T_A$. The structure of this theory is, however, far from transparent. We will next see that anything provable in $T_A$ can in fact be proved in a rather simpler extension of $T$ which can be seen to be sound directly on the basis of the soundness of $T$.

The statement that the theorems of $T_A$ can be recognized as true on the basis of the repeated adding of consistency statements is not to be taken as referring to any actual cognitive processes. What is meant insuch formulations is only, in acordance with the discussion of provability in §1.3, that it is a mathematical consequence of principles that we actually recognize as valid that an autonomous $T_a$ is an iterated consistency extension of $T$, and therefore sound, even if this consequence is so remote that we are not actually able to arrive at it, and an actual proof of a theorem of $T_a$ need not involve any iterative steps.

This leads to a further caveat: because of the unconstrained and qualitative character of the notion of an autonomous notation, it can be misleading to think of a proof in $T_a$ as a proof utilizing a series of consistency statements. To illustrate this, let us consider a more refined analysis of iterated consistency extensions in the case of PA. If we take the system of notations for ordinals smaller than $\varepsilon_0$ presented in §11.3 and regard it as a part of **O**, as also explained in §11.3, we obtain a subset of $A$ for any theory $T$. This is because we have assumed that $T$ extends PA, and as shown in §11.3, well-foundedness of $<_{\mathrm{K}}$ is provable in PA for these notations. This subset of $A$ can be considered a canonical system of autonomous notations for ordinals smaller than $\varepsilon_0$. For such ordinals $\alpha$, $T_\alpha$ will refer to the theory associated with the corresponding canonical notation, and $(\alpha)_{\mathbf{O}}$ to that notation itself.

Now if a sentence $\phi$ is provable in $T_\alpha$, we know that it can be proved on the basis of $\alpha$ iterated consistency statements, together with the axioms of PA, where each consistency statement asserts the consistency of the preceding theory, and nothing more. If $\phi$ is provable in $T_a$ for an autonomous notation $a$ with $|a| = \alpha$, no such conclusion can be drawn. We can see this by considering an autonomous variant of the notation used in Turing's completeness proof. For a $\Pi$-statement $\forall x\psi$, and an ordinal $\alpha < \varepsilon_0$ we define $e$ so that provably in PA,

> for every $n$, $\{e\}(n) = n_{\mathbf{O}}$ if $\psi(\overline{k})$ is true for every $k \leq n$, and otherwise $\{e\}(n) = (\alpha + n)_{\mathbf{O}}$.

This yields an autonomous notation $lim(e)$, and we can prove in PA that $T_{lim(e)}$ is either $T_\omega$ or $T_{\alpha+\omega}$, although if $\forall x\psi$ is undecidable in PA, we can't decide which theory it is. Now suppose we choose $\forall x\psi$ as "$T_\alpha$ is consistent". In $T_{suc(lim(e))}$, the consistency of $T_{lim(e)}$ is by definition provable, which means that $T_{suc(lim(e))}$ proves "if $T_\alpha$ is inconsistent, $T_{\alpha+\omega}$ (which extends $T_\alpha$) is consistent", and thus proves the consistency of $T_\alpha$. So for any $\alpha < \varepsilon_0$, an autonomous notation $a$ can be chosen with $|a| = \omega + 1$ such that $T_a$ proves the consistency of $T_\alpha$.

Thus $|a|$ is not really significant when we're talking about provability in $T_a$. But every iterated consistency extension that can be recognized as such on the basis of a more refined and informative ordinal analysis (which tells us how many iterated consistency statements are used in proving a statement) will be extended by $T_A$, and every theorem of $T_A$ can be justifed in terms of the soundness of iterated consistency extensions of $T$. This is a sufficient basis for the use of autonomous extensions made in the argument of this book.

## 13.3. Reflection and iterated consistency extensions

Consistency, as we know, is a weaker condition than soundness, and so is consistency of consistency extensions. That is, in general, for $T + Con_T$ to be consistent, it is not necessary that $T$ is sound. If $T$ is consistent but $T + Con_T$ is inconsistent, $\neg\, Con_T$ is provable in $T$ and is equivalent in $T$ to a false $\Sigma$-formula. Thus a sufficient condition for $T + Con_T$ to be consistent is that $T$ is $\Sigma$-sound. In fact $\Sigma$-soundness, like soundness, is inherited by consistency extensions: if $T$ is $\Sigma$-sound, then $T + Con_T$ is $\Sigma$-sound. This is because if $T + Con_T$ proves the false $\Sigma$-sentence $\psi$, $T$ proves the false $\Sigma$-sentence $\neg\, Con_T \vee \psi$. (This formula, it will be noted, is not really a $\Sigma$-sentence, but is equivalent to such a sentence by a logical transformation—a distinction that will usually be ignored in the following.)

It follows that if $T$ is $\Sigma$-sound, every iterated consistency extension of $T$ is $\Sigma$-sound, and thus consistent. (In fact we can also show, using the construction in the $\Pi$-completeness observation, that if every consistency extension of $T$ is consistent, $T$ is $\Sigma$-sound.) Applying this to autonomous extensions, we shall see that to formally prove every theorem of $T_A$, explicitly assuming $\Sigma$-soundness suffices.

Given $T$ and a $\Sigma$-formula $\phi$ defining the axioms of $T$, we extend $T$ to a theory $T^{\Sigma\,REF}$ obtained by adding the axiom "$T$ is $\Sigma$-sound". Here we will take "$T$ is $\Sigma$-sound" to be the formula

$$\forall x(Thm_T(x) \wedge \Sigma_1\text{-}sentence(x) \supset True\text{-}\Sigma_1(x))$$

which uses the restricted truth predicate $True\text{-}\Sigma_1$ introduced in §9.5. Instead of $\Sigma$-formulas, the above formulation refers to $\Sigma_1$-formulas, but we know it is provable in $T$ that every $\Sigma$-formula is equivalent in $T$ to a $\Sigma_1$-formula. We can now make the following

**Observation:** For every autonomous iterated consistency extension $T_a$ of $T$, $T^{\Sigma\,REF}$ extends $T_a$.

To prove this, we prove by induction on $a$ in $A$ that $T^{\Sigma\,REF}$ extends $T_a$ and proves the $\Sigma$-soundness of $T_a$.

The base case is immediate, since $T^{\Sigma\,REF}$ is defined as the extension of $T_0$ obtained by adding the axiom that $T_0$ is $\Sigma$-sound. For the induction step, suppose the statement is true of $a$ and that $T_a$ proves $\Omega(\overline{b})$. Since $T^{\Sigma\,REF}$ extends $T_a$, $T^{\Sigma\,REF}$ also proves $\Omega(\overline{b})$. We can therefore prove by $<_K$-induction in $T^{\Sigma\,REF}$ that $T_c$ is $\Sigma$-sound for every $c \leq_K b$, as follows. $T_0$ is $\Sigma$-sound, as postulated in $T^{\Sigma\,REF}$. To show that for any $a$, if $T_a$ is $\Sigma$-sound, $T_{suc(a)}$ is $\Sigma$-sound (where the axioms of $T_a$ and $T_{suc(a)}$ are given by the formulas $\phi_a$ and $\phi_{suc(a)}$ defined in the progression), we need only formalize in $T$ the argument given above to show that $\Sigma$-soundness is inherited by consistency extensions. For the case where $c$ is $lim(e)$, we first need to prove by $<_K$-induction that $T_{d'}$ extends $T_d$ whenever $d <_K d' \leq_K b$, and can then prove that if $T_{\{e\}(n)}$ is $\Sigma$-sound for every $n$, $T_{lim(e)}$ is $\Sigma$-sound, which concludes the induction. To show that $T^{\Sigma REF}$ extends $T_b$ we use $<_K$-induction (justified since $b$ is in **O**) to show that $T^{\Sigma REF}$ extends $T_c$ for every $c \leq_K b$. For the base case, $T^{\Sigma REF}$ extends $T_0$ by definition. For the successor induction step, if $T^{\Sigma REF}$ extends $T_c$ and $c \leq_K b$, it is provable in $T^{\Sigma REF}$ that $T_c$ is $\Sigma$-sound, hence consistent, and it follows that $T^{\Sigma REF}$ extends $T_{suc(c)}$. Finally for the limit induction step, if $T^{\Sigma REF}$ extends every $T_{\{e\}(n)}$, it extends $T_{lim(e)}$, whose axioms are the union of the axioms of the $T_{\{e\}(n)}$.

The statement "$T$ is $\Sigma$-sound" is a *reflection principle* for $T$: it states that every theorem of $T$ in some class of sentences is true. As we know, the statement that $T$ is consistent can also be formulated as a reflection principle, stating that $T$ is $\Pi$-sound. In fact $\Sigma$-soundness is equivalent to $\Pi_2$-soundness, that is, truth of every $\Pi_2$-theorem of $T$, for if $\forall x \exists y \phi(x, y)$ is provable in $T$, so is $\exists y \phi(\overline{n}, y)$ for every $n$, and if $T$ is $\Sigma$-sound, $\exists y \phi(\overline{n}, y)$ is true for every $n$, and so $\forall x \exists y \phi(x, y)$ is true.

What the discussion above has shown is that given a sound theory $T$, extending $T$ by a fairly weak soundness or reflection principle ($\Sigma$-reflection) gives a theory which formally extends every autonomous extension obtained by repeatedly adding an even weaker soundness or reflection principle (consistency). We will next consider, in the following chapter, the application of this pattern to the case where we extend $T$ by the strongest soundness principles that we can express in the language of arithmetic.

## Iterated consistency extensions and soundness

Note the central role played in this chapter by the assumption that $T$ is *sound*. The justification for taking every autonomous consistency extension of $T$ to be an iterated consistency extension was that $T$ is sound. When we consider what can be proved in autonomous iterated consistency extensions starting from a sound theory, we find that anything provable in such an extension is provable in a sound extension of $T$ obtained by adding the axiom "$T$ is $\Sigma$-sound". Thus we find that in this case, the potentially confusing business of repeatedly adding consistency statements will not yield any new conclusion that does not follow from a statement expressing a limited soundness of $T$, as long as we are only considering iterated consistency extensions that can be recognized as such on the basis of $T$.

While it is true that any iterated consistency extension of $T$ is $\Sigma$-sound if $T$ is $\Sigma$-sound, if we don't know $T$ to be sound, but only to be $\Sigma$-sound, it is not clear that anything in the above discussion tells us anything about what extensions of $T$ we can *prove* to be iterated consistency extensions on the basis of $T$ (together with some suitable uniform bit of set-theoretical reasoning). Further, there is no obvious candidate for a theory which extends every such consistency extension. In particular, merely accepting $T$ as $\Sigma$-sound gives us no basis for believing $T^{\Sigma\,REF}$ to be even consistent. For an example to show this, let $T$ be PA +"PA is not $\Sigma$-sound". Note that "PA is $\Sigma$-sound" can be formulated "For every $p$ and every $\Sigma_1$-formula $\exists x\phi$ there is a $k$ such that if $p$ is a proof of $\exists x\phi$ in PA then $\phi_x(\overline{k})$ is true", and thus as a $\Pi_2$-sentence, so "PA is not $\Sigma$-sound" can be taken to be a $\Sigma_2$-sentence. For this $T$, $T$ + "$T$ is $\Sigma$-sound" is clearly inconsistent, but $T$ itself is $\Sigma$-sound. This follows from the fact that $T$ is a conservative extension of PA with respect to $\Sigma$-sentences, that is, every $\Sigma$-sentence provable in $T$ is also provable in PA. For suppose $T$ proves a $\Sigma$-sentence $\exists x\phi$. Then PA $+\forall x\neg\phi$ proves "PA is $\Sigma$-sound", and hence proves "if PA proves $\exists x\phi$ then $\exists x\phi$", and so proves "PA does not prove $\exists x\phi$", that is, it proves its own consistency, hence is inconsistent, so PA proves $\exists x\phi$.

There may of course be many other approaches to the question of what can be proved on the basis of iterated consistency statements. What has been argued above is only that accepting $T$ as sound yields a convincing candidate for a theory $T'$ which is also sound and extends every iterated consistency extension of $T$ that can be recognized as such on the basis of the knowledge expressed in $T$, while accepting $T$ as $\Sigma$-sound yields no obvious such candidate, in spite of the formal role played by $\Sigma$-soundness in the argument.

Having extended $T$ by "$T$ is $\Sigma$-sound", we inevitably reflect that the process can be continued. For the soundness of $T$ implies that not only every consistency extension of $T$, but also the extension $T'$ obtained by adding "$T$ is $\Sigma$-sound" is sound. We could now consider consistency extensions of $T'$, but that would just entail repeating what has already been said. So we turn instead to the possibility of formulating stronger extensions that are still justified by the soundness of $T$.

CHAPTER 14

# ITERATED REFLECTION

## 14.1. Reflection principles

### Local and uniform $\Sigma$-reflection

In general we must expect that adding infinitely many new axioms $\phi(\overline{0}), \phi(\overline{1}), \phi(\overline{2})\ldots$ to a theory $T$ yields a theory weaker than that obtained by adding the single axiom $\forall x\phi$, unless $\forall x\phi$ is already a theorem of $T$. This is so since if $\forall x\phi$ could be proved from the instances $\phi(\overline{0}), \phi(\overline{1}), \phi(\overline{2})\ldots$, it could be proved from finitely many of them, which although sometimes the case is not to be expected.

This simple observation explains why there is a difference in logical strength between what are called *local* and *uniform* reflection principles, even though semantically they have the same content. We will illustrate this using the principle of $\Sigma$-reflection as formulated in Chapter 13, that is, as the sentence

$$(1) \qquad \forall x(\mathit{Thm}_T(x) \wedge \Sigma_1\text{-}\mathit{sentence}(x) \supset \mathit{True}\text{-}\Sigma_1(x))$$

This principle is also called the *uniform* reflection principle for $\Sigma_1$-formulas, and $T^{\Sigma\,REF}$, the theory obtained by adding (1) as a new axiom to $T$, is called an extension by uniform $\Sigma_1$-reflection of $T$.

The corresponding *local* reflection principle is obtained by adding instead every instance of the uniform reflection principle as a new axiom:

$$(2) \qquad \mathit{Thm}_T(\overline{n}) \wedge \Sigma_1\text{-}\mathit{sentence}(\overline{n}) \supset \mathit{True}\text{-}\Sigma_1(\overline{n})$$

Since

$$\mathit{True}\text{-}\Sigma_1(\overline{\phi}) \equiv \phi$$

is provable in PA for every $\Sigma_1$-sentence $\phi$, and since $\Sigma_1\text{-}\mathit{sentence}(x)$ computably defines the $\Sigma_1$-sentences in PA, adding every instance (2) of the uniform $\Sigma_1$-reflection principle as a new axiom to $T$ is equivalent to adding as a new axiom every formula

$$(3) \qquad \Box\phi \supset \phi$$

for $\Sigma_1$-sentences $\phi$. (Recall that if $\phi$ is a sentence, $\Box\phi$ stands for $Thm_T(\overline{\phi})$). (3) is the usual formulation of the local reflection principle. (1) and (3) give semantically equivalent extensions of $T$, in the sense that (1) is true if and only if all formulas (3) (with $\phi$ a $\Sigma_1$-sentence) are true. In accordance with the general observation above, we expect that extending $T$ by the local reflection principle will yield a formally weaker theory than that obtained by adding the uniform reflection principle. As will be shown below, this is indeed the case.

## Unlimited local reflection

$\Pi_1$-soundness is equivalent to consistency, while the property of $\Sigma_1$-soundness (equivalent to $\Pi_2$-soundness, as noted in §13.3) has been considered in the foregoing because it is inherited by consistency extensions. We can similarly define $\Sigma_n$-soundness and $\Pi_n$-soundness, and formulate corresponding uniform and local reflection principles for $\Sigma_n$ and $\Pi_n$ sentences, for every $n$. Inevitably, logicians have studied these principles in depth and established precise logical comparisons between them, and between the results of iterating different reflection principles ad infinitum. The reader is referred to the works by Schmerl, Beklemishev, and Smorynski cited in the bibliography for systematic presentations of these results. In this book, we will go directly from $\Sigma$- and $\Pi$-reflection to *unlimited* reflection, for our concern is with theories $T$ which we know to be not only $\Sigma$-sound but sound, that is, for which we know that every axiom and therefore every theorem of $T$ is true. For such $T$, the extension $T^{Ref}$ of $T$ by an *unlimited* local reflection principle, obtained by adding (3) as a new axiom for *every* sentence $\phi$, is also known to be sound. We will refer to $T^{Ref}$ as an extension of $T$ by local reflection, or a local reflection extension of $T$. As in the case of consistency extensions, $T^{Ref}$ is not unique, since we need to choose a $\Sigma$-definition of the axioms of $T$ in order to express the reflection schema.

We can see that unlimited local reflection does not yield all that strong an extension of $T$ by noting that every instance of (3) is either provable in $T$, which by Löb's theorem holds if and only if $\phi$ is provable in $T$, or else a consequence in $T$ of a true $\Pi$-sentence, namely $\neg\Box\phi$. This is in contrast to (1), which cannot be deduced in $T$ from any $\Pi$-sentence consistent with $T$. For suppose "$T$ is $\Sigma$-sound" is deducible in $T$ from a $\Pi$-sentence $\psi$. Then in $T + \psi$ we can argue as follows: if $\neg\psi$ is provable in $T$, $\neg\psi$ is true (since $T$ is $\Sigma$-sound), so $\neg\psi$ is not provable in $T$, or in other words $T + \psi$ is consistent. Thus $T + \psi$ proves its own consistency and so is inconsistent. It follows that (1) cannot be proved in $T^{Ref}$ if $T^{Ref}$ is $\Sigma$-sound, for it could then be proved from a conjunction of finitely many true $\Pi$-formulas, which would be inconsistent with $T^{Ref}$. Thus the unlimited local reflection principle does not prove the uniform $\Sigma_1$-reflection principle.

As an example of the more informative results that have been proved about the relative strength of reflection principles, we shall prove a stronger statement about the relation between $T^{Ref}$ and $T^{\Sigma REF}$. The consistency extension $T_\omega$ of $T$ (defined in §13.2) is the union of $T_0, T_1, T_2 \ldots$, where $T_0$ is $T$ and $T_{i+1}$ is $T_i + Con_{T_i}$. For $i = 0, 1, 2, \ldots$, $T^{Ref}$ and $T_i^{Ref}$ are equivalent theories, as can be seen by an inductive argument. For $i = 0$, $T^{Ref}$ and $T_i^{Ref}$ are identical. For the induction step, we can argue in $T^{Ref}$ that if $\phi$ is a theorem of $T_{i+1}$, $Con_{T_i} \supset \phi$ is a theorem of $T_i$, and thus $\phi$ follows (since by the induction hypothesis, $T^{Ref}$ extends $T_i^{Ref}$). It follows that $T^{Ref}$ is an extension of $T_\omega$, and in fact, $T^{Ref}$ is strictly stronger than $T_\omega$. This is shown by the instance $\Box\neg\, Con_{T_\omega} \supset \neg\, Con_{T_\omega}$ of local reflection, which is not provable in $T_\omega$, for if $T_\omega$ proves $\Box\neg\, Con_{T_\omega} \supset \neg\, Con_{T_\omega}$, $T_\omega + Con_{T_\omega}$ proves the consistency of $T + Con_{T_\omega}$, which implies that $T_\omega + Con_{T_\omega}$ is inconsistent, since it is provable in PA that $T + Con_{T_\omega}$ extends $T_\omega$. But although $T^{Ref}$ is strictly stronger than $T_\omega$, it does not prove the consistency of $T_\omega$. This is a consequence of the following

> **Theorem** (Goryachev): For every subtheory of $T^{Ref}$ obtained by extending $T$ with $n$ of the local reflection axioms, the consistency of the theory is provable in $T_{n+1}$.

It follows that $T_\omega$ and $T^{Ref}$ prove the same $\Pi$-sentences. Since the consistency of $T_\omega$ (and of the theories in any autonomous consistency sequence beginning with $T$) is provable using the uniform $\Sigma_1$-reflection principle, we see that $T^{Ref}$ does not even prove every $\Pi$-sentence provable in $T^{\Sigma REF}$.

Goryachev's theorem has an elegant proof using a completeness theorem for the modal logic GL (named after Gödel and Löb). For this, see Beklemishev [1995]. Here the theorem will be proved by direct use of the deducibility conditions.

For the proof of Goryachev's theorem we will make use of the fact that "$T_i$ is consistent" can be expressed as

$$\neg\Box\Box\ldots\Box\bot$$

with $i+1$ occurrences of the provability operator, interpreted as provability in $T_0$. We denote the above formula by $\neg\Box^{i+1}\bot$. For $i = 0$, it was noted in §12.2 that $Con_T$ is equivalent in PA to $\neg\Box\bot$. To prove the assertion for $i > 0$, we show by induction in PA that $T_i$ is equivalent to the theory $T_0 + \neg\Box^i\bot$. For $i = 1$, this is the observation just stated. For the induction step, we get from the induction hypothesis that the theory $T_i$ is inconsistent if and only if $T_0$ proves $\Box^i\bot$, so $T_i$ is consistent if and only if $\neg\Box^{i+1}\bot$, and $T_{i+1}$ is equivalent to $T_0 + \neg\Box^i\bot + \neg\Box^{i+1}\bot$ which is equivalent to $T_0 + \neg\Box^{i+1}\bot$.

Now for the proof of the theorem. The base case $n = 0$ is immediate. For the induction step, we need to show that for every $n > 0$, the following is provable in $T$:

(1) $$\Box\neg((\Box\phi_1 \supset \phi_1) \wedge \cdots \wedge (\Box\phi_n \supset \phi_n)) \supset \Box^{n+1}\bot$$

For this, we will need to use a slight strengthening of the deducibility condition (D3), namely

(D3)′ $\Box(\phi \supset \psi) \supset (\Box\phi \supset \Box\psi)$ is provable in $T$ for all sentences $\phi$ and $\psi$.

The proof of (D3)′ is essentially the same as the proof of (D3).

We treat the case $n = 1$ separately, to simplify the notation. In this case we get in $T$ from the assumption $\Box\neg(\Box\phi_i \supset \phi_1)$ the conclusions $\Box\Box\phi_1$ and $\Box\neg\phi_1$ and so also $\Box\Box\neg\phi_1$, and thus $\Box^2\bot$.

For the case $n > 1$, let $\psi$ be the formula $\Box\neg((\Box\phi_1 \supset \phi_1)\wedge\cdots\wedge(\Box\phi_n \supset \phi_n))$. By the deducibility conditions we can prove in $T$

(2) $$\psi \supset \Box(\phi_1 \supset (\Box\phi_2 \wedge \neg\phi_2) \vee \cdots \vee (\Box\phi_n \wedge \neg\phi_n))$$

From this it follows that $T$ proves

(3) $$\psi \supset (\Box\phi_1 \supset \Box((\Box\phi_2 \wedge \neg\phi_2) \vee \cdots \vee (\Box\phi_n \wedge \neg\phi_n)))$$

By the induction hypothesis $T$ proves

(4) $$\Box((\Box\phi_2 \wedge \neg\phi_2) \vee \cdots \vee (\Box\phi_n \wedge \neg\phi_n)) \supset \Box^n\bot$$

From (4) and (3), $T$ proves

(5) $$\psi \supset (\neg\Box^n\bot \supset \neg\Box\phi_1)$$

Since $\psi \supset \Box\psi$ is provable in $T$, we get from (5) that $T$ proves

(6) $$\psi \supset \Box(\neg\Box^n\bot \supset \neg\Box\phi_1)$$

We also have that $T$ proves

(7) $$\psi \supset \Box(\neg\Box\phi_1 \supset (\Box\phi_2 \wedge \neg\phi_2) \vee \cdots \vee (\Box\phi_n \wedge \neg\phi_n))$$

From (6) and (7) we get that $T$ proves

(8) $$\psi \supset \Box(\neg\Box^n\bot \supset (\Box\phi_2 \wedge \neg\phi_2) \vee \cdots \vee (\Box\phi_n \wedge \neg\phi_n))$$

From (6), by symmetry, $T$ proves

(9) $$\psi \supset \Box(\neg\Box^n\bot \supset \neg\Box\phi_i) \text{ for } i = 2, \ldots, n.$$

From (8) and (9), finally, we get that $T$ proves

(10) $$\psi \supset \Box^{n+1}\bot$$

**Unlimited uniform reflection**

Although the unlimited local reflection principle is true for a theory $T$, in the sense that every instance of

(1) $$\Box\phi \supset \phi$$

is true (for arithmetical sentences $\phi$), if and only if the theory is sound, we have seen that adding "$T$ is sound" as an axiom piecemeal in this way yields a theory that is formally rather weak. We would like to add instead an unlimited *uniform* reflection principle to $T$: "every sentence provable in $T$ is true". But here we come up against the difficulty that arithmetical truth is not arithmetically definable, so there is no unlimited uniform reflection principle corresponding to the uniform $\Sigma$-reflection principle. The best we can do while staying within the language of arithmetic is to add the principle of uniform $\Sigma_n$-reflection as an axiom for every $n$:

(2) $$\forall x(\mathit{Thm}_T(x) \wedge \Sigma_n\text{-}\mathit{sentence}(x) \supset \mathit{True}\text{-}\Sigma_n(x))$$

An extension $T^{REF}$ of $T$ obtained by adding (2) for every $n$ will be called an extension of $T$ by the unlimited uniform reflection principle, or a *uniform reflection extension* of $T$. In fact more often what is described by this term is not the theory obtained by adding (2) for every $n$, but that obtained by adding, for every formula $\phi$ with one free variable $x$, the new axiom

(3) $$\forall x(\Box\phi \supset \phi)$$

(2) and (3) yield equivalent extensions of $T$. To see this, we first note that since

(4) $$\forall x\,(\mathit{True}\text{-}\Sigma_n\,(\,\mathit{sub}(\mathit{num}(x), \overline{x}, \overline{\phi})) \equiv \phi)$$

is provable in PA for every $\Sigma_n$-formula $\phi$ with free variable $x$ (by §9.5) and

(5) $$\forall x \Sigma_n\text{-}\mathit{sentence}(\mathit{sub}(\mathit{num}(x), \overline{x}, \overline{\phi}))$$

is provable in PA for every such formula (by a formalization of the obvious argument), (2) implies (3) in $T$. In the other direction, we can prove (2) from (3) in PA by a formalization in PA of the following argument: If $\psi$ is a $\Sigma_n$-sentence provable in $T$ it follows that $\mathit{True}\text{-}\Sigma_n(\overline{\psi})$ is provable in $T$, since $\mathit{True}\text{-}\Sigma_n(\overline{\psi}) \equiv \psi$ is provable in $T$. Therefore, by (3), $\psi$ is $\Sigma_n$-true.

We get another equivalent formulation of the extension of $T$ by uniform reflection by adding instead the axiom

(6) $$\forall x \Box\phi \supset \forall x\phi$$

for every sentence $\forall x\phi$. (3) implies (6) by a simple logical step, while the other direction (Feferman [1962]) requires some logical sleight-of-hand, using Craig's construction from §8.6. Suppose the axioms of $T$ are defined by a formula $\exists y\psi(x, y)$, where $\psi$ is a $\Delta_0$-formula defining the primitive recursive relation $R$. Let the function $F$ be defined by $F(n) = \psi_1 \wedge \cdots \wedge \psi_n$, where $\psi_k$ is $(k)_1$ if $R((k)_1, (k)_0)$ and $\psi_k$ is the formula $0 = 0$ otherwise. By standard reasoning, the range $A$ of $F$ is primitive recursive, and it is provable in PA that $\phi$ is a theorem of $T$ if and only if there is a derivation of $\phi$ from the formulas in $A$. Let the primitive recursive predicate "$x$ is a derivation of $\phi$ from the formulas in $A$" be computably defined in PA by the formula $\pi(x, y)$, and consider the formula $\forall x\Box(\pi(x, \overline{\phi}) \supset \phi)$, where the provability operator refers to provability in $T$, using $\exists y\psi(x, y)$ to define the axioms of $T$. This formula is provable in PA, by a formalization of the following argument. If $n$ is in fact a proof of $\phi$ from the formulas satisfying $A$, then $\phi$ is a theorem of $T$, and so $\pi(\overline{n}, \overline{\phi}) \supset \phi$ is a theorem of $T$. If $n$ is not a proof of $\phi$ from the formulas satisfying $A$, then $\neg\pi(\overline{n}, \overline{\phi})$ is a theorem of $T$, and so again $\pi(\overline{n}, \overline{\phi}) \supset \phi$ is provable in $T$. Since $\forall x\Box(\pi(x, \overline{\phi}) \supset \phi)$ is thus provable in PA and therefore in $T$, (6) implies $\forall x(\pi(x, \overline{\phi}) \supset \phi)$, and thereby (3).

A derivation of $\phi$ from the formulas in $A$, with $A$ defined as above using some $\Sigma$-definition of the axioms of $T$, will be referred to as a *PR-proof of* $\phi$ in $T$. The usefulness of this notion derives from the fact that it is provable in PA that a formula is a theorem of $T$ if and only if it has a PR-proof in $T$, where the predicate "$k$ is a PR-proof of $\phi$ in $T$" is primitive recursive. This will be used below in the proof of the $\Pi_2$-completeness theorem.

## Reflection using a truth predicate

By extending the language of $T$ so as to allow us to express "$\phi$ is a true sentence in the language of $T$" we can formulate the assertion "$T$ is sound" as a single statement. We will consider here how this works in the case of arithmetical theories. Truth extensions of numerical theories generally will be considered in Chapter 15.

We can extend any arithmetical theory $T$ to a theory $T^{True}$ by adding the axioms of $T$ to those of the theory $\mathrm{PA}^{True}$ defined in §10.3. We can then express "$T$ is sound" in $T^{True}$ by

$$\forall x(\mathit{Thm}_T(x) \wedge \mathit{Sentence}_T(x) \supset \mathit{True}(x))$$

By the proof sketched in §10.3, it is provable in $\mathrm{PA}^{True}$ that every theorem of $T$ is true if every axiom of $T$ is true, so we can equivalently express "$T$ is sound" in $T^{True}$ by formalizing "every axiom of $T$ is true".

In $\mathrm{PA}^{True}$ we didn't need to include any axiom asserting the soundness of PA, since it is *provable* in $\mathrm{PA}^{True}$ from the general truth axioms and the axioms of PA that every theorem of PA is true, but this proof depends on the axioms of PA having a canonical description of the form "an axiom is one of the formulas $\phi_1, \ldots, \phi_n$ or an instance of one of the schemas $\Phi_1, \ldots, \Phi_m$". In the general case, when the axioms of $T$ can be any effectively enumerable set of sentences, it need not be provable in $T^{True}$ that all axioms of $T$ are true using only the axioms characterizing truth, the new induction axioms, and the axioms of $T$. For example, let $T$ have as axioms the axioms of PA together with $\phi_x(\overline{n})$ for every $n$, where $\phi$ formalizes "$x$ is not a derivation of $0 = s(0)$ in CA". $T^{True}$ without a soundness axiom for $T$ does not prove "every axiom of $T$ is true", since if it did, "CA is consistent" would be provable in $\text{PA}^{True}$. Of course, since in this case the formulas $\phi_x(\overline{n})$ are in fact provable in PA, the example leaves open the possibility that every theory $T$ (with only finitely many predicate symbols) can be suitably axiomatized so as to allow $T^{True}$ to prove "every axiom of $T$ is true", but this is not a question that we have any reason to pursue here. Thus $T^{True}$ will be assumed to contain, in addition to the axioms of $T$, the new instances of the induction schema, and the axioms characterizing the truth predicate, also an axiom formalizing "every axiom of $T$ is true".

Note that $T^{True}$, which we shall call a *truth extension* of $T$, like consistency extensions and reflection extensions is not unique. We must decide on some formula describing the axioms of $T$ in order to express "every axiom of $T$ is true", and we must also choose some predicate symbol *True* to add to the language. The latter point is of no significance when we are talking about truth extensions of a single theory $T$, but will need to be taken into account in a discussion of iterated truth extensions.

With this definition of $T^{True}$, we can note that $T^{True}$ is stronger than $T^{REF}$, for we can prove in $T^{True}$ that every theorem of $T^{REF}$ is true. In fact, as we shall next see, $T^{True}$ proves every statement provable in an autonomous sequence of extensions by unlimited reflection beginning with $T$.

## 14.2. Iterated uniform reflection

If the axioms of $T$ are defined by a $\Sigma$-formula $\phi$, those of $T^{REF}$ are defined by another $\Sigma$-formula $\phi^{REF}$, formalizing "$\phi(v_0)$, or $v_0$ is $\forall \boldsymbol{x}(\Box\psi \supset \psi)$ where $\psi$ is a formula with free variable $\boldsymbol{x}$". Here of course $\Box\psi$ is formulated using $\phi$ to define the axioms of $T$. $\phi^{REF}$ is a primitive recursive function of $\phi$, so we can define a (uniform) *reflection progression* just as we defined a consistency progression in §13.2, with only two modifications. The first is that $T_{suc(a)}$ is now a reflection extension of $T_a$ rather than a consistency extension. The second modification is just a small technicality: $T_{lim(e)}$ is defined so that $T_{lim(e)}$ extends not only $T_{\{e\}(n)}$ for every $n$, but also $T_0$. This of course makes no difference to $T_a$ for $a$ in $\mathbf{O}$, but for the proof of Feferman's result below we need to know that $T_{lim(e)}$ extends $T_0$ for every $e$.

The autonomous part of such a progression is defined in the same way as for consistency progressions. Thus, in the same sense and with the same justification as in the case of consistency progressions, we can take the autonomous reflection progressions to be those iterated extensions by reflection of $T$ that can be recognized as such on the basis of the mathematical knowledge contained in $T$.

That soundness is inherited by reflection extensions—in other words, that $T^{REF}$ is sound if $T$ is sound—is provable in $\mathrm{PA}^{True}$, and thus in $T^{True}$. Since it is provable in $\mathrm{PA}^{True}$ that every theorem of $T$ is true if every axiom of $T$ is true, we only need to show that the formalization of "$\forall x(\Box\phi \supset \phi)$ is true for every formula $\phi$ in the language of PA with free variable $x$" is provable in $\mathrm{PA}^{True}$ from the assumption "all axioms of $T$ are true". This follows by a straightforward formalization in $\mathrm{PA}^{True}$ of the obvious argument: letting $\psi$ be $\Box\phi \supset \phi$, the sentence $\forall x(\Box\phi \supset \phi)$ is true if and only if for every $n$, $\psi_x(\overline{n})$ is true. Since $\Box\phi$ is the formula $Thm_T(subseq(\langle num(x)\rangle, \langle\overline{x}\rangle, \overline{\phi}))$, $\psi_x(\overline{n})$ is the formula $Thm_T(subseq(\langle num(\overline{n})\rangle, \langle\overline{x}\rangle, \overline{\phi})) \supset \phi_x(\overline{n})$. Here the value of the variable-free term $subseq(\langle num(\overline{n})\rangle, \langle\overline{x}\rangle, \overline{\phi})$ is $\phi_x(\overline{n})$, so $\psi_x(\overline{n})$ is true if and only if $Thm_T(\overline{\theta}) \supset \theta$ is true, where $\theta$ is $\phi_x(\overline{n})$. That this formula is true again follows since "every axiom of $T$ is true" implies "every theorem of $T$ is true", so $Thm_T(\overline{\theta})$ implies $True(\overline{\theta})$.

That $T_{lim(e)}$ is sound if $T_0$ is sound and $T_{\{e\}(n)}$ is sound for every $n$ such that $\{e\}(n)$ is defined is also easily proved in $\mathrm{PA}^{True}$. (The situation is simpler than for $\Sigma$-soundness, because combining sound theories always yields another sound theory, whereas $\Sigma$-sound theories need not in general be logically compatible.) It follows that if proof by induction over $<_{\mathbf{K}}$ restricted to $\{b \mid b \leq_{\mathbf{K}} n\}$ can be carried out in $T^{True}$, it can be proved by induction in $T^{True}$ that $T_b$ is sound for every $b \leq_{\mathbf{K}} n$. As in the case of consistency progressions, if $\Omega(\overline{n})$ is provable in a theory $T^{True}$, proofs by induction on $<_{\mathbf{K}}$ restricted to $\{b \mid b \leq_{\mathbf{K}} n\}$ can be carried out in $T^{True}$ with respect to any formula $\phi(x)$ in the language of $T^{True}$. This again follows because we can take a proof of

$$\exists x(x \leq_{\mathbf{K}} \overline{n} \land P(x)) \supset \exists x(x \leq_{\mathbf{K}} \overline{n} \land P(x) \land \neg\exists y(y \leq_{\mathbf{K}} \overline{n} \land y <_{\mathbf{K}} x \land P(y)))$$

and replace $P(x)$ everywhere by $\phi(x)$ to obtain a proof of the relevant induction principle. Here we must note that this substitution will work only because the induction schema also covers formulas containing the new predicate *True*. Otherwise the uses of ordinary induction involving $P$ could not be replaced by induction involving $\phi(x)$.

Given this we can, analogously to the case of consistency sequences and $\Sigma_1$-soundness, make the following

> **Observation:** Every theory in an autonomous uniform reflection progression based on $T$ is extended by $T^{True}$.

We prove this by induction on autonomous notations. $T_0$ is by definition extended by $T^{True}$. Now assume $a$ is in the autonomous part and $\Omega(\overline{b})$ is provable in $T_a$. By the induction hypothesis, $T^{True}$ extends $T_a$, so $\Omega(\overline{b})$ is provable in $T^{True}$. It follows by induction in $T^{True}$, as indicated above, that $T_b$ is sound, and so by the Tarski equivalences the axioms of $T_b$ are provable in

$T^{True}$ (since they are defined by a $\Sigma$-formula, so that every axiom of $T_b$ can be proved in $T_0$ to be an axiom of $T_b$), showing that $T^{True}$ extends $T_b$.

## Iterated reflection extensions and soundness

As in the case of iterated consistency extensions, the epistemological conclusions arrived at above depend on our accepting $T$ as sound. If we don't know $T$ to be sound, there is no obvious reason why we should accept every autonomous extension by reflection, and there is no clear candidate for a simpler extension of $T$ that extends every extension by reflection that can be specified on the basis of $T$. And indeed, unlike the case of consistency extensions, there is no obvious justification for accepting even a one-step extension by reflection of $T$ except that $T$ is sound.

In the case of consistency extensions, we could formulate an arithmetically definable soundness condition—that of $\Sigma$-soundness—which is inherited by consistency extensions, but noted that merely knowing that $T$ satisfies this condition does not in any obvious way tell us what consistency extensions are acceptable on the basis of $T$. The situation is similar if we consider extensions obtained by adding "$T$ is $\Sigma$-sound", which as noted in §13.3 can be formulated as an $\Pi_2$-sentence. By an argument similar to that used in the consistency case, it follows that if $T$ is $\Sigma_2$-sound, then so is $T +$ "$T$ is $\Sigma$-sound", and we can show that the extension of $T$ by the principle "$T$ is $\Sigma_2$-sound" extends all autonomous extensions by iteration of "$T$ is $\Sigma$-sound".

When we go the whole hog and extend $T$ by the uniform reflection principle, there is a significant difference, in that in order to extend the union of all autonomous extensions by reflection, we need to introduce an extension of $T$ which is no longer arithmetical—here a truth extension, which is still numerical, but it could also have been ACA or a theory which allows the direct expression of inductive definitions. Epistemologically, this is not a great step, in the present exposition, since it has been emphasized that $T$ is assumed to be accepted as sound anyway, so that to introduce a truth extension is just to make this explicit. When we start considering iterated truth extensions in §15.1, the epistemological aspects will become progressively less clear-cut.

## 14.3. Feferman's completeness theorem

Feferman proved, in Feferman [1962], a far-reaching extension of Turing's completeness theorem for progressions based on unlimited uniform reflection:

> **Completeness theorem:** For any uniform reflection progression, there is a branch $B$ in $\mathbf{O}$ such that there is, for any true arithmetical sentence $\phi$, an $a$ in $B$ with $|a| < \omega^{\omega^{\omega+1}}$ for which $\phi$ is provable in $T_a$.

Thus there is a reflection sequence of length $\omega^{\omega^{\omega+1}}$ based on PA in which every true arithmetical sentence is provable. This depends on choosing a very special branch in **O**. By a theorem proved by Feferman and Spector, there are also branches of length $\omega_1^{CK}$, the maximal length possible, which don't even prove every true $\Pi_1$-sentence.

Feferman's completeness theorem, like Turing's, doesn't have any general epistemological significance for arithmetic. That is, there is no hint in the proof of the theorem of any way in which arithmetical truths in general can be formally derived from axioms that we recognize as valid. The part of Feferman's work that is directly relevant to the concerns of this book is rather that connected with reflection principles and autonomous progressions. However, apart from its generally interesting aspects, a sufficient reason for studying the proof of the completeness theorem is that a nagging sense that there is something mysterious going on is otherwise likely to remain. Here only part of the proof will be considered.

It should be clear in a general way from the discussion of Turing's much simpler result that a reflection sequence proving every true arithmetical sentence must make some rather remarkable use, at limit ordinals, of convoluted non-standard definitions of the union of the sets of axioms of earlier theories in the sequence. (Among the sequences of length $\omega^{\omega^{\omega+1}}$ it is easy to exhibit one that uses canonical definitions.) Feferman's proof of the completeness theorem rests on three main constructions. One is Turing's original completeness proof, another is Shoenfield's proof of the completeness of a certain infinite inference rule (called the recursive $\omega$-rule), and the third is a fairly technical combination of applications of the recursion theorem, by which Feferman greatly extends the range of Turing's argument.

Turing's construction was presented in Chapter 13. Feferman uses a slightly strengthened version: For any consistency progression, any $a$ in **O** and any true $\Pi_1$-sentence $\phi$, there is a $b \in \mathbf{O}$ with $a <_{\mathbf{O}} b$ and $|b| = |a|+\omega+1$ such that $\phi$ is provable in $T_b$. The proof is as in §13.2, except that instead of $\{e\}(n) = n_{\mathbf{O}}$ we define $\{e\}(n) = suc^n(a)$, where $suc^0(a) = a$ and $suc^{n+1}(a) = suc(suc^n(a))$. Note that in this notation, $n_{\mathbf{O}}$ as defined in §11.2 is $suc^n(0)$, and using the operation $x \oplus y$ defined in §11.2, we could instead define $suc^n(a)$ as $a \oplus n_{\mathbf{O}}$. We also need to note that Turing's completeness theorem applies to reflection extensions as well as to consistency extensions, since it only requires that the consistency of earlier theories in a progression can be proved in later theories.

Shoenfield's proof of the completeness of the recursive $\omega$-rule has several interesting aspects, but is not directly relevant to the questions considered in this book. To keep the exposition within reasonable bounds, the presentation here will therefore focus on Feferman's extension of Turing's proof to the case of $\Pi_2$-sentences, which should give the reader some appreciation of how a result of this kind can be obtained.

> **$\Pi_2$-completeness theorem** (Feferman): For any progression based on the uniform reflection principle and every true $\Pi_2$-sentence $\phi$, there is an $a$ with $|a| = \omega^2 + \omega + 1$ such that $\phi$ is provable in $T_a$.

The significance of the value $\omega^2 + \omega + 1$ will become clear at the end of the argument.

Primitive recursive functions play a large role in the proof of the $\Pi_2$-completeness theorem, for reasons shown by the following argument. Suppose $\forall x \exists y \psi(x, y)$ is a true $\Pi_2$-sentence. Then for every $n$ there is a smallest proof $f(n)$ in $T_0$ of $\exists y \psi(\overline{n}, y)$, by the $\Sigma$-completeness theorem. $f$ is computable, but may or may not be primitive recursive. Suppose $f$ is primitive recursive. Then (by the substitution property stated in §9.4) the formalization $\theta$ of "for every $n$, $f(n)$ is a proof in $T_0$ of $\exists y \psi(\overline{n}, y)$" is equivalent in $T_0$ to a $\Pi$-formula, and we can apply Turing's completeness theorem to conclude that $\theta$ is provable in some $T_a$, where $|a| = \omega + 1$. We can then use the uniform reflection principle for $T_0$ to prove $\forall x \exists y \psi(x, y)$ in $T_a$.

Unfortunately such a proof cannot be carried out in general, because $f$, although computable, is not in general primitive recursive. To see this, first recall that as shown in §14.1, for any $T$ there is a $\Pi_2$-sentence, namely (a suitable formulation of) "$T$ is $\Sigma$-sound", which cannot be deduced in $T$ from any $\Pi$-sentence consistent with $T$. Now suppose that for every true $\Pi_2$-sentence there is a primitive recursive $f$ as above. Then the theory $T = T_0^{\Sigma\ REF}$ has the property that for any true $\Pi_2$-sentence $\phi$ there is a true $\Pi$-sentence which implies $\phi$ in $T$, implying that $T$ is not $\Sigma$-sound.

Although the simple proof of the $\Pi_2$-completeness theorem envisaged above thus cannot be carried out, we can prove the theorem along similar lines using two rather more complicated functions that *are* primitive recursive:

> **Lemma:** There are primitive recursive functions $Q$ and $H$ such that for every $a$ in $\mathbf{O}$ and every sentence $\phi$, if $\phi$ is provable in $T_0$, $Q(\phi, a)$ is a proof in $T_0$ of the formalization of "$\phi$ is provable in $T_{H(\phi,a)}$", where $a <_{\mathbf{O}} H(\phi, a)$ and $|H(\phi, a)| = |a| + \omega * m$ for some $m$ (depending on $\phi$).

Note that there can't be any primitive recursive (or indeed computable) function $R$ such that for every theorem $\phi$ of $T_0$, $R(\phi)$ is a proof in $T_0$ of $\phi$, since we could then for example decide whether a $\Sigma_1$-formula $\phi$ is true or false by checking whether $R(\phi)$ is a proof of $\phi$ in $T_0$. The function $Q$, however, does not give us any method to decide whether $\phi$ is true, for even if $Q(\phi, 0)$ is indeed a proof in $T_0$ showing that $\phi$ is provable in $T_{H(\phi,0)}$, this doesn't tell us that $\phi$ is true as long as we don't know whether $H(\phi, 0)$ is in $\mathbf{O}$. $H(\phi, 0)$ is guaranteed to be in $\mathbf{O}$ only if $\phi$ is provable in $T_0$. So there is no obvious reason why there shouldn't be such primitive recursive functions $Q$ and $H$, but nor is it at all obvious how such functions might be defined.

Postponing the proof of the lemma, we first see how it can be used to prove the $\Pi_2$-completeness theorem, by a proof related to the unsuccessful argument attempted above. Suppose $\forall x \exists y \psi(x,y)$ is a true $\Pi_2$-sentence, and let $\phi(x)$ be $\exists y \psi(x,y)$. For every $n$, $\phi(\overline{n})$ is provable in $T_0$. We define primitive recursive functions $F$ and $G$ by the equations

$$F(0) = H(\phi(\overline{0}), 0)$$
$$F(n+1) = H(\phi(\overline{n}), F(n))$$
$$G(0) = Q(\phi(\overline{0}), 0)$$
$$G(n+1) = Q(\phi(\overline{n}), F(n))$$

$F(0) <_{\mathbf{O}} F(1) <_{\mathbf{O}} F(2) \ldots$ is a sequence of notations in $\mathbf{O}$, and for every $n$, $G(n)$ is a proof in $T_0$ that $\phi(\overline{n})$ is provable in $T_{F(n)}$. Let $\{e\}(n) = F(n)$ for every $n$. $lim(e)$ is then a notation for $\omega^2$, since each $F(n)$ is a notation for $\omega * m$ for some $m > 0$ depending on $n$ and the supremum of any sequence $\omega * m_1, \omega * m_1 + \omega * m_2, \ldots$ is $\omega^2$. Further, since it is provable in $T_0$ that $T_{lim(e)}$ extends every theory $T_{F(n)}$, we get from $G$ a primitive recursive $G'$ such that $G'(n)$ is a proof in $T_0$ that $\phi(\overline{n})$ is provable in $T_{lim(e)}$. (We just take the proof $G(n)$ and transform it into $G'(n)$ by adding an invocation of the proof in $T_0$ that everything provable in $T_{F(n)}$ is provable in $T_{lim(e)}$.) Since $G'$ is primitive recursive, we can apply Turing's completeness result (in the strengthened form given above) to conclude that there is a $b$ in $\mathbf{O}$ with $lim(e) <_{\mathbf{O}} b$ and $|b| = \omega^2 + w + 1$ such that $T_b$ proves "for every $n$, $G'(n)$ is a proof in $T_0$ that $\phi(\overline{n})$ is provable in $T_{lim(e)}$". Since $T_b$ proves the uniform reflection principle for both $T_0$ and $T_{lim(e)}$, we can conclude in $T_b$ that $\phi(\overline{n})$ is provable in $T_{lim(e)}$ for every $n$, and from this draw the conclusion $\forall x \phi$.

So what remains is to prove the lemma. We will introduce some ad hoc notation. $e_{d,k}$ means $s_1^2(e,d,k)$, so that $\{e_{d,k}\}(n) = \{e\}(d,k,n)$. Similarly, $e_d$ is defined as an abbreviation of $s_1^1(e,d)$. Recall that $\{e\}$ is defined using Kleene's $T$-predicate by $\{e\}(n) = U(\mu x T^1(e,x,n))$. We say that a number $m$ is the value of $\{e\}$ obtained at stage $k$, or more briefly just the *$k$-value* of $\{e\}$, if $k = \mu x T^1(e,x,n)$ and $m = U(k)$. The point of this definition is that the relation "$\{e\}$ has a $k$-value" is primitive recursive, and there is a primitive recursive function $U$ such that if $\{e\}$ has a $k$-value, that value is $U(k)$.

For the proof we need to define primitive recursive functions that involve deciding whether something is a proof in $T_0$. For this reason we will need PR-proofs in $T_0$, as defined in §14.1. By a "proof in $T_0$" will be meant, throughout the proof of the lemma, a PR-proof in $T_0$.

We begin the proof by invoking the recursion theorem to conclude that there is a primitive recursive function $E$ of two arguments such that, setting $e = E(\phi, a)$, it is provable in PA that for all $\phi, a, w, k, n$,

$$\{e_{w,k}\}(n) = \begin{cases} 0 & \text{if there is an } x < k \text{ such that } x \text{ is a proof in } T_0 \text{ of } \phi, \\ suc^n(a) & \text{if } k \text{ is the smallest proof in } T_0 \text{ of } \phi, \end{cases}$$

and otherwise (that is, if there is no $x \leq k$ such that $x$ is a proof in $T_0$ of $\phi$)

$\{e_{w,k}\}(n) = suc^n(lim(e_{w,k+1}))$ if $\{w_{k+1}\}$ has no $n$-value or has an $n$-value which is a proof in $T_0$ that $\phi$ is provable in $T_{lim(e_{w,k+1})}$, and otherwise $\{e_{w,k}\}(n) = suc(lim(e_{w,k}))$.

We will say that *the first case* applies if there is an $x \leq k$ such that $x$ is a proof in $T_0$ of $\phi$. The primitive recursive function $E$ is obtained through the recursion theorem as explained in §11.1.

To make some sense of the above, first consider what is required for $lim(e_{w,k})$ to be in **O**. There are two possibilities. First, if $a$ is in **O** and $k$ is the smallest proof in $T_0$ of $\phi$, $lim(e_{w,k})$ is a notation for $|a| + \omega$. Second, if the first case does not apply, $lim(e_{w,k+1})$ is a notation, and every value of $\{w_{k+1}\}$ is a proof in $T_0$ that $\phi$ is provable in $T_{lim(e_{w,k+1})}$, then $lim(e_{w,k})$ is a notation for $|\,lim(e_{w,k+1})| + \omega$.

Next we note that there is a primitive recursive function $G_1$ such that if $k$ is a proof of $\phi$ in $T_0$, $G_1(\phi, a, d, k)$ is a proof in $T_0$ that $\phi$ is provable in $T_{lim(e_{d,k})}$. This is a consequence of our having defined $T_{lim(e)}$ generally to be an extension of $T_0$. $G_1(\phi, a, d, k)$ is defined to be 0 if $k$ is not a proof of $\phi$ in $T_0$, and otherwise a formalization in $T_0$ of the argument "$k$ is a proof of $\phi$ in $T_0$ and $T_{lim(e_{d,k})}$ extends $T_0$, so $\phi$ is provable in $T_{lim(e_{d,k})}$". $G_1$ is primitive recursive since its value is obtained by a substitution operation involving $d$ and $k$ (while $\phi$ is just carried along and $a$ is not used).

Less obviously, there is a primitive recursive function $G_2$ such that if the first case does not apply and if $r$ is a proof in $T_0$ that $\{d\}^2$ is total, $G_2(\phi, a, d, k, r)$ is a proof in $T_0$ that $\phi$ is provable in $T_{lim(e_{d,k})}$. This is the proof $G_2(\phi, a, d, k, r)$ obtained by formalizing in $T_0$ the following reasoning:

> The first case does not apply (this is established by a computation). If $T_{lim(e_{d,k})}$ is inconsistent, anything is provable in the theory, and in particular $\phi$. If $T_{lim(e_{d,k})}$ is consistent, it cannot be the case that $\{e_{d,k}\}(n) = suc(lim(e_{d,k}))$ for any $n$, since the theory would then prove its own consistency and so be inconsistent. So we must have that for every $n$, if $\{d_{k+1}\}$ has an $n$-value, it is a proof in $T_0$ that $\phi$ is provable in $T_{lim(e_{d,k+1})}$. Since $\{d\}^2$ is total (here the proof $r$ is used to show this), it follows that there is in fact a proof in $T_0$, which is also a proof in $T_{lim(e_{d,k})}$, that $\phi$ is provable in $T_{lim(e_{d,k+1})}$. Since $T_{lim(e_{d,k})}$ contains the reflection principle for $T_{lim(e_{d,k+1})}$ as an axiom, it follows that $\phi$ can be proved in $T_{lim(e_{d,k})}$. (Note that we only need the local reflection principle for this proof.)

We set $G_2(\phi, a, d, k, r)$ to be 0 if the first case applies or if $r$ is not a proof in $T_0$ that $\{d\}^2$ is total. $G_2$ is primitive recursive because $\phi, a, d, k, r$ only enter into the proof $G_2(\phi, a, d, k, r)$ through simple substitutions and insertions.

We next define $\{d\} = \{G_3(\phi, a)\}$, where $G_3$ is primitive recursive, by the recursion theorem so that

$$\{d\}(k,i) = \begin{cases} 0 & \text{if } \phi \text{ has a proof in } T_0 \text{ smaller than } k, \\ G_1(\phi, a, d, k) & \text{if } k \text{ is the smallest proof in } T_0 \text{ of } \phi, \end{cases}$$

and if the first case does not apply,

$\{d\}(k,i) = G_2(\phi, a, d, k, r)$, where $r$ is a proof in $T_0$ that $\{d\}^2$ is total, that is, a proof of the sentence $\psi(\overline{d})$, where $\psi(x)$ is a formula formalizing "$\{x\}^2$ is total".

Here we need to convince ourselves, first that the recursion theorem does in fact allow us to define such a $d$, and second that $r$ can be defined as a primitive recursive function of $\phi, a$. For this, we consider a more general formulation. We wish to show that given an $h$ such that there is a proof in $T_0$ that $H$ is total, we can define $F = \{f\}^4$ so that for some primitive recursive $P$,

$F(a,b,c,d) = H(f,a,b,c,d,P(a,b))$, where $P(a,b)$ is a proof in $T_0$ that $\{s_2^2(f,a,b)\}^2$ is total.

For this, we first note that there is a primitive recursive function $R$ such that if $p$ is a proof in $T_0$ that $\{m\}^5$ is total, $R(m,p,n,a,b)$ is a proof in $T_0$ that $\{s_2^2(s_4^1(m,n),a,b)\}^2$ is total. We then find, using the recursion theorem, $f_1$ so that

$$f_1(u,a,b,c,d) = H(s_4^1(f_1,u),a,b,c,d,R(f_1,u,u,a,b))$$

Now let $q$ be a proof in $T_0$ that $\{f_1\}^5$ is total, and define $f = s_4^1(f_1,q)$ and $P(a,b) = R(f_1,q,q,a,b)$. We then have

$$\begin{aligned}\{f\}(a,b,c,d) &= \{s_4^1(f_1,q)\}(a,b,c,d) = \{f_1\}(q,a,b,c,d) \\ &= H(s_4^1(f_1,q),a,b,c,d,R(f_1,q,q,a,b))\end{aligned}$$

Here, since $q$ is a proof in $T_0$ that $\{f_1\}^5$ is total, $R(f_1,q,q,a,b)$ is a proof in $T_0$ that $\{s_2^2(s_4^1(f_1,q),a,b)\}^2$ is total.

We can now finally state that if $\phi$ is provable in $T_0$ then for $k$ smaller than or equal to the smallest proof of $\phi$ in $T_0$, $\{d\}(k,i)$ is a proof in $T_0$ that $\phi$ is provable in $T_{lim(e_{d,k})}$, and thus we can define $Q(\phi,a)$ as $\{d\}(0,0)$ and $H(\phi,a)$ as $lim(e_{d,0})$. $H$ is primitive recursive, since $e$ and $d$ are primitive recursive functions of $\phi$ and $a$. $Q$ is primitive recursive since $\{d\}(0,0) = G_2(\phi,a,d,0,r)$, where $G_2$ is primitive recursive and $d$ and $r$ are primitive recursive functions of $\phi$ and $a$. As indicated in the remarks following the statement of the lemma, an inductive proof shows that if $a$ is in **O** and $k$ is the smallest proof of $\phi$ in $T_0$, $lim(e_{d,i})$ is in **O** for $i \leq k$ and $|lim(e_{d,0})| = |a| + \omega * (k+1)$.

What the argument does is to define a sequence of notations "backwards", with earlier notations being defined in terms of later ones. Again it is only at limit ordinals that we can introduce a nonstandard definition of the axioms of a theory, which is why the final sequence has length $\omega * (k+1)$ instead of $k+1$. If $\phi$ is in fact provable in $T_0$, we get a reflection sequence by reading the sequence in the opposite direction. As in Turing's construction, the description of the axioms at limit ordinals actually describes the union of axioms of earlier theories in the sequence only if a certain $\Pi$-sentence is true, but this $\Pi$-sentence is now a particularly convoluted sentence carrying the information that it is provable in $T_0$ that $\phi$ is provable in the earlier theory in the sequence. This information can be formulated as a $\Pi$-sentence only because it can be "wrapped in" a $\Pi$-sentence of the form "every value of $\{f\}$ is a proof in $T_0$ that $\phi$ is provable in $T_k$". The way this is done, using the recursion theorem, is perhaps the most distinctive feature of the argument. If $\phi$ is not provable in $T_0$, the whole sequence makes no particular sense.

## 14.4. Accepting reflection principles

Is it reasonable to say that accepting the local or uniform reflection principle for a theory $T$ is implicit in our accepting the theory $T$ itself? This will depend on what "accepting $T$" amounts to. The discussion in this section will not be restricted to numerical theories $T$, but also applies to such theories as CA (presented in Chapter 10) and even to strong axiomatic set theories such as ZFC, Zermelo-Fraenkel set theory with the axiom of choice.

If by accepting $T$ we mean that we accept that all the axioms of $T$ are true, we will accept $T$ extended by a reflection principle in the same sense, since we can prove that every instance of the reflection principle is true if every axiom of $T$ is true. Here it must be kept in mind that accepting that the axioms of $T$ are true does not entail any particular view about the nature of mathematical truth, such as the view that mathematical truth is a matter of what is the case in a mathematical reality, or the view that mathematical truth is a matter of what we stipulate or prove or in some way find useful or pleasing to the intellect. To accept that the axioms of $T$ are true here only means to accept, in the same sense as we accept mathematical axioms or theorems in general, the mathematical statement "the axioms of $T$ are true", where "true" is a primitive or defined mathematical term, satisfying a set of basic conditions. Thus in the case of arithmetical sentences, as stated in §7.6, what is required of "true" is only that for all arithmetical sentences $\phi$, $\phi_1$, $\phi_2$,

if $\phi$ is atomic, $\phi$ is true if and only if $\phi$ is in $True_{VF}$,

$\phi_1 \lor \phi_2$ is true if and only if $\phi_1$ is true or $\phi_2$ is true,

$\neg\phi$ is true if and only if $\phi$ is not true, and

$\exists x\phi$ is true if and only if $\phi_x(\overline{n})$ is true for some $n$.

We may accept these principles, and the principle "the axioms of $T$ are true", either because they are themselves ordinary mathematical theorems (as is the case when we are talking about PA and arithmetical truth), or because we find them acceptable in the same way that we find other primitive mathematical concepts and axioms acceptable. The essential point in the present context is only that "the axioms of $T$ are true" mathematically implies "the axioms of $T^{REF}$ are true", so if we accept the axioms of $T$ as true, we will accept the axioms of $T^{REF}$ as true.

This point merits underlining because the impact and associations of the words "true" and "truth" are such that mathematicians sometimes shy away from the statement that the theorems of PA are true, even though this can be understood as a simple mathematical theorem about PA, proved using set-theoretical concepts and principles that the same mathematicians otherwise don't hesitate to use in their work.

Thus there are two cases that seem clear enough. First, if we don't accept the axioms of $T$ at all, we have no reason to accept a reflection principle for $T$. Second, if we accept the axioms of $T$ in the strong sense of accepting that they are true, we have the same grounds for accepting the uniform reflection principle for $T$ as we have for accepting $T$. But now let us consider on what grounds we might regard the reflection principle for a theory $T$ as dubious or not necessarily acceptable, even while accepting $T$ in a weaker sense, as part of the theoretical background or basis of mathematics.

First we need to set aside a particular kind of doubt about reflection principles. From a mathematical point of view, reflection principles may be seen as unsatisfactory because they are so easily come by. Consider the local reflection principle for PA applied to the statement "there are infinitely many twin primes", formalized as a sentence $\phi$. We can formulate this as

(1) For every $p$ and every $k$, if $p$ is a proof in PA of $\phi$, there is an $n > k$ such that $n$ and $n + 2$ are both primes.

If we present this statement to a mathematician as an assertion, he may take us to be holding out a promise of a mathematically interesting or illuminating connection between $p$, $k$ and $n$, showing how to get or extract a twin prime $n > k$ from $p$ and $k$. And indeed in the subject of proof theory, one does seek such mathematically interesting or illuminating general connections between proofs of existential statements and numbers instantiating those statements. However, in presenting (1) as a consequence of the reflection principle, which in turn is a mathematical consequence of "all axioms of PA are true", we are giving a proof of (1) which is trivial from a mathematical point of view, in that it does not establish any particular connection between proofs and primes, but only ascends to a higher level of mathematical abstraction in order to painlessly arrive at (1). This may not be very satisfying mathematically, although from a more philosophical point of view it is a striking illustration of the logical power of abstract concepts in mathematics. Thus from a mathematical point of view, (1) presented as an instance of reflection is perhaps not very striking or interesting, but this sort of disappointment with (1) is not what we're after in asking whether reflection principles can be regarded as dubious or not necessarily acceptable, even if the theory, like PA or CA, formalizes part of ordinary mathematics.

One relevant line of thought in this connection is to hold that mathematical statements can only be understood in conjunction with and in relation to some formal theory in which they are derivable. Thus for example (on this view) the statement "there is a non-measurable set of real numbers in the interval $[0, 1]$" is "true" in the sense that a formalization of this statement is derivable in a certain formal theory $T$ which formalizes concepts and principles used by mathematicians. It doesn't really make any sense to inquire into the acceptability of a mathematical statement of the form "if $\phi$ is provable in $T$ then $\phi$". All we can say is that we can get a new system $T'$ by adding this statement as a new axiom, and other statements will then be "true" in the sense of being provable in this new system. Such a view is often spoken of as "formalism", and without associating it with any particular philosopher or mathematician, it will be referred to below as "a formalist view".

A formalist view is never expressed using an example such as "$0 = 1$" rather than "there is a non-measurable subset of $[0, 1]$". That is, it is not said that if $0 = 1$ is provable in $T$, then $0=1$ is "true" in the sense of being provable in $T$, and that's all there is to it. Rather, everybody seems to accept that $0 = 1$ is understood independently of the formal theory $T$ as a false statement. There would be no conflict between this understanding of $0 = 1$ and accepting $0 = 1$ as "true" in the sense of being provable in $T$ if we did not assume any relationship between the arithmetical principles according to which $0 = 1$ is false and the theory $T$, but in fact we do assume such a relationship. In mathematically deriving formalizations of computational statements of various kinds, we want and expect the statement derived to be true as ordinarily understood. This is not restricted to computational statements. If a theory $T$ proves that a certain large number is composite, or that there are infinitely many twin primes, or that a certain theory is inconsistent, or that there is no polynomial-time algorithm for factorization, or that a certain system of differential equations has no stable solution, we draw conclusions from this (assuming $T$ to formalize a part of ordinary mathematics) that presuppose an understanding or interpretation or use of the theorem in question which is independent of the theory $T$, and which gives good sense to the question whether these consequences of $T$ are true. Arguing this point is not part of the inquiry to which this book is dedicated (but see Franzén [1987]), so in the remainder of the discussion it will simply be assumed that we are concerned with reflection principles for which we do think it makes sense to ask whether they are acceptable. In practice, this attitude is universal in the case of $\Pi_1$-reflection, or in other words, it seems pretty universally accepted that we may sensibly inquire whether we have any good grounds for accepting "$T$ is consistent".

Even if we do not lean towards a formalist view of mathematical statements, the above examples highlight a difference between two kinds of instances of local reflection, those in which we know $\phi$ to be false and those in which we do not. The statement "if it is provable in $T$ that there are infinitely many twin primes, then there are infinitely many twin primes" may seem acceptable merely on the basis that we expect to be convinced by any actual proof formalizable in $T$ that there are infinitely many twin primes. On the other hand, assent to "if it is provable in $T$ that $0 = 1$, then $0 = 1$" cannot be based on just on a readiness to accept future theorems proved in $T$, but requires us to assert that $T$ is consistent, which people are often reluctant to do. It may be argued that in fact accepting $T$ does not justify the latter conclusion. Mathematicians, according to this line of argument, accept $T$ only in the sense that they accept any mathematical statement as a theorem once it has been proved in $T$. Therefore, what is implicit in accepting $T$ is not the reflection principle for $T$, but only the reflection *rule*, whereby $\phi$ can be concluded from "$\phi$ is a theorem of $T$". In the formulation of logic used in this book, this means that we do not add the reflection axioms to $T$, but instead include a new rule for provability in $T$:

If $T' \Rightarrow \Box\phi$ then $T \Rightarrow \phi$

for sentences $\phi$, where $T'$ is some theory (perhaps $T$ itself) which we use to prove that $\phi$ is provable in $T$ (using the other rules). This rule is clearly *conservative* over $T$, that is, it does not allow us to prove anything not already provable in $T$, provided $T'$ is $\Sigma$-sound (although it may still be useful in practice, as commented on in §1.3).

This argument may be convincing if it is in fact the case that the theory $T$ is accepted only in the sense that a sentence is accepted as a theorem once we have *inspected* a proof in $T$ of the sentence. But formal theories such as PA and ZFC are accepted in mathematics in a stronger sense. We accept many theorems without having carried out or scrutinized any proof of the theorem, on the basis of the knowledge that it has been proved. In explaining just what it is that is assumed when we say that we regard a theorem as proved even though we haven't actually checked the proof, reference is often made to some formal theory in which a formal derivation of a formalization of the theorem in question is assumed to exist. It is difficult to see any justification for taking a theorem to be true on this basis which does not involve accepting the reflection principle as a hypothetical statement rather than only as a rule of inference. If we are not prepared to assert the hypothetical statement "if $\phi$ is provable in $T$ then $\phi$", on what basis do we conclude $\phi$ given only the information that $\phi$ is provable in $T$ (rather than a proof in $T$ which we have subjected to scrutiny)?

This point can be underlined by an illustration using the stronger reflection rule corresponding to the formulation (6) of the uniform reflection principle in §14.1: from "$\phi_x(\overline{n})$ is provable in $T$ for every $n$", conclude "$\forall x\phi$. This rule is *not* conservative over $T$, but can be used to prove that $T$ is consistent, by a special case of the argument showing that (6) implies (3). Given any derivation $k$ in $T$, either $k$ is a derivation of a contradiction, in which case anything and in particular "$k$ is not a derivation of a contradiction" is provable in $T$, or else $k$ is not a derivation of a contradiction in $T$, in which case again "$k$ is not a derivation of a contradiction" is provable in $T$. So for every $k$, "$k$ is not a derivation of a contradiction in $T$" is provable in $T$, so $T$ is consistent, by the reflection rule.

This consistency proof is a rather odd one, since one branch of the argument leading to the conclusion that $T$ is consistent assumes that $T$ is inconsistent (without deducing any contradiction from this assumption). If the reflection rule expresses trust in or acceptance of $T$, how can we accept a conclusion ("$T$ is consistent") which may be based on $T$ being inconsistent? This oddity is perfectly compatible with the use of the reflection rule, suggesting that the reflection rule is not in fact a statement of any trust in or acceptance of $T$, unlike the reflection principle as it is actually used and understood, namely in the form of an implication.

The claim here then is that if we accept a rule of inference by reflection, and apply that rule in cases when we haven't in fact checked the proof in $T$ (which is the only case when there is any point in applying the rule), this is justified only if we accept the general hypothetical statement "anything provable in $T$ is true". To the extent that we accept mathematical statements as theorems on the basis of taking them to be provable in $T$, even if we have not actually inspected any proof of them in $T$, we are appealing to the local reflection principle.

Note that the point of the above comments is not to assert the consistency or soundness of PA or CA or of any other theory. The comments concern only what we implicitly accept in accepting a theory $T$, in the sense that we accept a statement as a theorem on the basis of the knowledge that it is provable in $T$. If we are in fact disinclined to assert the local reflection principle, or even to assert that $T$ is consistent, we should be equally disinclined to accept a statement proved in $T$ as a theorem before having checked the proof to see if it is acceptable or proves anything at all. This is not how mathematicians or others who use mathematics actually proceed, and so there are good reasons to say that they implicitly accept at least a local reflection principle for those theories that they refer to in explaining theoretical standards of rigor or justifying methods of proof.

In the case when we accept the axioms of a theory as true, it is a simple matter to observe that iterated unlimited uniform reflection principles are equally acceptable. When we accept the theory only in some weaker or at least different sense, the matter is far from straightforward, and the argument above has only concerned the acceptability of local reflection, applied to statements that we regard as meaningful considered without reference to formal theories. No attempt will be made to argue that stronger reflection principles are necessarily implicit in accepting a theory, although such an argument could certainly be made. Inexhaustibility, the main topic of this book, centers on the case when we accept the axioms of a theory as true.

CHAPTER 15

# ITERATED ITERATION AND INEXHAUSTIBILITY

## 15.1. Beyond reflection

### Arithmetical soundness

As we saw in Chapter 14, anything that can be proved in an autonomous extension of an arithmetical theory $T$ by iterated reflection can also be proved in a truth extension of $T$ (using the same definition of the axioms of $T$). Truth extensions, like second order extensions of PA, extend not only the axioms, but the language of PA in a way which cannot be represented as an extension by definitions of the arithmetical language. Arithmetical truth is not arithmetically definable, and we cannot in general reduce quantification over sets to quantification over numbers. A theory $T$ which extends PA and may have new kinds of variables, or symbols for non-arithmetical functions or predicates, or both, is *arithmetically sound* if every theorem of $T$ in the language of PA is true.

Given an arithmetically sound theory $T$, we can extend $T$ to a stronger theory that is still arithmetically sound by adding a local or uniform reflection principle for $T$ restricted to *arithmetical* $\phi$. If $T$ extends $PA^{True}$, so that we can formalize in $T$ statements about arithmetical truth, we get a still stronger extension of $T$ by adding instead the axiom "$T$ is arithmetically sound". In general, however, extending an arithmetically sound theory $T$ by this axiom need not even yield a consistent theory. This situation is familiar from earlier similar cases—extending a consistent theory $T$ by the axiom "$T$ is consistent", or a $\Sigma$-sound theory $T$ by the axiom "$T$ is $\Sigma$-sound" may yield an inconsistent theory. We can use the same construction as in those cases to get an example of an arithmetically sound theory $T$ which cannot be consistently extended by adding the axiom "$T$ is arithmetically sound". Let $T$ be the extension of $PA^{True}$ obtained by adding as a new axiom "$PA^{True}$ is not arithmetically sound". $T$ is an arithmetically conservative extension of $PA^{True}$, that is, every arithmetical theorem $\phi$ of $T$ is provable in $PA^{True}$. For if $\phi$ is provable in $T$, we can prove in $PA^{True} + \neg\phi$ that $PA^{True}$ is arithmetically sound, and thus that $\phi$ is not provable in $PA^{True}$ (since $\phi$ is not true), and so that $PA^{True} + \neg\phi$ is consistent, which means that $\mathrm{PA}^{True}+\neg\phi$ is inconsistent by the second incompleteness theorem. Thus $T$ is arithmetically sound, but cannot be consistently extended by the axiom "$T$ is arithmetically sound".

Adding the new axiom "$\mathrm{PA}^{True}$ is arithmetically sound" to $\mathrm{PA}^{True}$ does yield a consistent, and indeed arithmetically sound theory. As in the parallel cases, our justification for holding the extension to be arithmetically sound must invoke a stronger soundness property of $\mathrm{PA}^{True}$ than that of being arithmetically sound, and the obvious justification is that $\mathrm{PA}^{True}$ is itself sound. Here, since soundness means that every axiom is true, we must extend the definition of "true" so as to apply not only to arithmetical sentences, but to any sentence in the language of $\mathrm{PA}^{True}$, with the truth predicate in that language having the meaning of "is a true arithmetical sentence". This extension can be carried out in the obvious way, by replacing the clauses for the case where $\phi$ is atomic in the definition of arithmetical truth given in §7.6 by two new clauses:

> For atomic arithmetical $\phi$, if $\phi$ is in $True_{Vf}$ it is true, and otherwise it is false.
>
> If $\phi$ is an atomic sentence $True(t)$ and the value of $t$ is a true arithmetical sentence, $\phi$ is true, and if the value of $t$ is not an arithmetical sentence or is a false arithmetical sentence, $\phi$ is false.

As the reader will no doubt appreciate from the pattern of Chapters 13 and 14, the stage is now set for establishing that anything provable in an autonomous sequence of iterated extensions of $\mathrm{PA}^{True}$ by an axiom "$T$ is arithmetically sound" can be proved in a truth extension of $\mathrm{PA}^{True}$, and the natural step after this is to consider autonomous sequences of iterated truth extensions. The subject of iterated extensions by the axiom "$T$ is arithmetically sound" will be left to the reader, while the topic of iterated truth extensions requires some further comments, and will lead to the final observations on which the discussion of inexhaustibility that concludes the book will be based.

In defining a truth extension of $\mathrm{PA}^{True}$, we could choose to retain the single non-arithmetical predicate symbol *True*, and add axioms corresponding to the definition above. There is in such a theory not necessarily anything paradoxical about a fixed point $\phi$ of the formula $\neg\, True(v_0)$. For example, if $\psi$ is $True(\overline{0=0})$, and $\phi$ is $True(\overline{\psi})$, $\psi$ is provable, and also $\phi$, but $\neg\, True(\overline{\phi})$ is provable, since $\psi$ is not an arithmetical sentence. Thus we can prove

$$\phi \equiv \neg\, True(\overline{\phi})$$

This is only apparently paradoxical, because $\phi$ is not a sentence for which $True(\overline{\phi})$ means "$\phi$ is true". A similar example is the pseudo-Liar: "This statement is a false arithmetical statement". The pseudo-Liar is simply false, since it is not an arithmetical statement. The use of a single predicate *True* in a truth extension of $\mathrm{PA}^{True}$ is however potentially confusing, and it makes for a more readily intelligible theory if we introduce two truth predicates $True_1$

and $True_2$ such that $True_1(x)$ means "$x$ is a true arithmetical sentence" and $True_2(x)$ means "$x$ is a true sentence in the language of arithmetic extended with $True_1$". This is also the pattern to be followed below in defining iterated truth extensions.

The use of $True_1$ and $True_2$ is often associated with the idea of a hierarchy of truth predicates as a way of dealing with paradoxes like the Liar, and this idea in turn is often associated with Tarski. In the present discussion no general doctrines about truth, and no ideas about how to come to grips with the semantic paradoxes, are at issue. We need to use the concept of truth applied to various classes of mathematical statements, and are free to define it whatever way we like, as long as the definition suits our purposes. The hierarchy of truth predicates makes its appearance in this context because it makes iterated truth extensions easier to handle.

### Iterated truth extensions

We know from Chapter 10 that $\mathrm{PA}^{True}$ is equivalent, as far as arithmetical theorems are concerned, to arithmetical analysis, ACA, in which arithmetical sets can be defined and induction can be used with formulas quantifying over arithmetical sets. As was mentioned in §5.2, there is a connection between inexhaustibility, as represented by the introduction of ever more comprehensive truth predicates, and a hierarchy of definable sets, starting with arithmetical sets. Instead of iterating the extension of a theory by adding a truth predicate, we could iterate the step leading from PA to ACA, in the following sense: given a second order theory $T$ with different categories of second order variables, used to quantify over different families of subsets of $\mathbf{N}$, we add a new category of second order variables, and new comprehension axioms expressing that these new variables range over sets definable by formulas in the language of $T$. Iterating this construction yields a theory in which we can talk about a smaller or larger portion of a hierarchy of sets, known as the *ramified* or *constructible* hierarchy, where sets later in the hierarchy can be defined by quantifying over natural numbers and the earlier sets. In the presentation followed here, we will not introduce different categories of second order variables, but instead use truth predicates to introduce iterated truth extensions of a numerical theory. This allows us to stay within the realm of first order extensions of PA, and also connects more directly with the discussion in earlier chapters.

First we define what is meant by a truth extension of a numerical theory. We presuppose that an infinite number of unary predicate symbols—say (using the identification of numbers and symbols given in §7.1) the predicate symbols $\langle 1, 1, 100n\rangle$ for $n = 1, 2, \ldots$—have been set aside to be used as truth predicates. We denote these $True_1, True_2, \ldots$. For simplicity, we assume that the language of the theory $T$ extends the primitive language of arithmetic only with predicate symbols, so that the set of terms is the same as in PA.

Given a $\Sigma$-formula $\phi$ defining the axioms of $T$ and given a truth predicate $True_i = True$ which is not in the language of $T$, we define the truth extension $T^{True}$ of $T$ as follows. The language of $T^{True}$ is the language of $T$, extended with the new predicate symbol *True*. The axioms of $T^{True}$ are the axioms of $T$ together with the universal closures of

$$vft(x_1) \wedge \cdots \wedge vft(x_n) \supset (True(subseq(\langle x_1, \ldots, x_n\rangle, \langle \overline{v}_1, \ldots, \overline{v}_n\rangle, \overline{\psi})) \\ \equiv p(val(x_1), \ldots, val(x_n)))$$

for every $n$-ary predicate $p$ in the language of $T$, where $\psi$ is $p(v_1, \ldots, v_n)$ and $vft(x)$ formalizes "$x$ is a variable-free term",

$$True(dis(x,y)) \equiv True(x) \vee True(y),$$
$$True(neg(x)) \equiv \neg\, True(x), \text{ and}$$
$$True(ex(x,y)) \equiv \exists w\, True(sub(num(w), x, y)).$$

Here *dis*, *neg*, and *ex* are the primitive recursive functions defining disjunction, negation, and existential quantification of formulas.

To these new axioms we add all induction axioms in the extended language, and the single axiom

$$\forall x(\phi(x) \supset True(x)),$$

stating that all axioms of $T$ are true. This axiom is needed even when we use a canonical definition of the axioms of $T$, if we have infinitely many predicate symbols with corresponding characteristic axioms (as we do in infinite sequences of truth extensions). Finally, there is an axiom stating that $True(x)$ implies that $x$ is a sentence in the language of $T$.

It follows from this description of the axioms of $T^{True}$ that they are canonically defined, relative to the given formula $\phi$ and the predicate $True_i$, by a formula $\phi^{True}$ which is a primitive recursive function of $\phi$ and $i$, and is a $\Sigma$-formula if $\phi$ is. In considering iterated truth extensions, the axioms of $T^{True}$ will be assumed to be defined by $\phi^{True}$.

Now consider a sequence $T_\alpha$, $\alpha < \lambda$, of iterated truth extensions, corresponding to consistency sequences and reflection sequences, beginning with some arithmetical theory $T$, where the axioms of $T_\alpha$ are given by the $\Sigma$-formula $\psi_\alpha$. $T_{\alpha+}$ is a truth extension of $T_\alpha$, which uses the formula $\psi_\alpha$ to define the axioms of $T_\alpha$ and as truth predicate uses $True_{\psi_\alpha}$, and for limit ordinals $\beta$, $T_\beta$ is the union of the theories $T_\alpha$ for $\alpha < \beta$.

A small complication appears at this point. The theory $T_0$ is an arithmetical theory, and so does not contain any truth predicate. At each successor ordinal, a new truth predicate is added to the language—it follows by a simple induction that $True_{\psi_\alpha}$ is not in the language of $T_\alpha$—and the language of $T_\alpha$ contains the truth predicates $True_{\psi_\beta}$ for $\beta < \alpha$. However, there is no guarantee that the set of these truth predicates is decidable for every $\alpha$, and if the language of a theory is not decidable, it is moot whether we are dealing with a theory in the sense of a possible codification of human knowledge. To eliminate this possibility, we stipulate that the language of every theory in the sequence in fact contains every truth predicate (although the only axioms involving truth predicates are those introduced at successor ordinals). We accordingly weaken the requirement on the new truth predicate *True* in the definition of "truth extension", and require only that the theory $T$ does not have any axioms containing the predicate *True*. We define the language $L$ to be the extension of the arithmetical language by all the truth predicates $True_{\psi_\alpha}$, and the *restricted language* of $T_\alpha$ to be the arithmetical language extended by the truth predicates $True_{\psi_\beta}$ for $\beta < \alpha$.

We can now, given a sequence of iterated truth extensions, give an inductive definition of "true" for sentences in the language $L$:

> For atomic arithmetical $\phi$, if $\phi$ is in $True_{Vf}$ it is true, otherwise $\phi$ is false.
>
> If $\phi$ is an atomic sentence $True_{\psi_\alpha}(\boldsymbol{t})$ and the value of $\boldsymbol{t}$ is a true sentence in the restricted language of $T_\alpha$, $\phi$ is true, while if the value of $\boldsymbol{t}$ is not a sentence in that language or is a false sentence in that language, $\phi$ is false.
>
> If $\phi_1$ or $\phi_2$ is true, $\phi_1 \vee \phi_2$ is true, and if $\phi_1$ and $\phi_2$ are both false, $\phi_1 \vee \phi_2$ is false.
>
> If $\phi$ is true, $\neg\phi$ is false, and if $\phi$ is false, $\neg\phi$ is true.
>
> If $\phi_x(\overline{n})$ is true for some $n$, $\exists x\phi$ is true, and if $\phi_x(\overline{n})$ is false for every $n$, $\exists x\phi$ is false.

From this definition, we can conclude the following:

> For atomic arithmetical $\phi$, $\phi$ is true if and only if $\phi$ is in $True_{Vf}$.
>
> If $\phi$ is an atomic sentence $True_{\psi_\alpha}(\boldsymbol{t})$, $\phi$ is true if and only if the value of $\boldsymbol{t}$ is a true sentence in the restricted language of $T_\alpha$.
>
> $\phi_1 \vee \phi_2$ is true if and only $\phi_1$ or $\phi_2$ is true.
>
> $\neg\phi$ is true if and only if $\phi$ is not true.
>
> $\exists x\phi$ is true if and only if $\phi_x(\overline{n})$ is true for some $n$.

The proof of this follows the pattern of the proof of the axioms of $\mathrm{PA}^{True}$ from the definition of arithmetical truth. Here we use induction on the well-founded relation $R$ between formulas defined by

> $R(\phi_1, \phi_2)$ if and only if $r(\phi_1) < r(\phi_2)$, where $r(\phi)$ is 0 for arithmetical $\phi$ and otherwise $\alpha+$ for the largest $\alpha$ such that $True_{\psi_\alpha}$ occurs in $\phi$, or $r(\phi_1) = r(\phi_2)$ and $\phi_1$ contains fewer occurrences of connectives and quantifiers than $\phi_2$.

As in the proof in §7.6, the crucial point is that we need to be able to show that $\phi$ is false if and only if $\phi$ is not true, and the crucial case in showing this is when $\phi$ is an atomic sentence $True_{\psi_\alpha}(t)$. Here we need an induction hypothesis by which the value of $t$, if it is a sentence in the language of $T_\alpha$, is true if and only if it is not false. This induction hypothesis is just what is yielded by using the relation $R$ for the proof, since if $\psi$ is a sentence in the language of $T_\alpha$, $r(\psi) < \alpha+ = r(\phi)$.

Using this definition of truth for sentences in $L$, we can now prove by a lengthy but conceptually simple induction that if every axiom of $T = T_0$ is true, every theorem of $T_\alpha$ is true for every $\alpha < \lambda$, in any sequence of iterated truth extensions of $T$. It follows in particular that every $T_\alpha$ is arithmetically sound.

## Autonomous truth extensions and second order extensions

As with consistency sequences and reflection sequences, recursive progressions give us a means of obtaining explicit sequences of iterated truth extensions, with the theories $T_\alpha$ being replaced by $T_a$ for notations $a$, so that the relation $R$ becomes arithmetically definable. If we now consider an *autonomous* sequence of iterated truth extensions starting with $T$, we can again define an extension $T'$ of $T$ which is stronger than every theory in the autonomous sequence. $T'$ is shown to be stronger by formalizing in $T'$ the argument given above to prove that every theory in a sequence of iterated truth extensions is sound if the first theory is sound. To carry out such a proof, we need to be able to define truth for sentences in $L$ in $T'$. This means that we must be able to somehow express inductive definitions in $T'$. In terms of the theories introduced in this book, we need to go from the numerical theory $T$ to a second order theory. The natural candidate here is a theory called $\Pi^1_1$-CA, which is obtained from ACA by strengthening the comprehension axioms to

$$\exists X \forall x (x \in X \equiv \phi)$$

with $\phi$ any formula of the form $\forall Y \psi$, where $\psi$ contains no second order quantifier and does not contain $X$. This yields a theory in which we can formalize the inductive definition of truth for sentences in the language $L$ given above, and also, for an autonomous sequence of iterated truth extensions, prove that every theorem of every $T_a$ is true. $\Pi^1_1$-CA is not strictly speaking stronger than every $T_a$, because $\Pi^1_1$-CA does not contain any axioms involving the truth predicates of $T_a$. However, we can *define* those truth predicates in $\Pi^1_1$-CA—$True_{\psi_a}(n)$ is defined to mean that $n$ is a true sentence in the restricted language of $T_a$—and then prove the corresponding truth axioms. In particular, it follows that every arithmetical theorem of $T_a$ is provable in $\Pi^1_1$-CA. Although much weaker than full classical analysis CA, $\Pi^1_1$-CA is a strong theory and one that is difficult to analyze proof-theoretically. Here we will only consider in informal terms some aspects of this theory regarded as an extension of PA, and more generally some aspects of extensions that prove every arithmetical theorem of every iterated autonomous truth extension of an arithmetical theory.

## Iterated truth extensions and soundness

As in the case of extensions by uniform reflection, the basis for accepting any autonomous iterated truth extension of $T$ is that $T$ is sound. The soundness of $T$ implies that every iterated truth extension of $T$ is also sound. However, the reflection whereby we arrive at these extensions is less straightforward, in that we need to introduce a new truth predicate at each step, and an infinite series of truth predicates as soon as we get beyond the first $\omega$ extensions. We can no longer use only a truth theory for $T$ to show that the extension principle preserves soundness, but need to introduce a longer or shorter series of new truth predicates to establish the soundness of a theory. It might be argued that this is a significant distinction which renders it moot whether every autonomous iterated truth extension can be said to be justifiable by reflection starting from $T$. However, what is essential in this context is only that the autonomous truth extensions can be reasonably held to represent an *upper bound* on what can be achieved by reflection on the basis of $T$, if we take reflection to consist in an open-ended series of observations "$T$ is sound", "$T$ + '$T$ is sound' is sound", and so on, where the "and so on" may extend into the transfinite and be taken "as far as we please". These particular autonomous extensions together take the idea of iterating the recognition that a theory is sound as far as it can be taken on the basis of an initial theory $T$.

Thus the extension of PA to $\Pi^1_1$-CA is not in the same sense an extension by reflection. Rather, it introduces a new category of mathematical objects, and principles for reasoning about them. Using these principles we can prove any arithmetical theorem provable in an iterated autonomous truth extension of PA. We could choose instead another kind of extension of PA of similar logical strength, such as one in which inductive definitions can be directly expressed, but it will still be an extension of PA by concepts and principles that do not arise from "pure reflection" in the indicated sense.

## Extensions beyond reflection

$\Pi^1_1$-CA can of course itself be extended by reflection principles or truth extensions to stronger theories that are still sound, given that $\Pi^1_1$-CA is sound (in the sense of a definition of truth for second order theories such as that outlined in §10.3). However, extending PA to $\Pi^1_1$-CA opens the door to a number of possible extensions by principles that go beyond reflection. In particular, we can extend a theory by introducing axioms about sets of higher type—meaning sets of sets of natural numbers, sets of sets of sets of natural numbers, and so on—and by introducing stronger comprehension principles for sets of a given type. The latter kind of extension has a natural limit, namely the introduction of unrestricted impredicative comprehension principles for sets of all types, as CA contains an unrestricted impredicative comprehension principle for sets of natural numbers. Repeatedly extending a theory by postulating the existence of sets of higher type does not however converge to any final theory, but can always be continued, just like extensions by reflection. The types are indexed by ordinals, and autonomous extensions by axioms for sets of higher type can be defined, analogous to the autonomous extensions that have been discussed in this book. Axiomatic set theories like ZF give powerful first order theories which prove everything provable in such iterated autonomous extensions. In this connection the term "reflection" reappears and takes on a new meaning. The *set-theoretic reflection principle* states, informally, that everything true of a hierarchical universe of sets is true of some proper initial part of that hierarchy. This principle leads to a further indefinite sequence of extensions of set theory, and furthermore, "axioms of infinity" have been formulated which can reasonably be argued to be stronger, as far as arithmetical theorems are concerned, than any such extension by set-theoretical reflection.

In the above brief sketch, nothing was said about the justification for taking the various further possible extensions of a theory to be arithmetically sound, and indeed when we come to axioms of infinity that are apparently arithmetically stronger than the set-theoretical principle of reflection, it is doubtful that anybody would hold it to be evident that the resulting theory is arithmetically sound, or even consistent. The point of the description above is to indicate that a contemplation of iterated extensions of a sound arithmetical theory by reflection principles will eventually lead to contemplation of an equally open-ended series of extensions by principles that are no longer justifiable in the same terms as reflection principles. And with this, it is time to return to the lowly consistency statements and to the central questions of this book.

## 15.2. Inexhaustibility

Having dealt with the technicalities, we are now in a position to consider directly the questions posed in Chapter 1. Quoting from that chapter:

> There is something rather strange about the impossibility of pinning down which axioms we recognize as correct. Shouldn't it be possible to somehow incorporate in those axioms the very principle that any collection of valid axioms can be extended to a stronger collection, and thereby set down in axiomatic form all of our current mathematical knowledge? And if this really can't be done, do we have some sort of ineffable or unformalizable insight that cannot be fully expressed in rules? If so, what are the limits of

> what can be proved using this insight, and how can those limits be described? In particular, just what is it that happens when we try to systematically and ad infinitum extend a theory by adding successive consistency statements as new axioms?

The discussion of these questions will refer to a formal theory $T$, which we will call a *fundamental theory* , in which we have attempted to set down the axioms and rules on which our arithmetical knowledge is based. "Arithmetical" rather than "mathematical", since the inexhaustibility phenomenon associated with Gödel's theorem specifically concerns our arithmetical knowledge. $T$ does not have to be a numerical theory, however, but can include any mathematical concepts and principles that we are prepared to use in arriving at arithmetical theorems, such as concepts and principles of analysis, set theory or the theory of inductive definitions. If $T$ *is* numerical, a canonical definition of the axioms of $T$ does not have to follow the pattern of PA, but must of course be one that we understand—for example, in a theoretical context one might choose to define the axioms of $T$ as "the arithmetical consequences of ZF". In accordance with the first "Gödelian tradition" commented on in §1.2, no precise description will be given of $T$. $T$ will be assumed to be at least as strong as PA, in the sense that the natural numbers and the arithmetical functions and relations can be defined in $T$ and theorems corresponding to the axioms of PA can be proved, and it will also be assumed that the theorems of $T$ can be effectively generated. On this basis, it is expected that results corresponding to those proved in the body of the book for first and second order extensions of PA will apply to $T$.

What will a fundamental theory contain, and how can a fundamental theory be justified? There is a very extensive tradition in the philosophy and foundations of mathematics in which these questions are argued and different answers explored. There are many differences of opinion about the extent to which various mathematical concepts and principles make good sense, and about the evidence for or acceptability of various rules and axioms. Whatever our views may be, it is very likely that unless we take a consistently formalist view of mathematics (in the sense explained in §14.4), we will find some concepts and principles unproblematic and others highly problematic, with yet others falling somewhere in between. The problem posed by Gödelian inexhaustibility concerns only fundamental theories that we recognize as *correct*, and thereby as consistent. By the incompleteness theorem this means, as Gödel puts it, that we have a mathematical insight not derivable from our axioms. No fundamental theory, it seems, can be truly fundamental, in the sense of containing only mathematical principles that we recognize as correct, and formally implying all of our mathematical insights. A fundamental theory which we recognize as correct is always subject to *reflection*, which leads from the recognition that the theory is correct to arithmetical conclusions not provable in the theory. So how can we take reflection into account when saying or describing what it is that we recognize as correct?

We will separate two aspects of this question, to be called the problem of description and the problem of formalization.

The *problem of description* is to determine to what extent we can make clear to ourselves, in whatever terms, formal or informal, how far we can get by reflection. If we have decided on some fundamental theory $T$ which we accept as correct, just what further axioms or principles can be justified by reflection?

In attacking the problem of description, we need to decide what we mean by "correct", and consider just how the basic principle of reflection (by which we conclude that a theory is consistent because correct) can be extended. An important aspect of the problem of description is that our concern is not with describing what some hypothetical idealized mathematician with a particular foundational or philosophical outlook (such as "the finitist", or "the Platonist", or "the predicativist") knows, or in what terms his knowledge can be described, but rather with the question what we ourselves can say about our own mathematical knowledge or insight. The significance of this aspect will become clear below.

The *problem of formalization* is to decide whether it is at all possible to program a computer so as to give it our own capacity for reflection. For vividness in discussing this second aspect, let us suppose that we wish to impart the sum total of our arithmetical knowledge to a group of robot explorers who will sail away through space as our mathematical representatives. These robots are equipped with the latest quantum computing nano-whatsits, and so we don't need to worry about their capacity for drawing formal conclusions from formal axioms. They will for example prove Gödel's theorem and every other theorem proved or referred to in this book, and they will be able to establish consequences of mathematical conjectures, and to generate all sorts of extensions of our ordinary formalizations of mathematics and derive theorems in those extensions. Our concern is only with what we can instruct the robots to take to be arithmetical *knowledge*. The difficulty posed by Gödelian inexhaustibility in this connection is that it doesn't seem to be possible to specify *formally* a set of basic principles or axioms expressing the sum total of our fundamental arithmetical insights, from which our explicit and implicit arithmetical knowledge is to be derived. For given any such formal specification of axioms which we recognize as correct, we will also recognize a stronger specification as correct. And how are we to convey our knowledge to the robots if not as a set of formal principles?

The discussion in the following will first consider various aspects of the problem of description, and then turn to the problem of formalization.

### Accepting theories as sound

What does it mean to say that $T$ is correct? The strongest relevant answer is that it means that $T$ is *sound*, in the sense that all axioms of $T$ are true, and the rules of $T$ can only lead from true axioms to true conclusions. While "correct" will be not be given any fixed interpretation in the following discussion, "sound" will only be used in this sense. As has been emphasized in earlier chapters, we need not associate the use of "true" with any particular philosophical view about the nature of mathematical truth, or about the place of mathematics in the world or in our thinking. What is essential is only that the axioms are taken to be true in a sense that conforms to the mathematically expressible sense of "true" exemplified in earlier parts of the book. Thus if we are talking about arithmetical sentences, what is essential is only that

if $\phi$ is atomic, $\phi$ is true if and only if $\phi$ is in $True_{Vf}$,

$\phi_1 \vee \phi_2$ is true if and only if $\phi_1$ is true or $\phi_2$ is true,

$\neg\phi$ is true if and only if $\phi$ is not true, and

$\exists x\phi$ is true if and only if $\phi_x(\overline{n})$ is true for some $n$.

If we are talking about sentences in the language of second order arithmetic, we will need, in mathematically characterizing truth for such sentences, to introduce also a predicate expressing that a formula containing free variables is true of a sequence of sets, as indicated in §10.3, and similarly in characterizing true sentences in the language of set theory.

Accepting the theory $T$ as correct in the strong sense of holding that it is sound can be justified on the basis of various doctrines, views, or inclinations. We may for example assert that the axioms of $T$ are true because we hold that they correctly describe a mathematical reality, or because they are in agreement with our conception of what mathematical reality is like, or because they are pleasing to the intellect and imagination, or because they are in agreement with mathematical intuition (whatever the source and nature of that intuition). In the present context, we need not consider whether any such justification is better than any other. The relevant point is that if $T$ is sound, it is consistent, and we can prove this implication mathematically. Indeed we can prove that if $T$ is sound, then $T + Con_T$ is also sound, so that soundness is inherited by consistency extensions.

Asserting that the axioms of $T$ are true is a way of endorsing or affirming the axioms of $T$, but it is not necessary to make use of a truth predicate in endorsing those axioms. In particular, if a theory $T$ has only finitely many axioms $\phi_1, \ldots, \phi_n$, a reasonable interpretation of the statement "$T$ is correct" is simply as the conjunction of those axioms. This is in accordance with ordinary usage—in response to the question "Is this statement (that there are more than five planets in the solar system) correct?", it would be a natural response to say "Yes, it is quite correct: there are indeed more than five planets in the solar system". It isn't necessary to introduce a truth predicate to affirm the correctness of the statement. Similarly we could affirm, at least for a finitely axiomatized theory, the conjunction of the axioms of the theory as a way of stating that the theory is correct.

While this is a very reasonable sense of "correct" applied to theories, it is another matter whether endorsing the axioms of a theory in this sense yields a sufficient basis for asserting that the theory is consistent. That no contradiction can be derived from the axioms of $T$ using the rules of logic depends on the properties of the specific rules used. In terms of truth, we can argue that the rules are logically valid, and that no false statement (in particular, no contradiction) can therefore be derived from true axioms using the rules. If we do not use a truth predicate, there is no obvious justification for concluding "no contradiction can be derived from $\phi_1, \ldots, \phi_n$ using the rules of logic" from "it is quite correct that $\phi_1, \ldots, \phi_n$". If we do draw this conclusion, some sort of inductive argument using specific properties of the rules and some relevant property of the axioms seems to be presupposed. And of course this is in accordance with the fact that the consistency of $T$ is not provable in $T$. Proving the consistency of $T$ on the basis of affirming the axioms of $T$ requires us to see the correctness of the theorems of $T$ as a *property* of those theorems, inherited from the axioms of $T$. And indeed, in the statement made above that we can prove mathematically that $T + Con_T$ is sound if $T$ is sound, it was implicitly assumed that we are prepared to reason about truth as a property of arithmetical sentences using induction, and not only to assert the defining conditions of the notion of truth.

## Iterating reflection principles

So suppose we accept $T$ as sound. We are then justified in asserting that $T$ is consistent, and indeed that any consistency extension $T + Con_T$ is also sound. Thus reflection leads us to accept as correct any consistency extension of $T$ that we recognize as a consistency extension. But our concern is with trying to describe just how far reflection will take us, and if we accept $T$ as sound, reflection will take us much further than consistency extensions.

First let us consider extensions of $T$ that do not extend the language of $T$. On the basis of reflection, we can conclude that $T$ extended by various formal reflection principles is still sound. We can distinguish between basic and iterated reflection principles. A basic reflection principle expresses "every theorem of $T$ in $\Gamma$ is true", where $\Gamma$ is some set of sentences in the language of $T$ for which truth is actually definable in the language of $T$ by some formula $\phi(x)$, at least extensionally, that is, in the sense that a sentence $\psi$ in $\Gamma$ is true if and only if $\phi(\overline{\psi})$ is true. Thus if we take $\phi(x)$ as the formula

$$(x = \overline{\psi}_1 \wedge \psi_1) \vee \cdots \vee (x = \overline{\psi}_n \wedge \psi_n)$$

this formula extensionally defines "$x$ is a true sentence in $\{\psi_1, \ldots, \psi_n\}$", yielding a local reflection principle for sentences in the set. More generally, depending on the theory, we may be able to define in the language of $T$ truth for various infinite sets of sentences and express a corresponding reflection principle which we can prove preserves soundness. But there is no strongest soundness-preserving reflection principle, for any such principle can be *iterated*, yielding a stronger soundness-preserving reflection principle. If $R(T)$ is a reflection principle for $T$, we can extend $T$ not only to $T + R(T)$ but to $T + R(T) + R(T + R(T))$. Indeed any such principle can be iterated ad infinitum (taking unions at limit ordinals), and as soon as we are able to prove that a particular sequence of theories is in fact a sequence of extensions by iterated soundness-preserving reflection principles, we can take the theories in the sequence to be justified by reflection.

There are thus several uncertainties regarding just what extensions of $T$ by reflection principles expressible in the language of $T$ can be justified on the basis of the soundness of $T$. In particular, which theories we can arrive at by iterating reflection principles depends on just what mathematical apparatus and principles having to do with infinite sequences we accept. We can bypass these uncertainties by taking a large step to a truth extension of $T$, regarded as justified by reflection. For if $T$ is a fundamental theory, we can expect that the mathematical machinery available to us for defining and handling infinite sequences of theories, in so far as we accept any such machinery as a means of establishing the truth of arithmetical statements, is included in the theory $T$. Further, the justification for accepting theories extended by reflection principles as correct is that $T$ is sound, so adding the soundness of $T$ as an axiom should allow us to formally prove the soundness of every iterated extension by reflection principles that we can prove to be such an extension on the basis of $T$, or indeed on the basis of $T$ extended by reflection principles. The verification of one version of this in the particular case of arithmetical theories $T$ has been carried out in the foregoing in terms of autonomous progressions and iterated uniform reflection principles, but the basic reasons for expecting such a result to hold are very general, as noted.

Thus we are led to truth extensions, considered as justified by reflection. A truth extension of $T$ is obtained by adding to the axioms of $T$ the defining axioms for the truth predicate together with an axiom stating that the axioms of $T$ are true, and also extending induction axioms and any other principles in $T$ for reasoning about properties of natural numbers to the new predicate. Is such a theory justified by reflection? Well, if we accept $T$ as sound, there is no apparent reason why we should not take the truth extension to be a correct theory, since it is obtained by adding axioms and principles that we affirm to a theory that we hold to be sound. Indeed, on the basis of our understanding of truth in the case of sentences of $T$, we can assert that the truth extension is also sound, now using a notion of truth for the sentences in the language of $T$ extended by a truth predicate, as illustrated for numerical theories in §15.1 above.

If we accept truth extensions as justified by reflection, the question arises how far we can iterate such extensions. In the case of extensions by the uniform reflection principle, the concept of truth for sentences in the language of $T$ was all we needed to add to $T$ in order to prove that every theory in an iterated reflection sequence is sound. In the case of iterated truth extensions, we need to introduce a concept of "true sentence of level $\alpha$", or some principle of inductive definition or of set existence by which such a concept can be defined. This would seem to be a greater, or a different kind of step compared to reflecting on our understanding of a given theory to produce a truth predicate for the theory. Thus it was suggested in §15.1 that the autonomous truth extensions of $T$ represent an "upper bound" on what can be justified by "pure reflection", in the sense of indefinitely iterated recognition of soundness. What this means is that in formulating a theory which is stronger than every theory occurring in an autonomous sequence of truth extensions of $T$, we need to introduce a new level of abstraction in $T$, not just based on reflection on the axioms of $T$ (unlike the truth predicate for $T$).

But now suppose we say that we accept, on the basis of reflection, any theory in an autonomous sequence of truth extensions starting with $T$, but cannot convince ourselves by reflection of the correctness of the union of all such theories. This is a rather strange thing to say. There may be good justification for holding that such a union represents an upper bound on what can be recognized as valid by "pure reflection" in the sense of iterated recognition of soundness, but if we grant that every autonomous truth extension of $T$ is correct, why not grant that the union of those extensions is correct? "Pure reflection" or not, by considering iterated pure reflection we are led to accept this union as correct. The only answer that immediately suggests itself is that we cannot really meaningfully speak of the union of these theories. But then it's entirely unclear why it should be any more meaningful to say that every theory in such a sequence is correct. It is common to seek to avoid the appearance of trying to have your cake and eat it too by casting such attempts to draw a line between the potentially evident and the non-evident or unintelligible in terms of what a hypothetical or idealized mathematician with a particular outlook would accept. Thus it might be said that "from a predicativist viewpoint", every theory in an autonomous sequence of truth extensions can be justified by reflection, but not their union. Here the "viewpoint" is to be understood to make sense of each of the theories taken in isolation, but not of statements referring to all theories in the sequence. Such descriptions may be quite useful and convincing as characterizations of some particular outlook, but the problem of description is not the problem of characterizing this or that restricted type of mathematical thinking or reasoning. We are trying to say what we ourselves find convincing on the basis of reflection. And if we find ourselves saying that we can see on the basis of reflection that every theory in an autonomous sequence is correct, we have no apparent grounds for avoiding the conclusion that the union of these theories is correct, even if this conclusion is obtained only on the basis of "second order reflection". It is pointless to attempt to say of ourselves that we can recognize every theory in the sequence in isolation as correct, but not their union. After all, there is no question of our being presented with every theory in the sequence in isolation. It's not like saying that we recognize each of our cousins when we meet him or her, but can't enumerate all the cousins when called upon to do so. The theories in autonomous sequences are something that we talk about in general terms, not something that we encounter in daily life. If we want to avoid being in the uncomfortable situation of making statements about sequences that we claim not to be able to refer to or make sense of, we can only affirm the soundness of various particular autonomous truth extensions, and nod sagely or smile mysteriously (somewhat in the spirit of Wittgenstein's *Tractatus*) when asked questions about all theories in such sequences.

But here we need to observe that it is not necessarily the case that we are able to describe every mathematical theory that we can define as either convincing or not convincing, intelligible or not intelligible, justified or not justified. It would be entirely reasonable to say that an extension of $T$ that proves every arithmetical theorem provable in any autonomous truth extension of $T$ is *fairly* clear and convincing, or *pretty* clear and *rather* convincing, while affirming the soundness of $T$ and various truth extensions of $T$ in more unhesitating terms. We may also feel inclined to say that we can keep track of short autonomous sequences of truth extensions, but get confused after a while and really cannot say with any confidence whether the theories are correct or not. Both of these are natural and common reactions when pondering sequences of extensions of theories by principles about sets of higher type, by truth predicates, or by other such towers of abstractions. This is in part because of the sheer confusion generated by considering large ordinals or complicated notations for ordinals.

If we do accept as correct some extension $T'$ of $T$ in which every arithmetical theorem in every autonomous truth extension of $T$ is provable, we will be led to similarly consider truth extensions of $T'$, although because of the potential for confusion in describing these theories we may well find it easier to consider corresponding theories in terms of set-theoretical hierarchies. Iteration will eventually lead to a discussion of all the usual set-theoretical concepts and principles discussed in the foundations of mathematics. The point of the above comments has been that we cannot expect, when considering the question just what is justifiable on the basis of reflection, an answer that is any more definite or clear-cut than in the case of the general question what mathematical abstractions make sense and what set-theoretical or other abstract principles are acceptable. The step from $T$ to $T + Con_T$ (where $T$ is a theory that we recognize as sound) is clear and unproblematic enough, but as we take larger strides in our iterations, all the uncertainty and potential confusion associated with general questions about evidence and justification in mathematics will creep up on us.

But now it must be noted that the above discussion was predicated on the assumption that we accept $T$ as correct in the strong sense of accepting the theory as *sound*. We also need to consider how far reflection will take us if we accept $T$ as correct only in some weaker sense. The situation then is rather different.

## Accepting theories as correct

Many or most mathematicians would hesitate to assert that the axioms of $T$ are true, if $T$ is for example the axiomatic set theory ZFC, which is commonly presented as a fundamental theory in the introductory chapters of mathematical textbooks, where the formalization in set theory of topology or analysis or mathematics as a whole is briefly described. Without considering whether or not we are justified in taking the axioms of ZFC to be true, let us suppose that we adopt instead a view of our fundamental theory $T$ as correct only in the weak sense of being *consistent*. If the consistency of $T$ is furthermore not regarded as evident, but as problematic or conjectural, as is often the case, there is of course no apparent way of arriving at new principles by reflection. If we do regard $T$ as evidently correct in this sense, we have, by the second incompleteness theorem, a mathematical insight not derivable in $T$, but by the same theorem that is as far as reflection starting with $T$ can take us. We don't know that $T + Con_T$ is consistent just on the basis of the knowledge that $T$ is consistent. So if we accept $T + Con_T$ as correct in the same weak sense, this can only be on the basis of a separate insight regarding that theory. It is not a consequence of Gödel's theorem that any such new insight is available. In other words, there is no inexhaustibility phenomenon here, or at least none that Gödel's theorem brings to light.

Now suppose we recognize $T$ as correct in the sense of being $\Sigma$-sound, but not in any stronger sense. Then indeed we recognize both $T$ and $T + Con_T$ as consistent, and more generally, we know that any theory in a consistency sequence beginning with $T$ is also $\Sigma$-sound. Thus we have a principle for obtaining ever stronger correct extensions of $T$. However, just what extensions we can recognize to be such is unclear. Further, the $\Sigma$-soundness of $T$ does not guarantee that $T$ + "$T$ is $\Sigma$-sound" is correct even in the weak sense of being consistent.

This is a rather strange situation. We have the arithmetical knowledge that $T$ is $\Sigma$-sound, but we don't know that this knowledge is even compatible with our fundamental theory $T$. Then in what context does this bit of arithmetical knowledge fit? If the fundamental theory $T$ is not seen to be compatible with our actual arithmetical knowledge, it seems that $T$ cannot be a fundamental theory. In the case of a theory $T$ which we accept as sound, inexhaustibility shows $T$ to be not as fundamental as we intended, in the sense that it can be extended to a stronger theory that we still accept as sound. The situation in this case is different: $T$ fails to be a truly fundamental theory, not because we see that it can be extended, but because we don't see that it is even compatible with our current arithmetical knowledge. Thus the remedy would seem to be not to extend $T$ but to restrict it. Again unlike the case of a sound $T$, where we can say quite generally that the obvious way to extend $T$ is to add a reflection principle as a new axiom, we cannot say in general just how $T$ should be restricted to be compatible with our arithmetical knowledge.

An example of this situation might be illuminating, but it seems that there isn't any to be had. That is, we can't find any example of a fundamental theory $T$ which is presented as evidently $\Sigma$-sound while it is not claimed to be evident that "$T$ is $\Sigma$-sound" is compatible with $T$. According to the line of thought presented above, no such example is to be expected.

We are faced with a similar hypothetical situation if we consider other possible interpretations of "correct" weaker than soundness. Consider, as an extreme, accepting $T$ as arithmetically sound. If $T$ is not an arithmetical theory, we cannot conclude that "$T$ is arithmetically sound" is compatible with $T$ just on the basis of the arithmetical soundness of $T$. And if we use "$T$ is arithmetically sound" to arrive at arithmetical reflection principles for $T$, this principle, although not arithmetical, is one that we use in arriving at arithmetical conclusions, and so it should be seen to be compatible with a fundamental theory.

It would seem, on the basis of the above argument, that the answer to the question how far one can get by reflection starting with a theory that is not accepted as sound can only be "we cannot say". Whenever we have a theory $T$ which we recognize as correct, and find through reflection that there is an open-ended series of extensions of $T$ which are also correct, we do so on the basis of the perceived correctness of $T$, which implies some soundness property of $T$ inherited by the extensions. How far such a series of extensions can take us depends on what principles are available for defining sequences of theories, and what the relevant soundness properties are. In the case of sound theories $T$, the uncertainties that this implies could be sidestepped by considering a truth extension of $T$, which is also sound and is stronger than any such iterated extension by soundness principles that can be formulated in the language of $T$ and which can be proved to be such using $T$. When $T$ is not known to be sound, autonomous sequences based on $T$ have no obvious justification, and we don't even know that $T$ is compatible with the assertion that $T$ is correct. In particular, $T$ cannot be part of a fundamental theory.

As with other arguments presented above, the above remarks have been substantiated only for a certain range of theories and interpretations of "correct". For example, if $T$ is a $\Sigma_n$-sound first or second order extension of PA, "$T$ is $\Sigma_n$-sound" is not necessarily compatible with $T$. This is typical of what we can expect when justifying an open-ended series of extensions by reflection by invoking a soundness property of $T$ that is weaker than unrestricted soundness.

## Reflecting robots

By the Gödelian inexhaustibility argument, we cannot define an effectively enumerable set of arithmetical sentences that we recognize as yielding as logical consequences those and only those arithmetical truths that are potentially evident to us. As many commentators have pointed out, in particular Gödel himself (*Collected Works*, Volume III, p. 309), this does not imply that there exists no effectively enumerable set of arithmetical sentences that in fact yields those and only those arithmetical truths that are potentially evident to us. All that follows is that if there is such a set, we cannot recognize it for what it is. However, to speak of this as a possibility is to suppose that there is such a thing as "the set of arithmetical truths that are potentially evident to us". A justification for the idea that there is some such definite human capacity for discovering arithmetical truth is lacking. In particular, an investigation of extensions by reflection of evident theories suggests that what we can make evident to ourselves on the basis of reflection is in no way determinate. There are several dimensions of vagueness, definiteness, uncertainty, confidence, confusion, and vacillation connected with reflection, just as there are in connection with other questions about meaning and evidence in mathematics.

It follows that if we are to program our mathematical robots to faithfully mirror our own mathematical understanding and insight, we must program them to exhibit similar vagueness, definiteness, uncertainty, confidence, confusion, and vacillation. We probably don't know how to do this, but the problems involved are no different from the problems of teaching the robots to mirror our own vagueness, definiteness, uncertainty, confidence, confusion, and vacillation in other matters. It is therefore not a consequence of Gödelian inexhaustibility that we possess some unmechanizable or ineffable capacity or insight.

This is not to say that we cannot program our robotic mathematical ambassadors with reflection principles. The robot, we may assume, pushes a green button to produce the next interesting theorem, using his current store of rules and axioms. Next to it is a blue button; when this is pushed, the store of axioms is extended by a reflection principle.

This description invites a comment: the robot can iterate the reflection principle only finitely many times. But if it is a valid reflection principle, an $\omega$-times iterated application of the principle also yields a valid extension of the basic axioms. How is this to be taught to the robot? No matter what reflection principle is added by the blue button, the robot will miss the stronger principle.

It is true that such a robot is not a faithful simulation of human mathematical behavior. If we accept a set of axioms and a reflection principle for those axioms, we will also most likely accept to the same degree an $\omega$-iteration of that reflection principle. Hence we introduce a more complicated model. The robot has its green button for producing the next theorem, but it also has a pink button for picking one of a range of effectively enumerated reflection principles $R_0, R_1, R_2, \ldots$, another pink button for picking one of a range of effectively enumerated set-theoretical extensions of the axioms $S_0, S_1, S_2, \ldots$, and a little white button for choosing an ordinal notation $n$. The white button may be pressed in conjunction with one of the $R_i$ whenever $n$ has been proved to be a notation, meaning that the reflection principle will be iterated up to the ordinal denoted by $n$, and the $S_i$ buttons may be pressed to add a set-theoretical principle. The theorems proved so far are stored together with information about what principles were used in proving them.

Because of the autonomy condition for pressing the white button for a given $n$, we know that the totality of sentences provable by the robot through any combination of button pushing is effectively enumerable. Indeed there is no reason why the robot should not be able to define this total set. There are no provisions, however, for allowing the robot to prove anything beyond this set, for example by a further reflection principle, although it can of course prove all sorts of theorems about the theory defined by the total set. This does not mean that there is anything missing in the robot's reflective abilities. For it is by no means the case that the robot unhesitatingly affirms (when interviewed in between button pushing) the validity of all its current principles, or the truth of all theorems proved so far. Instead it mirrors human behavior in this regard, and will express varying degrees of confidence in principles $R_i$ and $S_i$, and may indeed occasionally revise its set of proved theorems to contain only those not proved using a certain principle, and thenceforth or for a while no longer use that principle in proving theorems. Of course just what factors (external and internal) influence the robot in making these decisions or displaying such vacillation is unclear, and we may or may not be able to produce a convincing simulacrum of human mathematicians, philosophers, or logicians in this respect. The claim made in this present discussion is only that for all we know, we have met the robots and they are us. It is misleading in this context to concentrate on cases when we go from a theory that is clearly valid to an equally clearly valid consistency extension. Continued reflection leads us into deep and muddy waters, and so we can only be emulated by deeply muddled robots.

## 15.3. Overview and summary

In 1900, David Hilbert expressed a conviction that there are no unsolvable problems in mathematics:

> Take any definite unsolved problem, such as the question as to the irrationality of the Euler-Mascheroni constant C, or the existence of an infinite number of prime numbers of the form $2^n + 1$. However unapproachable these problems may seem to us, and however helpless we stand before them, we have, nevertheless, the firm conviction that their solution must follow by a finite number of purely logical processes. ... This conviction of the solvability of every mathematical problem is a powerful incentive to the worker. We hear within us the perpetual call: There is the problem. Seek its solution. You can find it by pure reason, for in mathematics there is no *ignorabimus*.[1]

Hilbert didn't explain the phrase "definite unsolved problem", but it is no doubt significant that he invoked in illustration two famous open *arithmetical* problems. It is the natural attitude of a mathematician that arithmetical problems cannot be solved by fiat (for example by *postulating* that there are infinitely many primes of the form $2^n + 1$) but only by a mathematical proof, the validity of which we will recognize when we see it. We may be more or less optimistic about the prospects of solving a particular problem, either in the near future or at any time, and Hilbert's affirmation (the "*non ignorabimus*") is naturally read as expressing and seeking to instill a general optimism: if we (mathematicians) try hard enough, we will eventually crack any mathematical problem.

The work on formalizing mathematics in the late nineteenth and early twentieth century made it possible to interpret the non ignorabimus as a theoretical conjecture. In the case of computational arithmetical problems, for example problems concerning the factors of large numbers, we know that however difficult or indeed impossible they may be to settle in practice, there are in theory a "finite number of purely logical processes" by which they can be solved. Since arithmetical problems such as those mentioned by Hilbert can be formulated in formalized theories with precise formal axioms and rules of inference, it might be conjectured that a *complete* theory of this kind can be formulated, one in which every arithmetical statement can be either proved or disproved, at least in principle.

Gödel's incompleteness theorem established that this is not the case. There is no single set of mathematical axioms from which the solutions to all arithmetical problems follow by a finite number of purely logical processes. In particular, given a formal theory $T$ of arithmetic, $T$ cannot prove the arithmetical formalization of "$T$ is consistent".

Several questions arise. First, we may wonder whether incompleteness should be a matter of practical concern to mathematicians. Is it possible that some of the famous open arithmetical problems in mathematics can in fact not be settled on the basis of our current mathematical axioms, so that Hilbert's non ignorabimus requires optimism not only with regard to human ingenuity

[1] In presenting his famous list of unsolved mathematical problems, see Browder [1976].

in finding proofs, but also about the ability of mathematicians to come up with new, valid and convincing axioms or rules of reasoning? The current answer to this question is that we simply don't know whether any open arithmetical problem which mathematicians are trying to solve is formally unsolvable in current mathematics. There is no particular reason to believe that any such incompleteness exists, although the possibility cannot be dismissed out of hand.

As another practical question, we may ask whether the arithmetization of "$T$ is consistent" has any mathematically interesting consequences, that is, if it can be profitably put to use as a new axiom, added to those of $T$. Quite a lot of work has been done in this direction, extracting combinatorial consequences of possible mathematical interest from consistency statements and from strengthenings of consistency statements such as reflection principles.

From the point of view of the special field of mathematical logic, many natural technical questions arise concerning extensions of theories obtainable by repeated addition of consistency statements or reflection principles, and there is a wealth of results about such extensions. These results do not however connect directly with questions about how incompleteness might affect the work of mathematicians in general.

Looking at the incompleteness theorem from a different angle and applying it specifically to formalizations of what we regard as our mathematical knowledge, it's striking that the particular instance of incompleteness exhibited in Gödel's theorem is in the nature of an *inescapable oversight*. Naturally $T$ is consistent, given that it embodies part of our mathematical knowledge, but we cannot consistently affirm this in $T$ itself. Hence the phenomenon of the apparent inexhaustibility of our arithmetical knowledge: whatever axioms we put down in a theory $T$, we can always add: "and of course $T$ is consistent". This is the aspect of the incompleteness theorem that forms the chief topic of this book.

It is probably a more widespread point of view regarding the second incompleteness theorem that it is in the nature of an epistemological *obstacle* to proving, or even to having confidence in, the consistency of our mathematical theories. While it is clear enough that the incompleteness theorem revealed an obstacle to Hilbert's original plan to prove the consistency of such theories as CA using only finitary methods, there is no immediately apparent reason why the impossibility of proving the consistency of a theory $T$ in $T$ itself should prompt any doubts about the consistency of $T$, or why any doubts that we may have about the consistency or soundness of $T$ should be strengthened by the incompleteness theorem. It has been argued in the foregoing that the presentation of the second incompleteness theorem as an epistemological obstacle is misleading because it *plays down* the character of the obstacle. A theory $T$ not only cannot prove its own consistency, it cannot truly assert its own consistency. Even if we are all set to postulate rather than prove, we cannot formulate a theory that truly affirms its own consistency. This is perfectly compatible with there simply being no reasonable doubt about the consistency of the theory, so that the impossibility of proving the consistency of $T$ in $T$ itself is not an obstacle to knowledge, but an obstacle to the formalization of knowledge.

We can formulate the phenomenon of inexhaustibility as an informal principle or rule of reasoning: whenever we have a valid theory $T$ of arithmetic, we can extend $T$ to a stronger valid theory by adding a formalization of "$T$ is consistent". Indeed we can go further and add any formal principle which can be seen to follow from the fact that $T$ expresses part of our mathematical knowledge. We will call this the informal reflection principle.

Like informal principles in general, the informal reflection principle has no well-defined set of formal consequences. Instead it has *applications*, some of which are formal principles, and the range of such applications is to a large extent indefinite. We are not adept at implementing informal principles (whether mathematical, physical, economic, political, philosophical, etc.) on computers, since this would involve programming computers with a capacity for association, judgment and endless philosophical or scientific discussion such as we ourselves make use of in applying the principles. This has been known to lead to arguments to the effect that there is something essentially non-algorithmic about the informal reflection principle. The point of view presented here is that no such conclusion can be drawn. While it may conceivably be the case that the great (verbal, mathematical, musical) buzzing produced by humanity can only be produced by beings who are not describable in algorithmic terms, the impossibility of squeezing informal principles into a formal mold is not proof of any unmechanizable insight underlying those principles, any more than the impossibility of drawing a line between "cheap" and "not cheap", or "fast" and "not fast" shows that the essence of cheapness or speediness can only be perceived through unmechanizable intuition. The informality of the informal reflection principle is of course not just a matter of vagueness, but it is connected with an inescapable elasticity and indefiniteness in the concept of knowledge. To require of an algorithm that it produce a maximal $n$ such that $n$ cents is cheap for a car, and then triumphantly point out that the human mind, but not the algorithm, will realize that $n + 1$ cents is then also cheap for a car, is not a convincing way of arguing for the non-mechanizability of human thought. An argument which requires an algorithm to produce a definite set of all knowable arithmetical truths is really no better.

It remains, however, a meaningful and interesting question whether we can characterize or at any rate gain some understanding of how far the informal reflection principle can advance our arithmetical knowledge. Here we are not trying to establish any connection between mathematical practice and the consequences of formal reflection principles, but to understand in more general and abstract terms the theoretical limits of informal reflection, whether those limits are mathematical or conceptual.

Since we are not asking about the limits of mathematical knowledge in general, but specifically about the informal reflection principle, we need to stipulate a starting point in the form of a formal theory $T$ embodying part of our mathematical knowledge. How far can $T$ be extended on the basis of the informal reflection principle?

Here we need to decide on what formal reflection principles can be taken to follow from the informal reflection principle. This question is complicated by the fact that any formal reflection principle leads to stronger extension principles by iteration, that is by repeatedly applying the same formal reflection principle to $T$, to $T$ with the formal reflection principle added, and so on. This "and so on" can be taken into the transfinite, but when we ask how far, we again encounter the question just how far our mathematical knowledge extends. The reason is that we may need to invoke mathematical knowledge neither given by the reflection principle, nor contained in $T$, to *prove* that a certain theory is obtainable from $T$ by such iteration.

To be able to say anything about how far we can get by reflection, we therefore also need to specify some basis in our mathematical knowledge for identifying theories resulting from iterated formal reflection. Here the concept of *autonomous progressions* provides a useful tool. The theories in autonomous progressions based on some reflection principle and starting with a theory $T$ are, or can reasonably be taken to be, those that can be recognized as iterated extensions of $T$ by the formal reflection principle in question precisely on the basis of $T$ and of that principle. Thus we can ask, as a way of exploring the range of a formal reflection principle derived from the informal principle, just what can be proved in an autonomous progression based on that principle and beginning with $T$.

When we pursue this line of thought formally, for theories $T$ containing formal arithmetic and axiomatized in the ordinary style, and for various formal reflection principles, a pattern emerges. Anything provable in an autonomous progression based on $T$ is also provable in an extension of $T$ by a single stronger formal reflection principle. There is a break or bump in this pattern when we choose as our formal reflection principle "every theorem of $T$ is true", which involves the introduction of a sequence of new truth predicates. The break or bump consists in the fact that it is no longer clear that a stronger formal principle which when added to $T$ yields every arithmetical theorem in an autonomous progression can be described as a formal reflection principle implied by the informal principle. The union $T'$ of the theories in an autonomous progression of such truth extensions would therefore seem to be the best available candidate for a limit on what can be achieved by reflection starting from $T$. On the other hand, why cannot the informal reflection principle itself be iterated? After all, we describe $T'$ as a limit for what can be achieved by formal reflection principles based on the informal reflection principle, and so $T'$ is itself arrived at by reflection on the informal reflection principle, or by a second order reflection. So if we are seeking the limits of our own potential mathematical knowledge, we need to allow the whole procedure to continue on the basis of $T'$. The alternatives to this would seem to be to either disavow the apparent results about $T'$ as not in fact established or intelligible, which leads to various difficulties, or to cast the whole investigation in terms of what can be established on the basis of reflection given some particular limited outlook on what is mathematically acceptable, which means that we are not pursuing the question of what we ourselves can recognize as valid on the basis of reflection. As we continue the process of reflection, we are faced with all of the questions concerning mathematical evidence otherwise associated with the development of set theory.

The upshot of the present investigation is that our arithmetical knowledge is indeed inexhaustible in the sense indicated by Gödel: we cannot set down axioms which we recognize as exhaustively expressing our arithmetical knowledge. But as in the analogous case of our inability to count up to a number that represents the absolutely greatest number we can count to, this does not imply that there is some well-defined body of arithmetical truths, "our arithmetical knowledge", which cannot be fully described in formal terms. Our potential or implicit mathematical knowledge, in the sense of the range of arithmetical truths that are formal consequences of principles that we recognize as valid, is determinate only to the extent that those principles are determinate. The process of reflecting on iterated Gödelian extensions of theories is itself one of several paths that lead us from evident axioms to an indefinite range of mathematical principles that are more or less evident at different times, and from different points of view.

# REFERENCES

L. D. BEKLEMISHEV [1995], *Iterated local reflection versus iterated consistency*, ***Annals of Pure and Applied Logic***, vol. 75, pp. 25–48.

L. D. BEKLEMISHEV [1997], *Notes on local reflection principles*, ***Theoria***, vol. 58 part 3, pp. 139–146.

P. BERNAYS [1935], *On platonism in mathematics*, ***Philosophy of mathematics: Selected readings*** (P. Benacerraf and H. Putnam, editors), 2nd edition, Cambridge 1983.

F. E. Browder (editor) [1976], ***Mathematical developments arising from Hilbert problems***, Proceedings of Symposia in Pure Mathematics, no. 28, American Mathematical Society.

S. FEFERMAN [1960], *Arithmetization of metamathematics in a general setting*, ***Fundamenta Mathematicae***, vol. 49, pp. 35–92.

S. FEFERMAN [1962], *Transfinite recursive progressions of axiomatic theories*, ***The Journal of Symbolic Logic***, vol. 27, no. 3, pp. 259–316.

S. FEFERMAN [1984], *Kurt Gödel: Conviction and caution*, ***Philosophia Naturalis***, vol. 21, no. 2–4, pp. 546–562.

S. FEFERMAN [1993], *What rests on what? The proof-theoretic analysis of mathematics*, ***Proceedings of the 15th international Wittgenstein symposium***, Philosophy of Mathematics, Part I, Verlag Hölder-Pichler-Tempsky, Vienna, pp. 147–171.

S. FEFERMAN AND G. HELLMAN [1999], *Challenges to predicative foundations of arithmetic*, ***Between logic and intuition: Essays in honor of Charles Parsons*** (Gila Sher and Richard L. Tieszen, editors), Kluwer Academic, Dordrecht & Boston, pp. 317–339.

S. FEFERMAN AND C. SPECTOR [1962], *Incompleteness along paths in progressions of theories*, ***The Journal of Symbolic Logic***, vol. 27, no. 4, pp. 383–390.

S. Feferman, et al. (editor) [1995], ***Kurt Gödel: Collected works. Vol. III***, Oxford.

T. FRANZÉN [1987], *Provability and truth*, ***Acta universitatis stockholmiensis***, Stockholm Studies in Philosophy, vol. 9, Almqvist & Wiksell International, Stockholm.

P. HÁJEK AND P. PUDLÁK [1993], ***Metamathematics of first order arithmetic***, Springer-Verlag, Berlin & New York.

G. H. HARDY [1929], *Mathematical proof*, ***Mind***, vol. XXXVIII, no. 149.

J. HARRISON [1995], *Metatheory and reflection in theorem proving: A survey and critique*, ***SRI CRC-053***, Cambridge.

P. PUDLÁK [1999], *A note on applicability of the completeness theorem to human mind*, ***Annals of Pure and Applied Logic***, vol. 96, pp. 335–342.

W. V. O. QUINE [1969], ***Set theory and its logic***, second ed., Harvard University Press.

U. R. SCHMERL [1979], *A fine structure generated by reflection formulas over Primitive Recursive Arithmetic*, ***Logic colloquium '78*** (M. Boffa, D. van Dalen, and K. McAloon, editors), North Holland, Amsterdam.

J. R. SHOENFIELD [1967], ***Mathematical logic***, Addison-Wesley, Reading, Massachusetts, reprinted by the Association for Symbolic Logic, 2001.

S. G. SIMPSON [1999], ***Subsystems of second order analysis***, Perspectives in Mathematical Logic, Springer-Verlag.

C. SMORYNSKI [1977], *The incompleteness theorems*, ***Handbook of mathematical logic*** (J. Barwise, editor), North-Holland, Amsterdam.

R. M. SMULLYAN [1992], ***Recursion theory for metamathematics***, Oxford University Press, New York.

R. M. SMULLYAN [1993], ***Gödel's incompleteness theorems***, Oxford University Press, New York.

A. TARSKI [1944], *The semantic conception of truth and the foundations of semantics*, ***Philosophy and phenomenological research 4***.

E. ZERMELO [1908], ***Investigations in the foundations of set theory I***, Translation in ***From Frege to Gödel***, (van Heijenoort, editor), Harvard University Press, 1971.

# INDEX

LECTURE NOTES IN LOGIC

## General Remarks

This series is intended to serve researchers, teachers, and students in the field of symbolic logic, broadly interpreted. The aim of the series is to bring publications to the logic community with the least possible delay and to provide rapid dissemination of the latest research. Scientific quality is the overriding criterion by which submissions are evaluated.

Books in the Lecture Notes in Logic series are printed by photo-offset from master copy prepared using LaTeX and the ASL style files. For this purpose the Association for Symbolic Logic provides technical instructions to authors. Careful preparation of manuscripts will help keep production time short, reduce costs, and ensure quality of appearance of the finished book. Authors receive 50 free copies of their book. No royalty is paid on LNL volumes.

Commitment to publish may be made by letter of intent rather than by signing a formal contract, at the discretion of the ASL Publisher. The Association for Symbolic Logic secures the copyright for each volume.

The editors prefer email contact and encourage electronic submissions.

## Editorial Board

## Editorial Policy

1. Submissions are invited in the following categories:

| | |
|---|---|
| i) Research monographs | iii) Reports of meetings |
| ii) Lecture and seminar notes | iv) Texts which are out of print |

Those considering a project which might be suitable for the series are strongly advised to contact the publisher or the series editors at an early stage.

2. Categories i) and ii). These categories will be emphasized by Lecture Notes in Logic and are normally reserved for works written by one or two authors. The goal is to report new developments quickly, informally, and in a way that will make them accessible to non-specialists. Books in these categories should include

– at least 100 pages of text;
– a table of contents and a subject index;
– an informative introduction, perhaps with some historical remarks, which should be accessible to readers unfamiliar with the topic treated;

In the evaluation of submissions, timeliness of the work is an important criterion. Texts should be well-rounded and reasonably self-contained. In most cases the work will contain results of others as well as those of the authors. In each case, the author(s) should provide sufficient motivation, examples, and applications. Ph.D. theses will be suitable for this series only when they are of exceptional interest and of high expository quality.

Proposals in these categories should be submitted (preferably in duplicate) to one of the series editors, and will be refereed. A provisional judgment on the acceptability of a project can be based on partial information about the work: a first draft, or a detailed outline describing the contents of each chapter, the estimated length, a bibliography, and one or two sample chapters. A final decision whether to accept will rest on an evaluation of the completed work.

3. Category iii). Reports of meetings will be considered for publication provided that they are of lasting interest. In exceptional cases, other multi-authored volumes may be considered in this category. One or more expert participant(s) will act as the scientific editor(s) of the volume. They select the papers which are suitable for inclusion and have them individually refereed as for a journal. Organizers should contact the Managing Editor of Lecture Notes in Logic in the early planning stages.

4. Category iv). This category provides an avenue to provide out-of-print books that are still in demand to a new generation of logicians.

5. Format. Works in English are preferred. After the manuscript is accepted in its final form, an electronic copy in LaTeX format will be appreciated and will advance considerably the publication date of the book. Authors are strongly urged to seek typesetting instructions from the Association for Symbolic Logic at an early stage of manuscript preparation.

# Lecture Notes in Logic

From 1993 to 1999 this series was published under an agreement between the Association for Symbolic Logic and Springer-Verlag. Since 1999 the ASL is Publisher and A K Peters, Ltd. is Co-publisher. The ASL is committed to keeping all books in the series in print.

Current information may be found at `http://www.aslonline.org`, the ASL Web site. Editorial and submission policies and the list of Editors may also be found above.

Previously published books in the *Lecture Notes in Logic* are:

1. *Recursion theory.* J. R. Shoenfield. (1993, reprinted 2001; 84 pp.)
2. *Logic Colloquium '90; Proceedings of the Annual European Summer Meeting of the Association for Symbolic Logic, held in Helsinki, Finland, July 15–22, 1990.* Eds. J. Oikkonen and J. Väänänen. (1993, reprinted 2001; 305 pp.)
3. *Fine structure and iteration trees.* W. Mitchell and J. Steel. (1994; 130 pp.)
4. *Descriptive set theory and forcing: how to prove theorems about Borel sets the hard way.* A. W. Miller. (1995; 130 pp.)
5. *Model theory of fields.* D. Marker, M. Messmer, and A. Pillay. (1996; 154 pp.)
6. *Gödel '96; Logical foundations of mathematics, computer science and physics; Kurt Gödel's legacy. Brno, Czech Republic, August 1996, Proceedings.* Ed. P. Hajek. (1996, reprinted 2001; 322 pp.)
7. *A general algebraic semantics for sentential objects.* J. M. Font and R. Jansana. (1996; 135 pp.)
8. *The core model iterability problem.* J. Steel. (1997; 112 pp.)
9. *Bounded variable logics and counting.* M. Otto. (1997; 183 pp.)
10. *Aspects of incompleteness.* P. Lindstrom. (1997, 2nd ed. 2003; 163 pp.)
11. *Logic Colloquium '95; Proceedings of the Annual European Summer Meeting of the Association for Symbolic Logic, held in Haifa, Israel, August 9–18, 1995.* Eds. J. A. Makowsky and E. V. Ravve. (1998; 364 pp.)
12. *Logic Colloquium '96; Proceedings of the Colloquium held in San Sebastian, Spain, July 9–15, 1996.* Eds. J. M. Larrazabal, D. Lascar, and G. Mints. (1998; 268 pp.)
13. *Logic Colloquium '98; Proceedings of the Annual European Summer Meeting of the Association for Symbolic Logic, held in Prague, Czech Republic, August 9–15, 1998.* Eds. S. R. Buss, P. Hájek, and P. Pudlák. (2000; 541 pp.)
14. *Model Theory of Stochastic Processes.* S. Fajardo and H. J. Keisler. (2002; 136 pp.)

15. *Reflections on the Foundations of Mathematics; Essays in honor of Solomon Feferman.* Eds. W. Seig, R. Sommer, and C. Talcott. (2002; 444 pp.)

16. *Inexhaustibility; a non-exhaustive treatment.* T. Franzén. (2004; 255 pp.)

17. *Logic Colloquium '99; Proceedings of the Annual European Summer Meeting of the Association for Symbolic Logic, held in Utrecht, Netherlands, August 1–6, 1999.* Eds. J. van Eijck, V. van Oostrom, and A. Visser. (2004; 208 pp.)

* 9 7 8 1 5 6 8 8 1 1 7 5 8 *

T - #0517 - 101024 - C0 - 229/152/14 - PB - 9781568811758 - Gloss Lamination